Lecture Notes in Statistics

Edited by D. Brillinger, S. Fienberg, J. Gani, J. Hartigan, and K. Krickeberg

26

Robust and Nonlinear Time Series Analysis

Proceedings of a Workshop Organized by the Sonderforschungsbereich 123 "Stochastische Mathematische Modelle", Heidelberg 1983

Edited by J. Franke, W. Härdle and D. Martin

Springer-Verlag
New York Berlin Heidelberg Tokyo 1984

Editors

Jürgen Franke
Wolfgang Härdle
Fachbereich Mathematik, Universität Frankfurt
Robert Mayer Str. 6–10, 6000 Frankfurt, FRG

Douglas Martin
Department of Statistics, GN-22, University of Washington
Seattle, WA 98195, USA

ISBN 3-540-96102-X Springer-Verlag Berlin Heidelberg New York Tokyo
ISBN 0-387-96102-X Springer-Verlag New York Heidelberg Berlin Tokyo

Library of Congress Cataloging in Publication Data. Main entry under title: Robust and nonlinear time series analysis. (Lecture notes in statistics; 26) Bibliography: p. 1. Time-series analysis–Congresses. 2. Robust statistics–Congresses. I. Franke, J. (Jürgen) II. Härdle, Wolfgang. III. Martin, D. (Douglas) IV. Sonderforschungsbereich 123–"Stochastische Mathematische Modelle." V. Series: Lecture notes in statistics (Springer-Verlag); v. 26.
QA280.R638 1984 519.5'5 84-23619

Printing and binding: Beltz Offsetdruck, Hemsbach / Bergstr.
2146/3140-543210

PREFACE

Classical time series methods are based on the assumption that a particular stochastic process model generates the observed data. The most commonly used assumption is that the data is a realization of a stationary Gaussian process. However, since the Gaussian assumption is a fairly stringent one, this assumption is frequently replaced by the weaker assumption that the process is wide-sense stationary and that only the mean and covariance sequence is specified. This approach of specifying the probabilistic behavior only up to "second order" has of course been extremely popular from a theoretical point of view because it has allowed one to treat a large variety of problems, such as prediction, filtering and smoothing, using the geometry of Hilbert spaces.

While the literature abounds with a variety of optimal estimation results based on either the Gaussian assumption or the specification of second-order properties, time series workers have not always believed in the literal truth of either the Gaussian or second-order specification. They have none-the-less stressed the importance of such optimality results, probably for two main reasons: First, the results come from a rich and very workable theory. Second, the researchers often relied on a vague belief in a kind of continuity principle according to which the results of time series inference would change only a small amount if the actual model deviated only a small amount from the assumed model. Somehow second-order theory was held in particular esteem because in such theory one does not assume a Gaussian distribution.

Unfortunately blind faith in the existence of such a continuity principle turns out to be totally unfounded. This fact has been made clear with regard to the behavior of maximum-likelihood estimates based on the Gaussian assumption in the classical setting of independent observations. The seminal work of Hampel, Huber and Tukey formed the foundation for over a decade of work on robust methods for the independent observations setting. These methods have been designed to perform well in a neighborhood of a nominal parametric model. The most frequently treated case is where the nominal model is Gaussian and the neighborhood contains outlier-producing heavy-tailed distributions.

"Doing well" in a neighborhood can be described in terms of three distinct concepts of robustness, namely: *efficiency robustness*, *min-max robustness*, and *qualitative robustness*. These are in historical order of inception. Qualitative robustness, introduced by Hampel (1971), is a fundamental continuity condition, which is in fact a basic principle of statistics. Efficiency robustness is a simple and accessible concept, whose importance was made quite clear by J.W. Tukey (1960). An estimate is efficieny robust if it has high efficiency both at the nominal model and at a strategically-selected finite set of "nearby" distributions. Min-max robustness was introduced by Huber (1964) in the problem of estimating location, with asymptotic variance as the loss function. In general, a min-max robust estimate is one for which the maximum mean-squared error of an estimate, either for a finite sample size or asymptotically, is minimized over a neighborhood of the nominal model. For further details on robustness see Huber (1981).

In general an estimate which has all three robustness properties is preferred. For some problems, typically in the asymptotic framework such estimates exist. However, this is not always possible, and often one will have to settle for qualitative robustness (or its data-oriented analogue, *resistance*, a term coined by Tukey - see Mosteller and Tukey, 1977), along with some sort of a verification of efficiency robustness. The latter is obviously required since some rather ridiculous estimates, e.g., identically constant estimates, are qualitatively robust.

It is only in relatively recent years that attention has turned to robustness in the context of the time-series setting. As in the independent observations setting, classical time-series procedures based on normality assumptions lack robustness, and the same is true of most optimal linear procedures for time series based on second-order specifications. For example, the least-squares estimates of autoregressive parameters are quite non-robust, as are smoothed periodogram spectrum estimates.

In the time-series setting it is important to realize that there are a variety of outlier types which can occur and which have different consequences. Two particular broad types of outliers stand out: (i) *innovation outliers*, which are associated with heavy-tailed distributions for the innovations in perfectly-observed linear processes, and (ii) *additive outliers*. See, for example, Fox (1972) and Martin (1981). Additive outliers, or more general forms of replacement type contamination (see Martin and Yohai, 1984 b), cause by far the most serious problems for the classical time-series estimates. A striking example of how extremely small outliers can spoil spectrum estimates is provided by Kleiner, Martin and Thomson (1979), who show how to obtain robust spectrum estimates via robust prewhitening techniques.

Since qualitative robustness is a fundamental concept, it should be noted that Hampel's definition of qualitative robustness does not really cover the time-series setting, where a variety of definitions are possible. Fortunately, this issue has recently been dealt with by a number of authors: Papantoni-Kazakos and Gray (1979), Cox (1981), Bustos (1981), Boente, Fraiman and Yohai (1982), and by Papantoni-Kazakos in this volume. Boente, Fraiman and Yohai (1982) not only provide results which relate the various definitions of qualitative robustness, they also provide an intuitively appealing new definition which is based on Tukey's data-oriented notion of resistance.

In addition to the issue of qualitative robustness for time series, a variety of other problems concerning robust estimation for time series have begun to receive attention. For a recent survey of results on robustness in time series and robust estimates of ARMA models, see Martin and Yohai (1984 a).

Heavy-tailed distributions, and other diverse mechanisms for model failure (e.g., local nonstationarity, interventions - see Box and Tiao, 1975), which give rise to outliers are not the only kind of deviation from a nominal model having a severe adverse impact on time-series estimates. Deviations from an assumed second-order specification, either in terms of covariance sequence or spectrum, can also cause problems. This issue is important, for example, when constructing linear filters, smoothers and predictors. Some details may be found in the article in this volume by Franke and Poor, and in the many references therein.

With regard to deviations from an assumed second-order model, a particularly potent kind of second-order deviation is that where one assumes exponentially decaying correlations (e.g., as in the case of stationary ARMA processes), and in fact the correlations decay at a slower rate.Here we have unsuspected long-tailed correlations as a natural counterpart of long-tailed distribution functions. In this regard, the article by Graf and Hampel in this volume is highly pertinent.

In both the time series and the non-time series settings nonparametric procedures form a natural complement to robust procedures. One reason for this, in the nonparametric regression setting with independent observations for example, is the following. Many data sets have a "practical" identification problem: there are a few unusual data points which on the one hand may be regarded as outliers, and on the other hand may be due to nonlinearity. The data may or may not be sufficient to say which is the case. We shall often wish to use both nonparametric and robust regression. Then if the sample size is large enough, we may be able to decide whether or not nonlinearity is present, and whether or not a few data points are outliers.

In the time-series setting the situation is similar. It is known that many times series arising in practice exhibit outliers, either in isolation or in patches, and that certain nonlinear models generate sample paths which manifest patches of outliers (Subba Rao, 1979; Nemec, 1984). Thus, we will need to have both good robust procedures for estimating linear time-series models, and good nonparametric procedures for estimating nonlinear time-series models (see Watson, 1964, for early work in this area). And of course, we may need to combine the two approaches, as does Robinson in his article in this volume, and Collomb and Härdle, 1984).

It may also be noted that robust estimates which guard against outliers are always nonlinear, and that many Gaussian maximum-likelihood estimates of time-series models are also nonlinear. Thus nonlinearity is also a common theme which bridges both robust and nonparametric approaches to modelling and analysis of time series.

In order to provide a forum for discussions and exchange of ideas for researchers working on robust and nonlinear methods in time-series analysis, or on related topics in robust regression, a workshop on "Robust and Nonlinear Methods in Time Series Analysis" took place at the University of Heidelberg in September 1983, organized under the auspices of the SFB 123. This volume contains refereed papers, most of which have been presented at the workshop. A few particularly appropriate contributions were specifically invited from authors who could not attend the meeting.

We take the opportunity to thank Dr. H. Dinges and the speaker of the SFB 123, Dr. W. Jäger, for encouraging and supporting this workshop. The meeting could not have taken place without the generous support of the Deutsche Forschungsgemeinschaft. Finally, we would like to thank all participants for their enthusiastic cooperation, which made the workshop a lively and successful meeting.

The Editors

References

Boente, G., Fraiman, R. and Yohai, V.J. (1982). "Qualitative robustness for general stochastic processes," Technical Report No. 26, Department of Statistics, University of Washington, Seattle, WA.

Box, G.E.P. and Tiao, G.C. (1975). "Intervention analysis with applications to economic and environmental problems," J. Amer. Stat. Assoc. 70, 70-79.

Bustos, O.H. (1981). "Qualitative robustness for general processes", Informes de Matematica, Serie B-002/81, Instituto de Matematica Pura e Aplicada, Brazil.

Collomb, G. and Härdle, W. (1984). "Strong uniform convergence rates in robust nonparametric time series analysis: kernel regression estimation from dependent observations", submitted to Annals of Stat.

Cox, D. (1981). "Metrics on stochastic processes and qualitative robustness", Technical Report No. 3, Department of Statistics, University of Washington, Seattle, WA.

Hampel, F.R. (1971). "A general qualitative definition of robustness", Ann. Math. Stat. 42, 1887-1896.

Huber, P.J. (1964). "Robust estimation of a location parameter", Ann. Math. Stat. 35, 73-101.

Huber, P.J. (1981). Robust Statistics. Wiley, New York, NY.

Kleiner, B., Martin, R.D. and Thompson, D.J. (1979). "Robust estimation of power spectra", J. Royal Stat. Soc. B41, 313-351.

Martin, R.D. (1981). "Robust Methods for time series", in Applied Time Series II, D.F. Findley, ed. Academic Press, New York, NY.

Martin, R.D. and Yohai, V.J. (1984 a). "Robustness for Time Series and Estimating ARMA models", in Handbook of Statistics, Vol. 4, edited by Brillinger and Krishnaiah, Academic Press, New York, NY.

Martin, R.D. and Yohai, V.J. (1984 b). "Influence function for time series", Technical Report No.51, Department of Statistics, University of Washington, Seattle, WA.

Mosteller, F. and Tukey, J.W. (1977). Data Analysis and Regression, Addison-Wesley, Reading, MA.

Nemec, A.F.L. (1984). "Conditionally heteroscedastic autoregressions", Technical Report No. 43, Department of Statistics, University of Washington, Seattle, WA.

Papantoni-Kazakos, P. and Gray, R.M. (1979). "Robustness of estimators on stationary observations", Ann. Probab. 7, 989-1002.

Subba Rao, T. (1979). Discussion of "Robust estimation of power spectra" by Kleiner, Martin and Thomson, J. Royal Stat. Soc., B, 41, 346-347.

Tukey, J.W. (1960). "A survey of sampling from contaminated distributions" in Contributions in Probability and Statistics, I. Olkin, ed., Stanford University Press, Stanford, CA.

Watson, G.S. (1964). "Smooth regression analysis", Sankhya, 26, A, 359-372.

CONTENTS

ON THE USE OF BAYESIAN MODELS IN TIME SERIES ANALYSIS

Hirotugu Akaike
The Institute of Statistical Mathematics
4-6-7 Minami-Azabu
Minato-ku
Tokyo 106
Japan

Abstract

The Bayesian modeling allows very flexible handling of time series data. This is realized by the explicit representation of possible alternative situations by the model. The negative psychological reaction to the use of Bayesian models can be eliminated once we know how to handle the models.
In this paper performances of several Bayesian models developed for the purpose of time series analysis are demonstrated with numerical examples. These include models for the smoothing of partial autocorrelation coefficients, smooth impulse response function estimation and seasonal adjustment. The concept of the likelihood of a Bayesian model is playing a fundamental role for the development of these applications.

Introduction

The basic characteristic of the Bayesian approach is the explicit representation of the range of possible stochastic structures through which we look at the data. The concept of robustness can be developed only by considering the possibilities different to the one that is represented by the basic model. In this sense the Bayesian modeling is directly related to the concept of robustness which is one of the two subjects discussed in this workshop.

In the area of time series analysis, and also in other areas of statistics, the practical application of Bayesian models is not fully developed yet. This is obviously due to historically well-known difficulty of the selection of prior distributions. This is the difficulty that made distinguished statisticians as R.A. Fisher and J. Neyman consider the procedure not quite fully fledged at their time.

Basically, there is a psychological reaction against the use of the Bayesian model that describes the parameter of the basic model as a random variable. This is amplified by the apparent arbitrariness of the assumption of the prior distribution. However, this psychological reaction can easily be controlled once we recognize that even the basic parametric model is seldom an objectively confirmed structure. If we adopt the predictive point of view that specifies the objective of statistics as the construction of a predictive distribution, the distribution of some future observation defined as a function of the presently available data, it becomes obvious that the justification of the use of a probabilistic reasoning, including Bayesian modeling, can be found only in what it produces. Only when the final output is found to be useful then the assumption underlying the reasoning is acceptable.

In the present paper performances of several Bayesian models for time series analysis will be illustrated with numerical examples. The first example is the smoothing of partial autocorrelation coefficients. The second is the estimation of a smooth impulse response function, with application to Series J of Box-Jenkins (1970). The third is the spectrum estimation with the assumption of smoothness in the frequency domain. The fourth is the seasonal adjustment, where a comparison is made of the BAYSEA procedure developed by Akaike and Ishiguro (1980) and a new Bayesian procedure based on the model that assumes an AR model generated from an innovation process with smoothly changing mean. This last example demonstrates a potential for the seasonal adjustment of a time series with significant stochastic variation of the seasonal component.

1. Smoothing of partial autocorrelation coefficients

In 1969 the present author introduced a criterion called FPE (Final Prediction Error) for the determination of the order of an autoregressive (AR) model. The criterion was later generalized to an information criterion (Akaike, 1974)

$$\text{AIC} = -2\ (\text{log maximum likelihood}) + 2\ (\text{number of parameters})$$

for the evaluation of models with the parameters determined by the method of maximum likelihood. In discussing the asymptotic behavior of the sequence of FPE's for increasing orders Bhansali and Downham (1977) suggested the use of criteria that corresponded to the generalizations of AIC obtained by replacing the multiplicative factor 2 of the number of parameters by arbitrary α's ($\alpha > 0$).

The work by Bhansali and Downham and the later work by Hannan and Quinn (1979), both on the problem of order determination, are essentially based on the assumption of finiteness of the true order. However, in practical applications it is often more natural to assume the true order to be infinite. This means the necessity of considering very high order AR models. Obviously, if the ordinary method of maximum likelihood is applied to an AR model with a high order, relatively to the available data length, the resulting estimate of the power spectrum exhibits significant sampling variability.

As a practical solution to this problem Akaike (1979) suggested the use of $\exp(-0.5\ \text{AIC}(k))$ as the likelihood of the k-th order model and to take the average of models with the weight proportional to $\exp(-0.5\ \text{AIC}(k))p(k)$, where $p(k)$ denotes a properly chosen prior weight of the model. One particular choice of $p(k)$ proposed in the paper, was $p(k)=1/(1+k)$. The averaging of the models was realized by taking the average of the sets of the partial autocorrelation coefficients that specify the AR models. The performance of the procedure, as evaluated in terms of the one-step ahead prediction error variance, was reasonable for the examples discussed by Bhansali and Downham.

Table 1 reproduces the result of the simulation experiment reported in Akaike (1979). Now the mean squared error of one-step prediction is evaluated analytically, instead of the evaluation by sampling experiment in the original paper, to produce the statistics. The statistics are based on 1000 samples each of length 100. Experiment IV is about the ARMA model of an EEG record and is newly added to the table.

Table 1 Statistics of mean squared prediction error

Procedure	Mean	St. dev.	NLM*	Mean	St. dev.	NLM*
	Experiment I			Experiment II		
AIC	1.032	.047	3.2	1.077	.048	7.4
AIC4	1.017	.023	1.7	1.094	.047	9.0
BAYES	1.022	.029	2.2	1.066	.038	6.4
FBAYES	1.026	.027	2.6	1.071	.031	6.9
	Experiment III			Experiment IV		
AIC	1.364	.216	31.0	1.193	.097	17.6
AIC4	1.484	.112	39.4	1.204	.077	18.6
BAYES	1.365	.178	31.1	1.168	.088	15.5
FBAYES	1.413	.211	34.6	1.122	.100	11.5

* NLM = $100 \times \log_e$ (Mean)

Experiment I : $x(n) = 0.55x(n-1) + 0.05x(n-2) + w(n)$

Experiment II : $x(n) = w(n) + 0.5w(n-1) + 0.5w(n-2)$

Experiment III : $x(n) = 0.50x(n-1) - 0.06x(n-2) + 0.45x(n-15) + w(n)$

Experiment IV :
$$x(n) = 3.812x(n-1) - 6.022x(n-2) + 4.648x(n-3) -1.165x(n-4) - 0.825x(n-5) + 0.742x(n-6) -0.192x(n-7) + w(n) - 3.000w(n-1) +3.457w(n-2) - 1.603w(n-3) -0.036w(n-4) + 0.200w(n-5)$$

The procedures denoted by AIC and AIC4 are realized by choosing the AR models with minimum AIC and AIC4, respectively, where AIC4 is defined by replacing the multiplicative factor 2 by 4 in the definition of AIC. BAYES denotes the Bayesian type procedure described in the preceding paragraph and FBAYES denotes a new Bayesian procedure to be discussed in Section 5. The mean squared one-step prediction error of each fitted model can easily be obtained by using the computational procedures of the state covariance matrix given in Akaike (1978). Form table 1 we can see that the maxima of the means of mean squared prediction errors are attained only by AIC and AIC4. This suggests the non-robustness of the ordinary model selection approach compared with the present Bayesian approach that considers possible alternative situations explicitly.

2. Estimation of smooth response functions

The concept of smoothness prior was first introduced by Shiller (1973). It is based on the data distribution specified by the structure

$$y(n) = \sum_{m=o}^{M} a_m x(n-m) + w(n) \quad ,$$

where $y(n)$ and $x(n)$ denote the sequences of the output and input of the system, respectively, and $w(n)$ denotes a white noise, assumed to be Gaussian with mean O and variance σ^2. The smoothness prior is defined by assuming a homogeneous spherical Gaussian distribution of $\Delta^d a$, the d-th difference of the sequence a_m.

By assuming $a_m = O$ for $m > M$ the density of the prior distribution may simply be represented by

$$p(a|d,\tau) = \left(\frac{1}{2\pi\tau^2}\right)^{(M+1)/2} |R_d| \exp\left(-\frac{1}{2\tau^2} a'R_d'R_d a\right) \quad ,$$

where $a = (a_o, a_1, \ldots, a_M)'$, τ^2 denotes the common variance of the differences $\Delta^d a_m$ and R_d is a matrix whose (i,j) element is defined by $R(i,j) = e_{j-i}$, for $j = i, i+1, \ldots, i+d$ and O, otherwise, where e_k denotes the coefficient of B^k in the polynomial expansion of $(1-B)^d$. The data distribution is given by

$$p(y|a,\sigma) = \left(\frac{1}{2\pi\sigma^2}\right)^{N/2} \exp\left(-\frac{1}{2\sigma^2} ||y-Xa||^2\right)$$

where $y = (y(1), y(2), \ldots, y(N))'$ and X denotes a matrix with its (i,j) element defined by $X(i,j) = x(i-j)$. We adopt the parametrization $\tau = \sigma/c$, where c is a scaling constant. The estimate of the response sequence is given by the posterior mean of a and the necessary computation reduces to the minimization of $|| y^*-X^*a||^2$, where

$$y = \begin{bmatrix} y \\ o \end{bmatrix} \quad \text{and} \quad X = \begin{bmatrix} X \\ cR_d \end{bmatrix} \quad .$$

The very basic problem here is how to choose the lag length M, the order of differencing d and the scaling constant c. For the variance parameter we assume the ignorance prior density $1/\sigma$, proposed by Jeffreys (1946).

We define the likelihood of each model by

$$\ell(d,c,M) = \iint p(y|a,\sigma,M)p(a|d,\sigma c^{-1})\frac{1}{\sigma}da\,d\sigma \ .$$

In analogy to the definition of AIC we use

$$\begin{aligned} \text{ABIC} &= (-2)\log \ell(d,c,M) \\ &= N\log\{S(c)\} + \log|X'X+c^2R_d'R_d| - \log|c^2R_d'R_d| \end{aligned}$$

as the criterion of fit of the model to the data, where $S(c) = \min\|y^*-X^*a\|^2$. A model with a smaller value of ABIC is considered to be a better model.

To show the performance of the smoothness prior model thus determined the result of its application to Series J of Box and Jenkins (1970) is illustrated in fig. 1. Instead of choosing an M by minimizing the ABIC the estimate was obtained by assuming a uniform prior distribution over M = 1 to 25 and using $\ell(d,c,M)$ as the likelihood of each model. The selection of d and c were realized by maximizing the sum of the likelihoods. It was assumed that $a_0 = 0$. Along with the Bayesian estimate is given the least squares estimate of the response function for M = 25. Since we already know that the data show very low coherence at the higher frequency band it is almost certain that the spiky response shown in the least squares estimate is due to the high frequency noise that contaminated the response.

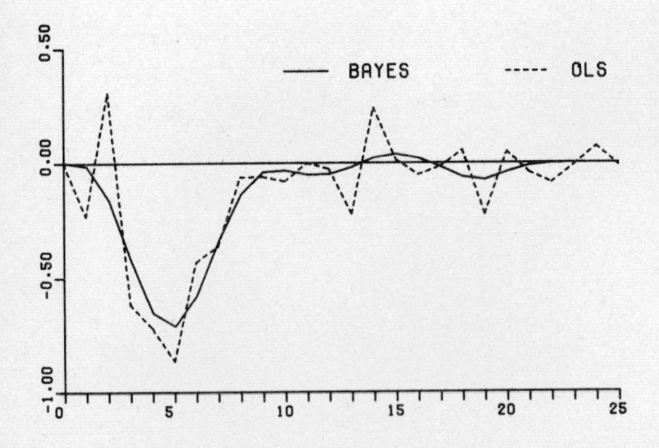

FIG.1 IMPULSE RESPONSE ESTIMATES BY THE SMOOTHNESS PRIOR AND THE ORDINARY LEAST SQUARES. (SERIES J BOX-JENKINS (1970))

3. Smoothness in the frequency domain

The concept of smoothness is applicable also to the frequency domain characteristic of a time series. The smoothness of a power spectral density function p(f) may be measured by

$$S = \int \left(\frac{d}{df} \log p(f)\right)^2 Q(f)df,$$

where Q(f) is a properly chosen weight function and the integration extends over $-1/2 < f < 1/2$. For an AR process defined by

$$x(n) = \sum_{m=1}^{M} a_m x(n-m) + w(n) \quad ,$$

where w(n) is a white noise with mean 0 and variance σ^2, the power spectral density function p(f) is given by

$$p(f) = |A(f)|^{-2}\sigma^2, \text{ where } A(f) = \sum a_m \exp(-i2\pi fm)$$

$(m = 0,1,\ldots,M;\ a_o = -1)$. The smoothness S is then defined by

$$S = 8\pi^2 \int \frac{|\sum m a_m \exp(-i2\pi fm)|^2}{|A(f)|^2} Q(f)df.$$

In ordinary applications the estimate of the power spectrum in its low power region usually represents only the effects of the background noise, rather than the essential characteristic of the signal under consideration. In such circumstances it will be a reasonable choice to put Q(f) = p(f). (Obviously this choice makes Q(f) dependent on p(f).) With this choice of Q(f), and by ignoring the constant multiplication factor, we get

$$S = \sum_{m=1}^{M} m^2 a_m^2.$$

Consider the fitting of an AR model by assuming the likelihood function

$$f(x|a,\sigma^2) = \left(\frac{1}{2\pi\sigma^2}\right)^{N/2} \exp\left(-\frac{1}{2\sigma^2} \sum_{n=1}^{N} \left(x(n) - \sum_{m=1}^{M} a_m x(n-m)\right)^2\right),$$

where $x = (x(1), x(2), \ldots, x(N))$, $a = (a_1, a_2, \ldots, a_M)$ and the first M observations $x(-M+1), x(-M+2), \ldots, x(0)$ are treated as given constants. The prior distribution of a can be defined by

$$p(a|c) = \left(\frac{1}{2\pi\tau^2}\right)^{M/2} \exp\left(-\frac{1}{2\tau^2} S\right).$$

Again for the convenience of numerical manipulations we use the parametrization $\tau = \sigma/c$. With this choice of the prior distribution the Bayesian estimate of the AR coefficient can be obtained by the same computational procedure for the smooth impulse response function estimation, if only y is replaced by x and R_d by a diagonal matrix with the m-th diagonal element equal to m.

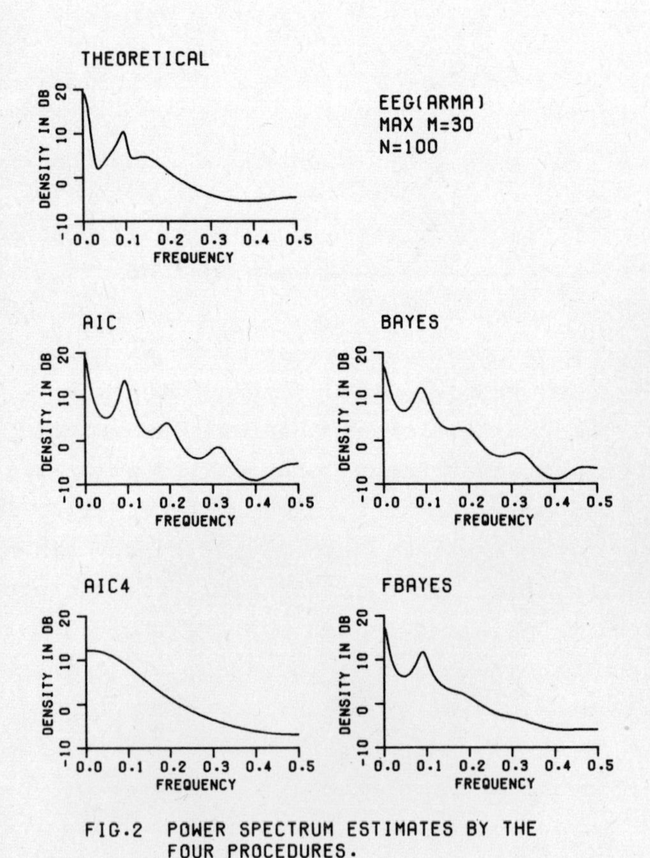

FIG.2 POWER SPECTRUM ESTIMATES BY THE FOUR PROCEDURES.

The procedure denoted by FBAYES in table 1 was obtained by this model. From the table we can see that the procedure is performing quite well in the experiment IV where the ARMA process that simulates an EEG spectrum is approximated by AR models. To see the differences among the procedures the estimates of the power spectrum obtained by the four procedures are illustrated in fig. 2, along with the theoretical power spectrum, for one typical example. For these estimates the highest order was put equal to 30. The result of FBAYES was obtained by putting $M = 30$ and properly choosing c by maximizing the likelihood of the Bayesian model defined analogously to $\ell(d,c,M)$ of the smooth response function model.

4. Seasonal adjustment by Bayesian modeling

The seasonal adjustment procedure BAYSEA (Akaike 1980 b, Akaike and Ishiguro, 1980) is realized by assuming a smoothness prior for the trend and seasonal components of a time series. One significant advantage of this procedure is that it can easily handle the missing value problem. The future values cf the trend and seasonal components can also be estimated similarly. However, it is obvious that the simple model used to define the procedure may not be quite sufficient for this purpose. This should be reflected in the large expected error of prediction. Also it is generally known that in the seasonal adjustment of a monthly series the adjusted values for the last six months or so usually undergo significant revisions when new observations are added. This again should be reflected in the large expected errors of the estimates.

The Bayesian modeling is particularly suited for the evaluation of the uncertainties of the estimated components. The posterior standard deviations serve for this purpose. Some examples are given in fig.'s 3 and 4. The dotted lines show the 2 s.d. range of each component represented by the solid line. We can see a clear representation of what we usually expect of the variabilities of the estimated components. The discontinuous jumps of the 2 s.d. lines are due to the one year shift of the four year span of data used for the adjustment.

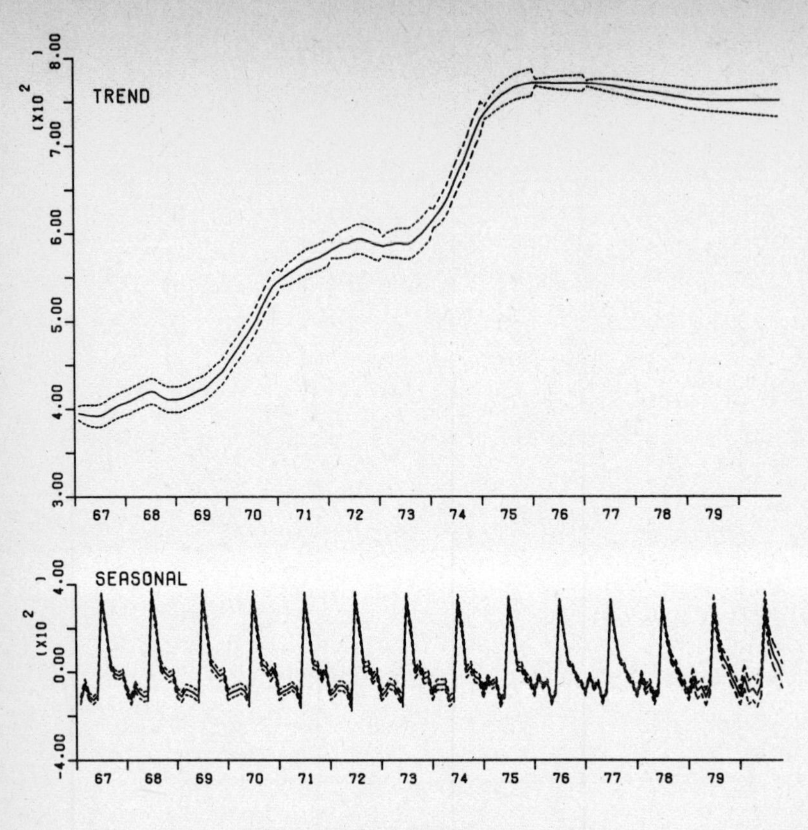

FIG.3 2 S.D. RANGES OF THE TREND AND SEASONAL COMPONENTS OBTAINED BY BAYSEA

The result illustrated in fig. 4 was obtained under the assumption that the data for the last 12 months were unavailable. The inflated 2 s.d. range of the trend component clearly demonstrates that the simple linear extrapolation of the trend based on the particular model lacked necessary predictive power in this case. Another model which assumed locally constant trend produced much narrower 2 s.d. range for this set of data.

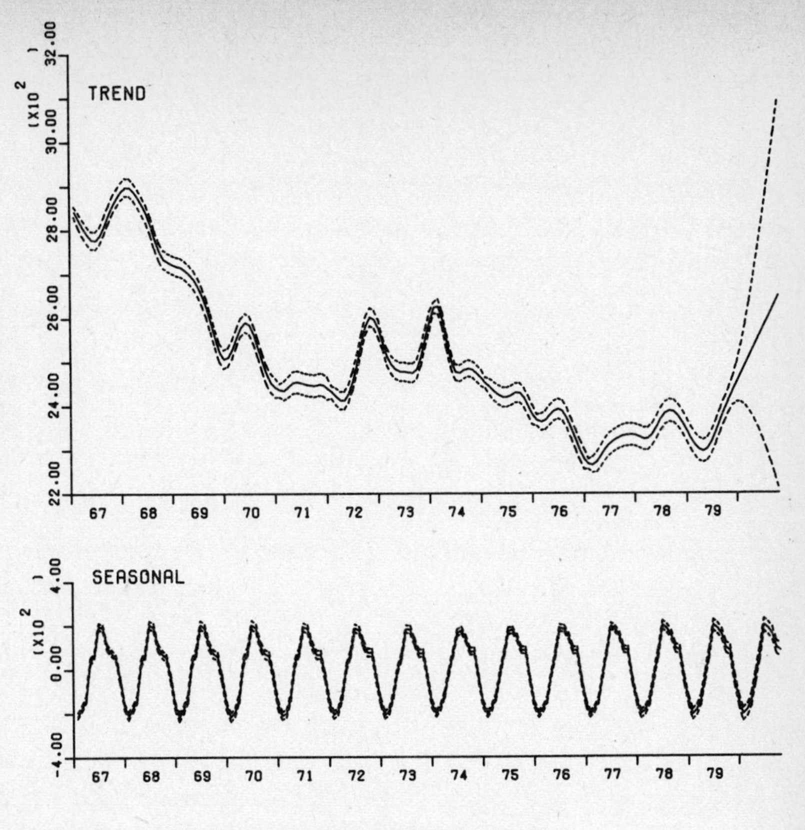

FIG.4 EFFECT OF MISSING OBSERVATIONS ON THE 2 S.D. RANGES

5. AR model with smoothly changing mean of the innovation

Some economic time series give the impression that the variation of a time series around a local base line is controlled by the recent temporal variation of the series. This suggests the use of a simple time series model defined by

$$x(n)-t(n) = \sum_{m=1}^{M} a_m \Delta^d x(n-m)+w(n) ,$$

where Δ^d denotes the d-th order difference and t(n) denotes the smoothly changing level of the base line. Obviously this may be viewed as an AR model defined by the innovation w(n) with smoothly changing mean t(n).

By the present model the smoothness requirement can be satisfied by assuming a smoothness prior for t(n). We may also assume some prior distribution for the coefficients a_m, like the smoothness prior in the frequency domain discussed in Section 3. The same computational procedure can be used for this model as the one used for the preceeding smoothness priors.

Fig.'s 5A and 5B show the three components of a time series, obtained by BAYSEA and the present procedure, called TRAR. By TRAR the coefficients a_m were determined by maximizing the likelihood of the corresponding Bayesian model defined with a as its parameter. The order M of the AR-part was set equal to 13 to allow for the AR-part respond to the significant yearly pattern of the seasonality. The order of differencing was d = 2. The first 15 data points were used to define the initial values for the TRAR procedure and the result is missing for this part of the data.

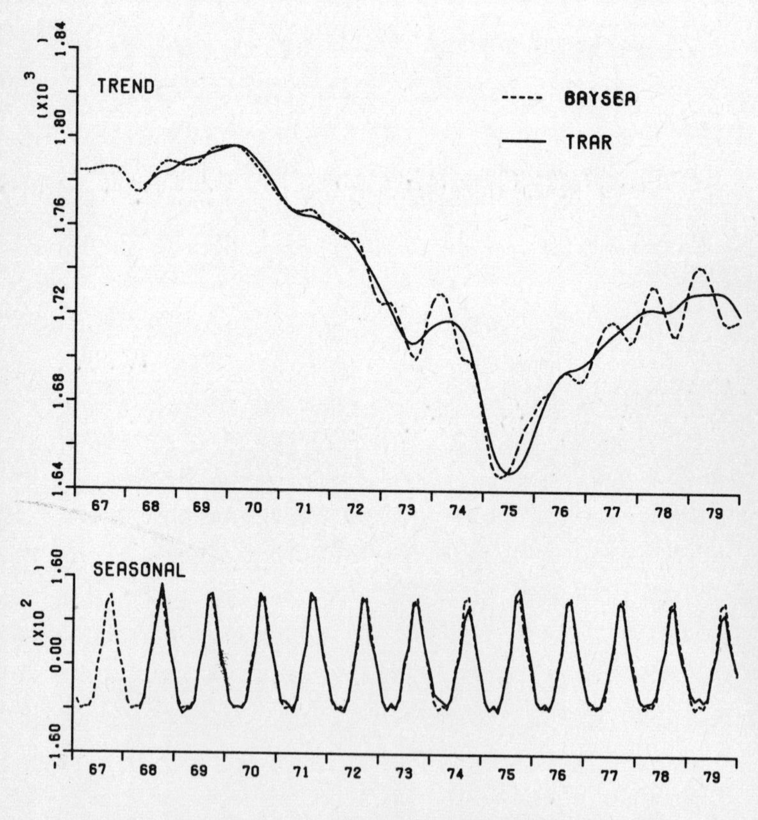

FIG.5A THE TREND AND SEASONAL COMPONENTS OBTAINED BY BAYSEA AND TRAR

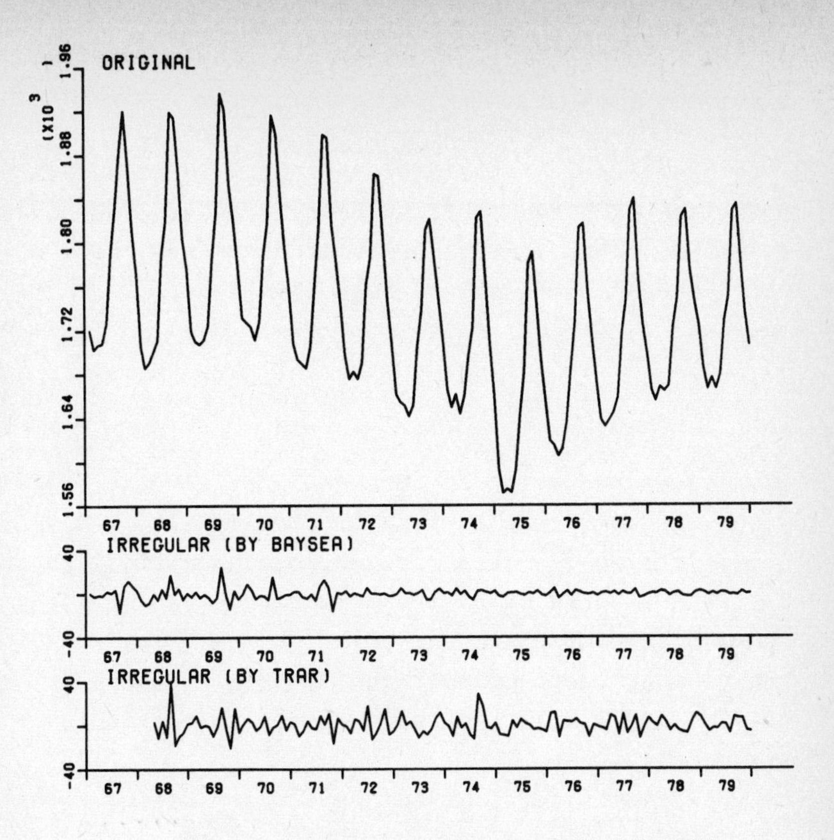

FIG.5B THE ORIGINAL DATA AND THE IRREGULAR COMPONENTS OBTAINED BY BAYSEA AND TRAR

It can be seen that TRAR is producing a much more natural estimate of the trend. The irregular behavior of the trend obtained by BAYSEA is obviously due to the presence of significant stochastic variation in the seasonal component. It has been observed that the conventional Census Method II X-11 variant produces even more irregular trend components than BAYSEA (Akaike and Ishiguro, 1983). Thus the present result suggests that TRAR may find interesting applications in the area of seasonal adjustment. The stochastic modeling of the seasonal component also allows simple realization of the prediction of the seasonal component.

It is generally recognized that seasonal adjustment generates very unnatural behavior of the irregular component, or the additive white noise. Although the power spectrum of an irregular component usually looks fairly flat, probably the structure of the series is very far from being Gaussian, i.e. the adjustment tends to produce a highly non-linear distortion of the series. If this is the case the seasonally adjusted series is not quite suitable for the analysis of linear relations between several time series. Hopefully the linear structure of the TRAR procedure might reduce this difficulty.

6. Discussion

Time series analysis is a branch of statistics characterized by the extensive use of some complex models. The models are usually developed either through some analytical characterization of the process or by the structural information based on the knowledge of the subject area. The Bayesian modeling allows us to supplement conventional models with the information that broadly limits the range of possible structures of the basic model.

The unlimited possibility of developing different Bayesian models for a particular time series might produce the impression that the whole procedure is quite arbitrary. However, even the choice of a non-Bayesian model is always based on a highly subjective judgement. If this view is accepted then what is the cause of the uneasy feeling of the Bayesian approach?

The conventional approach is realized by a collection of some specific procedures of which performances can be checked by repeated applications. The Bayesian approach will become more acceptable, if, instead of stressing the uniqueness of the model for a particular data set, its procedural aspect is stated clearly so that the merit and demerit can be evaluated before its application. The claim of optimality of the Bayesian approach under the assumption of the Bayesian set up is unfounded, as the assumption of the model itself is the basic question. It is only through the comparison of various possible models that we can gain some confidence in a particular model. ABIC is an example of the criterion developed for this purpose.

In the conventional approach we often consider the use of some test statistic when a particular deviation of the data from the assumed model is suspected. By the Bayesian approach the situation is described by taking the possible alternatives into the model explicitly. In this sense the robustness is built-in to a Bayesian procedure. It is only the proper choice of alternatives that provides the procedure with the desired type of robustness.

Obviously the development of new basic time series models is fundamental for the progress of time series analysis. Nevertheless, the flexibility of the modeling and the convenience in practical applications of the Bayesian modeling can significantly contribute to the improvement of the practice of time series analysis. The increasing number of successful applications shows that the Bayesian modeling is now fully operational.

Appendix

Orginal data (y(n), n = 1,2,...,100) used to produce the result of fig. 1

——> n

0.88	3.41	4.85	2.95	0.75	-1.84	-2.53	-2.35	-1.15	-0.85
0.66	2.61	2.63	0.54	-0.45	-1.15	-0.46	0.43	0.51	-0.17
-0.52	0.49	1.53	1.92	-0.02	0.14	-1.57	-2.10	-2.09	-1.08
1.14	1.50	0.57	0.72	2.09	0.50	-0.84	-2.88	-3.74	-2.43
-2.00	-1.70	-1.00	0.12	-1.11	-0.84	-1.37	-1.35	-3.37	-2.66
-2.12	-1.02	-1.92	-1.52	-1.17	-1.63	-1.29	-3.27	-3.97	-2.99
-2.32	-4.29	-4.12	-2.95	-2.27	-0.47	0.03	-0.21	-1.06	-0.91
0.50	-1.03	-2.19	-3.12	-3.08	-1.39	-0.34	-0.11	0.04	0.34
-0.92	0.16	-0.93	-1.62	-0.62	-2.19	-2.79	-0.21	1.41	0.00
-0.02	0.08	0.55	-0.96	-0.20	-2.16	-2.61	-2.62	-2.37	-2.21

Acknowledgements

The author is grateful to Ms. E. Arahata for preparing the graphical outputs presented in this paper. He is also particularly indebted to his late wife Ayako for her help in preparing the original manuscript of this paper. This work was partly supported by the Ministry of Education, Science and Culture, Grant-in-Aid No. 58450058.

References

Akaike, H. (1974) A new look at the statistical model identification. IEEE Transactions on Automatic Control, AC-19, 716-723.

Akaike, H. (1978) Covariance matrix computation of the state variable of a stationary Gaussian process. Ann. Inst. Statist. Math., 30, B. 499-504.

Akaike, H. (1979) A Bayesian extension of the minimum AIC procedure of autoregressive model fitting. Biometrika 66, 237-242.

Akaike, H. (1980a) On the identification of state space models and their use in control. In "DIRECTIONS IN TIME SERIES", D.R. Brillinger and G.C. Tiao eds., Institute of Mathematical Statistics: Hayward, California. 175-187.

Akaike, H. (1980b) Seasonal adjustment by a Bayesian model. J. Time Series Analysis, 1, 1-13.

Akaike, H. (1982) The selection of smoothness prior for distributed lag estimation. Technical Summary Report No. 2394, Mathematics Research Center, University of Wisconsin-Madison. (To be included in the de Finetti volume on Bayesian statistics edited by A. Zellner and P.K. Goel.)

Akaike, H. and Ishiguro, M. (1980) BAYSEA, A Bayesian seasonal adjustment program. Computer Science Monographs, No. 13, The Institute of Statistical Mathematics, Tokyo.

Akaike, H. and Ishiguro, M. (1983) Comparative study of the X-11 and BAYSEA procedures of seasonal adjustment (with discussions). Proceedings of the Conference on Applied Time Series Analysis of Economic Data, U.S. Department of Commerce, Bureau of the Census. 17.53.

Bhansali, R.J. and Downham, D.Y. (1977) Some properties of the order of an autoregressive model selected by a generalization of Akaike's FPE criterion. Biometrika, 64, 547-551.

Box, G.E.P. and Jenkins, G.M. (1970) Time Series Analysis, Forcasting and Control. Holden Day, San Francisco.

Jeffreys, H. (1946) An invariant form for the prior probability in estimation problems. Proc. Royal Society of London, Ser. A, 186, 453-461.

Hannan, E.J. and Quinn, B.G. (1979) The determination of the order of an autoregression. J. Royal Statist. Soc., B. 41, 190-195.

Shiller, R.J. (1973) A distributed lag estimator derived from smoothness priors. Econometrica, 41, 775-788.

ORDER DETERMINATION FOR PROCESSES WITH INFINITE VARIANCE

R.J. Bhansali
University of Liverpool, England

1. INTRODUCTION

Several studies, see, e.g., Granger and Orr (1972), have provided evidence to suggest that many observed time series may be realizations of processes whose innovations follow an infinite variance stable law. Such processes are said to have innovation outliers (Fox, 1972) and their correlation and partial correlation functions are not defined. Nevertheless, Nyquist (1980), Rosenfeld (1976) and Granger and Orr (1972), amongst others, report that the model identification procedure of Box and Jenkins (1970) is still useful for these processes. As is well-known, an alternative approach to model identification is via the use of an order determining criterion. A number of different criteria have recently been suggested. However, little is currently known analytically concerning the behaviour of these criteria for processes with infinite variance. Therefore, in this paper, their behaviour when the innovations follow an infinite variance stable law is examined empirically by means of a simulation study. The finite sample properties of the order of an autoregressive model selected by the FPE_{α} criterion of Bhansali and Downham (1977) and of the order of a moving average model selected from the autoregressive estimates of the inverse correlations (see Bhansali, 1983) are investigated. The motivation for considering these criteria is also explained in these references.

2. THE ASYMPTOTIC BEHAVIOUR OF THE ESTIMATES

Let $\{x_t\}$ be a real-valued, discrete-time, autoregressive process

$$\sum_{u=0}^{m} a_m(u)x_{t-u} = \varepsilon_t, \quad a_m(0) = 1 ,$$

where the ε_t's are independent identically distributed and follow a symmetric stable law with characteristic exponent δ and the location and scale parameters set equal to 0 and 1 respectively and the polynomial $A_m(z) = 1+a_m(1)z+...+a_m(m)z^m$ does not vanish for $|z| \leq 1$. Also let $\hat{\underset{\sim}{a}}_m = \hat{a}_m(1),...,\hat{a}_m(m)$ ' denote the least-squares estimator of $\underset{\sim}{a}_m = a_m(1),...,a_m(m)$ ' based on a realization of T consecutive observations of x_t. Consistency of $\hat{\underset{\sim}{a}}_m$ for $\underset{\sim}{a}_m$ is established by Kanter and Steiger (1974), who show, also, that for any

$$T^{1/\theta}(\hat{\underset{\sim}{a}}_m - \underset{\sim}{a}_m) = o_p(1) , \tag{2.1}$$

while Hannan and Kanter (1977) show that the above result holds almost surely. Some further information about the asymptotic covariance structure of $\hat{\underset{\sim}{a}}_m$ is given by Yohai and Maronna (1977). The work of these authors suggests that asymptotically the variance of the $a_m(u)$ decreases as δ decreases and vice versa. Another suggestion is that for $\delta<2$ a consistent estimator of m may be obtained by minimizing the FPE_α criterion with any fixed $\alpha>1$. For an intuitive explanation, note that in the Gaussian case ($\delta=2.0$) the asymptotic distribution of the order selected is derived by writing the differences

$$V_{p-m}=T\{\hat{\sigma}^2(m)-\hat{\sigma}^2(p)\}/\hat{\sigma}^2(m) \approx T \sum_{i=m+1}^{p} \hat{a}_i^2(i) \quad (p > m) ,$$

where $\hat{\sigma}^2(m)$ is the least-squares estimate of the variance of ε_t, as a sum of p - m independent χ_1^2 random variables, see Akaike (1970). But, if $\delta<2$, (2.1) suggests that V_p converges to 0 in probability as $T \to \infty$.

Suppose instead that x_t is a moving average process of order h and, as in Bhansali (1983), let the $\hat{\beta}_{h,k}(j)$ and $\hat{h}_k(\alpha)$, respectively, denote the estimates of the moving average coefficients and the order obtained from the autoregressive estimates of the inverse correlation function. No analytical results are currently available converning the properties of these estimates when $\delta<2$. However, the well-known duality between the moving average and autoregressive models, see, e.g., Chatfield (1979), suggests that the results discussed in the last paragraph would also apply to these estimates, and, for any fixed $\alpha>1$, $\hat{h}_k(\alpha)$ may provide a consistent estimator of the order.

3. PLAN OF THE EXPERIMENTS

A stretch of T observations following the models

$$1:\ x_t = \varepsilon_t \qquad\qquad 2:\ x_t = 0.5x_{t-1}+\varepsilon_t$$

$$3:\ x_t = 0.55_{t-1}+0.05_{t-2}+\varepsilon_t; \qquad 4:\ x_t = 0.5x_{t-1}-0.3x_{t-2}+\varepsilon_t;$$

$$5:\ x_t = 0.1x_{t-1}+0.8x_{t-2}+\varepsilon_t; \qquad 6:\ x_t = \varepsilon_t+0.3\varepsilon_{t-1};$$

$$7:\ x_t = \varepsilon_t-0.3\varepsilon_{t-1}+0.5\varepsilon_{t-2}; \qquad 8:\ x_t = \varepsilon_t+1.8\varepsilon_{t-1}+0.9\varepsilon_{t-2};$$

with T = 96 and 500, was simulated. For convenience, these models are henceforth referred to as Experiments 1,2,...,8, respectively.

In these experiments, $\{\varepsilon_t\}$ is a computer-produced sequence of independent, symmetric stable variates with location and seale parameters set equal to 0 and 1, respectively. For each T, four different values of δ were selected, namely δ= 0.8, 1.5, 1.7 and 2.0. The algorithm of Chambers et al (1976) was used for generating the ε_t's. For each (Experiment, T, δ) configuration, the pth order estimates of the autoregressive parameters were determined for every integral p up to L by recursively solving the Yule-Walker equations. We set L = 24 for T = 96 and L = 30 for T = 500. The order of the autoregressive model selected by minimizing the FPE_α criterion was determined. The 'autoregressive' estimates of the moving average coefficients and the order were also computed; see Bhansali (1983) for computational details. Four different values of α were used, namely, α=2,3,4 and log T. The total number of simulations for each experiment and each choice of (T,δ,α) values was 100. The computing was done on the 1906A/7600 computer of the University of Manchester.

4. THE SIMULATION RESULTS

The frequency distributions of the autoregressive order selected with α=2,3 and log T are shown in Table 1, and of the moving average order selected with α= log T are shown in Table 2, along with the asymptotic frequencies. The latter were obtained from the results of Bhansali and Downham (1977) and Bhansali (1983) and apply when δ=2.0. To save space, the results for the other values of α are not shown, but are consistent with those presented.

The simulation results strongly suggest that even when the ε_t's follow an infinite variance stable law, the choice of α = log T would provide a consistent estimator of the order.

On the other hand, when the value of α is fixed, the simulated frequencies of selecting the correct order increase as decreases and are not well-approximated by the asymptotic. An implication is that the asymptotic results obtained by assuming that the ε_t's have finite variance may not apply when their distribution follows an infinite variance stable law.

For δ = 0.8, the simulation results support the suggestion made in Section 2 that a consistent estimator of the order may be obtained with any fixed $\alpha > 1$. However, for δ = 1.5 and 1.7 and α = 2, the simulated frequencies of selecting the correct order seem too small to be consistent with this suggestion. Nevertheless, with a finite T, no one choice of α is always optimal, since this choice appears to depend simultaneously on the value of T and the nature of the generating process.

On the whole, the results suggest that an order determining criterion may be used for model selection even when the innovations follow an infinite variance stable law; the suggestions made by Bhansali (1983) for the choice of α also apply to this case.

The applicability of the asymptotic results described in Section 2 for a finite T was investigated further by examining the simulated sampling distributions of the statistic V_{p-m} and the means and variances of the estimated coefficients.

Figure 1 shows the chi-square Q - Q plots of V_1 and V_2 for Experiment 2 with T = 96 and δ = 0.8. The plots lie entirely below the 45 degree line. Thus, the simulated distributions are more concentrated near zero than the corresponding χ^2 distributions, which accords with the asymptotic theory.

The simulated means and variances of the estimated coefficients are shown in Tables 3 and 4 along with the asymptotic, which were calculated from the results of Walker (1961) and apply when $\delta = 2.0$.

The consistency property of the estimates even when $\delta < 2.0$ is clearly apparent. However, the result (2.1) concerning the order of consistency is not strongly supported. For Experiment 3, the simulated variances increase monotonically with δ and behave in accordance with the asymptotic theory. However, for Experiment 8, they decrease monotonically as δ increases and behave in exactly the opposite way to that suggested by the asymptotic theory. On the other hand, for Experiments 6 and 7, and with $T = 500$ for Experiment 4, the simulated variances increase with δ for all $\delta < 2.0$ but then decrease as δ increases to 2.0. Thus, here the simulation results are in partial agreement with the asymptotic theory, but, for Experiments 2 and 5, the agreement is rather poor, though the disparity is not as wide as for Experiment 8.

A unifying explanation which accounts for the differing behaviour of the simulated variances just described may not be easily given. It would, however, be unrealistic to expect a perfect agreement between the simulation results and the asymptotic theory for all processes and all values of T. The seemingly wild behaviour occurs for Experiment 8 and to a lesser extent for Experiment 3. For these two experiments, the roots, μ_j, of the corresponding characteristic polynomials are close to the unit circle. The simulations indicate that for such processes the asymptotic results described in section 2 may not apply even with T as large as 500. This finding is perhaps not surprising. The proofs of Yohai and Maronna (1977), e.g., depend crucially on the property that the estimated roots are strictly less than one.

REFERENCES

Akaike, H. (1970). Statistical predictor indentification. *Ann. Statist. Math.*, **22**, 203-17.

Bhansali, R.J. (1983). Estimation of the order of a moving average model from autoregressive and window estimates of the inverse correlation function. *J. Time Series Analysis*, **4**, 137-162.

Bhansali, R.J. and Downham, D.Y. (1977). Some properties of the order of an autoregressive model selected by a generalization of Akaike's FPE criterion. *Biometrika*, **64**, 547-51.

Box, G.E.P. and Jenkins, G.M. (1970). *Time Series Analysis: Forecasting and Control*. San Francisco: Holden Day.

Chambers, J.C., Mallows, C.L. and Stuck, B.W. (1976). A method for simulating stable random variables. *J. Amer. Statist. Assoc.*, **71**, 340-344.

Chatfield, C. (1979). Inverse Autocorrelations. *J. Royal Statist. Soc.*, A, **142**, 363-377.

Fox, A.J. (1972). Outliers in time series. *J. Royal Statist. Soc.. B*, **34**, 350-363.

Granger, C.W.J. and Orr, D. (1972). "Infinite Variance" and research strategy in time series analysis. *J. Amer. Statist. Assoc.* **67**, 275-285.

Hannan, E.J. and Kanter, M. (1974). Regression and autoregression with infinite variance. *J. Appl. Prob.*, **14**, 411-415.

Kanter, M. and Steiger, W.L. (1974). Regression and autoregression with infinite variance. *Adv. Appl. Probl.*, **6**, 768-783.

Nyquist, H. (1980). On the identification and estimation of stochastic processes generated by stable distributions. Research Report, Department of Statistics, University of Umea, Sweden.

Rosenfeld, G. (1976). Identification of time series with infinite variance. *Applied Statistics*, **25**, 147-153.

Walker, A.M. (1961). On Durbin's formula for the limiting generalized variance of a sample of consecutive observations from a moving-average process. *Biometrika*, **48**. 197-199.

Yohai, V.J. and Maronna, R.A. (1977). Asymptotic behaviour of least-squares estimates for autoregressive processes with infinite variance. *Ann.Statist.*, **5**, 554-560.

TABLE 1

THE FREQUENCY OF AUTOREGRESSIVE ORDER SELECTED, AOS, BY THE FPE_{α} CRITERION

T	Experiment	AOS / δ	α = 2					α = 3					α = log T				
			0	1	2	3	>3	0	1	2	3	>3	0	1	2	3	>3
96	2	0.8	1	90	4	0	5	1	92	4	0	3	1	94	2	0	3
		1.5	0	82	8	1	9	1	90	5	1	3	0	92	5	2	1
		1.7	0	78	7	6	9	0	92	5	2	1	0	97	2	0	1
		2.0	0	72	7	5	16	0	94	2	1	3	0	98	1	1	0
		Asymptotic	0	71	11	6	12	0	88	7	2	3	0	100	0	0	0
	3	0.8	1	88	5	0	6	1	95	1	0	3	1	97	0	0	2
		1.5	0	79	9	4	8	0	90	7	0	3	1	96	1	0	2
		1.7	0	75	11	5	9	0	89	5	4	2	0	96	3	0	1
		2.0	0	71	10	3	16	0	90	6	1	3	0	98	2	0	0
		Asymptotic	0	0	71	11	18	0	0	88	7	5	0	0	100	0	0
	4	0.8	1	0	90	1	8	1	0	95	1	3	1	3	92	1	3
		1.5	1	3	79	5	12	1	7	87	2	3	1	11	85	2	1
		1.7	0	4	74	9	13	0	6	88	3	3	1	15	81	2	1
		2.0	0	2	68	12	18	0	8	75	8	9	0	15	80	4	1
		Asymptotic	0	0	71	11	18	0	0	88	7	5	0	0	100	0	0
	5	0.8	1	0	75	12	12	2	0	84	10	4	2	0	89	8	1
		1.5	0	0	68	18	14	0	0	88	7	5	0	0	94	5	1
		1.7	0	0	75	10	13	0	0	87	7	6	0	0	96	3	1
		2.0	0	0	72	8	20	0	0	89	5	6	0	0	98	2	0
		Asymptotic	0	0	71	11	18	0	0	88	7	5	0	0	100	0	0
500	2	0.8	0	84	5	2	9	0	91	3	1	5	0	94	3	0	3
		1.5	0	77	3	4	15	0	89	3	2	6	0	97	1	1	1
		1.7	0	80	4	4	12	0	91	3	3	3	0	98	1	1	0
		2.0	0	74	10	6	10	0	90	5	3	2	0	99	1	0	0
		Asymptotic	0	71	11	6	12	0	88	7	2	3	0	100	0	0	0
	3	0.8	0	72	16	0	12	0	88	5	1	6	0	95	1	1	3
		1.5	0	51	24	6	19	0	75	15	3	7	0	95	3	1	1
		1.7	0	46	30	8	16	0	72	21	3	4	0	93	6	1	0
		2.0	0	48	30	8	14	0	70	26	3	1	0	92	7	1	0
		Asymptotic	0	0	71	11	18	0	0	88	7	5	0	0	100	0	0
	4	0.8	0	0	85	3	12	0	0	89	3	8	0	0	96	1	3
		1.5	0	0	72	8	20	0	0	88	5	7	0	0	97	1	2
		1.7	0	0	71	9	20	0	0	89	7	4	0	0	98	2	0
		2.0	0	0	73	10	17	0	0	90	7	3	0	0	99	1	0
		Asymptotic	0	0	71	11	18	0	0	88	7	5	0	0	100	0	0
	5	0.8	0	0	81	4	15	0	0	83	4	13	0	0	90	5	5
		1.5	0	0	74	8	18	0	0	87	4	9	0	0	94	4	2
		1.7	0	0	69	14	17	0	0	83	9	8	0	0	95	5	0
		2.0	0	0	72	12	16	0	0	87	7	6	0	0	100	0	0
		Asymptotic	0	0	71	11	18	0	0	88	7	5	0	0	100	0	0

TABLE 2

THE FREQUENCY OF MOVING AVERAGE ORDER SELECTED, MOS, BY THE $FPEA_\alpha$ CRITERION WITH $\alpha = \log T$

Experiment	MOS \ δ	T = 96: 0.8	1.5	1.7	2.0	T = 500: 0.8	1.5	1.7	2.0	Asymp.
1	0	93	89	85	88	93	95	94	98	100
	1	3	8	13	10	3	3	4	2	0
	2	2	2	2	2	1	1	2	0	0
	3	0	0	0	0	0	0	0	0	0
	>3	2	1	0	0	3	1	0	0	0
6	0	8	11	14	9	1	1	0	0	0
	1	85	84	82	87	94	93	96	98	100
	2	2	3	3	4	2	4	3	2	0
	3	1	0	0	0	0	0	0	0	0
	>3	4	2	1	0	3	2	1	0	0
7	0	1	1	1	1	0	0	0	0	0
	1	2	2	3	0	0	0	0	0	0
	2	92	92	91	89	96	97	98	100	100
	3	2	3	4	7	1	1	2	0	0
	>3	3	2	1	3	3	2	0	0	0
8	0	1	0	0	0	0	0	0	0	0
	1	4	4	1	1	0	0	0	0	0
	2	90	91	97	96	93	97	98	99	100
	3	1	1	1	1	4	3	2	1	0
	>3	4	4	1	2	3	0	0	0	0

TABLE 3

Means and variances of $\hat{a}_1(1)$ for Experiment 2 and of $\hat{a}_2(1)$ for Experiments 3, 4 and 5

Experiment	δ	Means: T=96	Means: T=500	Variances × 10^3: T=96	Variances × 10^3: T=500
2	0.8	-0.481	-0.492	5.24	1.87
	1.5	-0.486	-0.492	6.73	1.50
	1.7	-0.491	-0.492	6.69	1.42
	2.0	-0.494	-0.496	7.53	1.62
	Asymptotic	-0.5	-0.5	7.81	1.50
3	0.8	-0.538	-0.544	6.64	1.95
	1.5	-0.542	-0.543	8.56	1.97
	1.7	-0.545	-0.543	9.04	2.06
	2.0	-0.548	-0.546	8.11	2.14
	Asymptotic	-0.55	-0.55	10.39	2.00
4	0.8	-0.484	-0.495	5.97	1.64
	1.5	-0.488	-0.494	8.04	1.81
	1.7	-0.492	-0.493	7.87	1.92
	2.0	-0.495	-0.495	6.92	1.82
	Asymptotic	-0.5	-0.5	9.48	1.82
5	0.8	-0.128	-0.102	13.39	1.88
	1.5	-0.110	-0.098	4.01	0.65
	1.7	-0.111	-0.097	4.22	0.62
	2.0	-0.112	-0.101	4.94	0.72
	Asymptotic	-0.1	-0.1	3.75	0.72

TABLE 4

Means and variances of $\hat{\beta}_{k,1}(1)$ for Experiment 6

and of $\hat{\beta}_{k,2}(1)$ for Experiments 7 and 8

		Means		Variances × 10^3	
Experiment	δ	T=96	T=500	T=96	T=500
6	0.8	0.286	0.296	4.25	2.01
	1.5	0.294	0.295	9.02	2.30
	1.7	0.298	0.294	10.76	2.51
	2.0	0.294	0.295	10.29	2.16
	Asymptotic	0.3	0.3	9.48	1.82
7	0.8	-0.289	-0.303	3.89	1.28
	1.5	-0.277	-0.301	6.41	1.71
	1.7	-0.272	-0.301	7.11	1.94
	2.0	-0.276	-0.301	9.20	1.76
	Asymptotic	-0.3	-0.3	7.81	1.50
8	0.8	1.535	1.727	79.40	14.72
	1.5	1.473	1.692	48.45	12.95
	1.7	1.443	1.675	34.37	9.84
	2.0	1.400	1.639	20.40	7.17
	Asymptotic	1.8	1.8	1.98	0.38

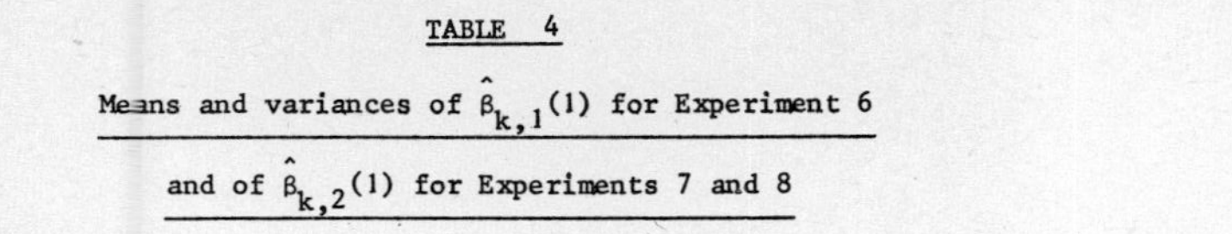

Fig. 1: Q-Q plots for V_1 and V_2 with Experiment 2.

T = 96; ... , Plot for V_1; ——— · ——— Plot for V_2.

ASYMPTOTIC BEHAVIOUR OF THE ESTIMATES BASED ON RESIDUAL AUTOCOVARIANCES FOR ARMA MODELS

Oscar Bustos, Ricardo Fraiman and Victor J. Yohai

Instituto de Matemática Pura e Aplicada, Rio de Janeiro
Universidad de Buenos Aires
Universidad de Buenos Aires and Centro de Estudios
Macroeconomicos de Argentina, Buenos Aires

ABSTRACT

In a recent paper Bustos and Yohai introduce the class of estimates based on residual autocovariances (RA-estimates) for the parameters of an ARMA model. They show using a Monte Carlo study that this class contains estimates which are highly efficient when the observations correspond to a perfectly observed Gaussian ARMA model and robust under the presence of outliers. In this paper we show the consistency and asymptotic normality of a class of estimates containing the RA-estimates.

1. Introduction

Suppose that z_t, $1 \leq t \leq T$ are observations corresponding to an stationary and invertible autoregressive-moving average process with p autoregressive parameters and q moving average parameters (ARMA(p,q)), i.e., we can write

$$(z_t-\mu_o) - \varphi_{o1}(z_{t-1}-\mu_o) - \dots - \varphi_{op}(z_{t-p}-\mu_o) = \tag{1.1}$$
$$= U_t - \theta_{o1}U_{t-1} - \dots - \theta_{oq}U_{t-q},$$

where the U_t's are i.i.d. random variables with common distribution F. Bustos and Yohai (1983) define a class of estimates based on the residual autocovariances (RA-estimates). The basic idea is to replace the sample autocovariances of the residuals which appear in a convenient form of the least squares (LS) estimate by robust autocovariances. The class of RA-estimates includes a subclass which is asymptotically equivalent to the M-estimates, see Denby and Martin (1979) and Lee and Martin (1982), and in particular to the LS-estimates too. Bustos and Yohai (1983) show by Monte Carlo simulation, that some conveniently chosen RA-estimates have a good performance in efficiency under a perfect observed gaussian model and also they are robust under the presence of additive outliers. They also show that the RA-estimates compare favorably with respect to the M-estimates and to the GM-estimates, see Denby and Martin (1979), Bustos (1982) and Martin (1981) for a definition of the GM-estimates. Martin and Yohai (1984) give theoretical support to these results. They show that given a GM-estimate for the AR(1) model, there exists a RA-estimate with the same asymptotic variance under a Gaussian model but with smaller supremum of a conveniently defined influence curve, which may be considered an extension of Hampel's (1974) influence curve for i.i.d. observations.

Bustos and Yohai (1983) give a heuristic derivation of the covariance matrix of the asymptotical normal distribution of the RA-estimates, but no formal proof is given. This asymptotic covariance matrix is the same as that of the LS-estimate except in a scalar factor independent of the ARMA(p,q) model. As a result of this, the tuning of the constants which define the estimate for obtaining a given degree of relative efficiency with respect to the LS-estimate under normal innovations, is independent of p and q. This property is not shared by the GM-estimates.

In this paper we prove the consistency (Secion 3) and the asymptotic normality (Section 4) of a class of estimates which include the RA-estimates. We call this class General RA-estimates (GRA-estimates). In Section 2 we give some notation and introduce the GRA-estimates.

2. General RA-estimates.

Given $\underset{\sim}{\beta} = (\beta_1,\ldots,\beta_h) \in R^h$, $\underset{\sim}{\beta}(B)$ denotes the polynomial operator $\underset{\sim}{\beta}(B) = 1 - \beta_1 B - \ldots - \beta_h B^h$, where 1 is the identity operator and B the backward shift operator, i.e., $Bz_t = z_{t-1}$. Therefore (1.1) may be written as

$$(2.1) \qquad \underset{\sim}{\varphi}_o(B)(z_t - \mu_o) = \underset{\sim}{\theta}_o(B)U_t ,$$

where $\underset{\sim}{\varphi}_o = (\varphi_{o1},\dots,\varphi_{op})$, $\underset{\sim}{\theta}_o = (\theta_{o1},\dots,\theta_{oq})$.

We denote by $\underset{\sim}{\alpha} = (\underset{\sim}{\varphi},\underset{\sim}{\theta})$ the set of all autoregressive moving average parameters and by $\underset{\sim}{\nu} = (\underset{\sim}{\varphi},\underset{\sim}{\theta},\mu)$ the set of all the parameters. $\underset{\sim}{\alpha}_o$ and $\underset{\sim}{\nu}_o$ denote the corresponding sets of true parameters.

Define

$$R^{*h} = \{\beta \in R^h : \beta(B) \text{ have all the roots with absolute value } > 1\}.$$

R^{*h} is a bounded and open set; we denote by $\bar{R}^{*h}$ its closure.

Since z_t is stationary $\underset{\sim}{\varphi}_o \in R^{*p}$ and since it is invertible $\underset{\sim}{\theta}_o \in R^{*q}$.

Given $\underset{\sim}{\beta} \in R^{*h}$ we can define the inverse operator $\underset{\sim}{\beta}^{-1}(B)$ (sse Anderson (1971)) by

$$\underset{\sim}{\beta}^{-1}(B) = \sum_{i=0}^{\infty} c_i(h,\underset{\sim}{\beta})B^i. \tag{2.2}$$

Given $\underset{\sim}{\beta} \in R^{*h}$, $\underset{\sim}{\delta} \in R^j$ define

$$\underset{\sim}{\beta}^{-1}(B)\underset{\sim}{\delta}(B) = \sum_{i=0}^{\infty} g_i(j,h,\underset{\sim}{\delta},\underset{\sim}{\beta})B^i. \tag{2.3}$$

Clearly

$$c_i(h,\underset{\sim}{\beta}) = g_i(j,h,\underset{\sim}{0},\underset{\sim}{\beta}) \quad \forall\ j. \tag{2.4}$$

The coefficients $g_i(j,h,\underset{\sim}{\delta},\underset{\sim}{\beta})$ can be obtained recursively by $g_o(j,h,\underset{\sim}{\delta},\underset{\sim}{\beta}) = 1$ and

$$g_i = \delta_i - \sum_{k=0}^{\max(i-1,h)} g_k\beta_{i-k}, \quad \text{with } \delta_i = 0 \text{ for } i \geq j. \tag{2.5}$$

It is easy to prove that the functions g_i are continuously differentiable for $\underset{\sim}{\beta} \in R^{*h}$. Moreover, given $C^j \subset \mathbb{R}^j$ and $C^h \subset \mathbb{R}^{*h}$, compact sets, there exist, $A > 0$, $0 < b < 1$ such that

$$\sup\{|g_i(j,h,\underset{\sim}{\delta},\underset{\sim}{\beta})|\ \underset{\sim}{\delta} \in C^j,\ \underset{\sim}{\beta} \in C^h\} \leq Ab^i \tag{2.6}$$

$$\sup\{\frac{|\partial g_i(j,h,\underset{\sim}{\delta},\underset{\sim}{\beta})|}{\partial\gamma}\ \underset{\sim}{\delta} \in C^j,\ \underset{\sim}{\beta} \in C^h\} \leq Ab^i \tag{2.7}$$

where $\gamma = \beta_1,\dots,\beta_h,\delta_1,\dots,\delta_j$.

Given $\underset{\sim}{\nu} = (\underset{\sim}{\varphi},\underset{\sim}{\theta},\mu)$, define the residuals of order k by

$$(2.8) \qquad U_t^{(k)}(\underset{\sim}{\nu}) = \sum_{i=0}^{k} g_i(p,q,\underset{\sim}{\varphi},\underset{\sim}{\theta})(z_{t-i}-\mu), \qquad 1 \le k \le \infty$$

and the estimated residuals

$$(2.9) \qquad \hat{U}_t(\underset{\sim}{\nu}) = U_t^{(t-i)}(\underset{\sim}{\nu}).$$

Bustos and Yohai (1983) define a class of estimates based on the residual autocovariances (RA-estimates). According to this proposal the estimation of $\underset{\sim}{\varphi}$, $\underset{\sim}{\theta}$, μ and of a scale parameter σ of the U_t's is obtained as follows: let $\eta: R^2 \to R$, $\psi: R \to R$ and $\pi: R \to R$ be given. Put $\underset{\sim}{\lambda} = (\underset{\sim}{\varphi},\underset{\sim}{\theta},\mu,\sigma)$, then the autocovariances $\gamma_j(\underset{\sim}{\lambda})$ are defined by

$$(2.10) \qquad \gamma_j(\underset{\sim}{\lambda}) = \sum_{t=j+1}^{T} \eta\left(\frac{\hat{U}_t(\underset{\sim}{\nu})}{\sigma}, \frac{\hat{U}_{t-j}(\underset{\sim}{\nu})}{\sigma}\right).$$

The function $\underset{\sim}{R}_T(\underset{\sim}{\lambda}) = (R_{T,1}(\underset{\sim}{\lambda}),\ldots,R_{T,p+q+2}(\underset{\sim}{\lambda}))$: $R^p \times R^{*q} \times R \times R \to R^{p+q+2}$ is defined by:

$$(2.11) \quad \begin{cases} R_{T,i}(\underset{\sim}{\lambda}) = \dfrac{1}{T-T_{o,i}} \displaystyle\sum_{j=0}^{T-T_{o,i}} \ell_j(p,\underset{\sim}{\varphi})\gamma_{i+j}(\underset{\sim}{\lambda}), & i=1,\ldots,p. \\ R_{T,p+i}(\underset{\sim}{\lambda}) = \dfrac{1}{T-T_{o,p+i}} \displaystyle\sum_{j=0}^{T-T_{o,p+i}} \ell_j(q,\theta)\gamma_{i+j}(\underset{\sim}{\lambda}), & i=1,\ldots,q. \\ R_{T,p+q+1}(\underset{\sim}{\lambda}) = \dfrac{1}{T-T_{o,p+q+1}} \displaystyle\sum_{j=T_{o,p+q+1}+1}^{T} \psi\left(\dfrac{\hat{U}_t(\underset{\sim}{\nu})}{\sigma}\right) & \\ R_{T,p+q+2}(\underset{\sim}{\lambda}) = \dfrac{1}{T-T_{o,p+q+2}} \displaystyle\sum_{j=T_{o,p+q+2}+1}^{T} \pi\left(\dfrac{\hat{U}_t(\underset{\sim}{\nu})}{\sigma}\right) & \end{cases}$$

where $T_{o,i} = i$, $1 \le i \le p$, $T_{o,p+i} = p+i$ $1 \le i \le q$, $T_{o,p+q+1} = T_{o,p+q+2} = 0$. The RA-estimates are defined as a solution of

$$(2.12) \qquad \underset{\sim}{R}_T(\underset{\sim}{\lambda}) = 0$$

using as $\ell_j(p,\underset{\sim}{\varphi})$ and $\ell_j(q,\underset{\sim}{\theta})$ the functions $c_j(p,\underset{\sim}{\varphi})$ and $c_j(q,\underset{\sim}{\theta})$ respectively. The class of GRA-estimates is defined by (2.12) too, but $\ell_j(p,\underset{\sim}{\varphi})$ and $\ell_j(q,\underset{\sim}{\theta})$ are not restricted to be $c_j(p,\underset{\sim}{\varphi})$ and $c_j(q,\underset{\sim}{\theta})$.

The class of M-estimates is defined by $\hat{\underset{\sim}{\nu}}$, $\hat{\sigma}$ which satisfy:

(i) $\hat{\underset{\sim}{\nu}}$ minimizes $\sum_{t=p+1}^{T} \rho\left(\frac{|\hat{U}_t(\underset{\sim}{\nu})|}{\hat{\sigma}}\right)$, where $\rho: [0,\infty) \to R$

(ii) $\underset{\sim}{R}_{T,p+q+2}(\hat{\underset{\sim}{\nu}},\hat{\sigma}) = 0.$

It may be shown that if $\eta(u,v) = \psi(u)v$, with $\psi = \rho'$, the corresponding RA-estimate is asymptotically equivalent to the corresponding M-estimate. In particular the LS-estimate corresponds to $\eta(u,v) = u \cdot v$. The M-estimates are not qualitatively robust, even in the case of bounded ψ. Bustos and Yohai (1983) show through a Monte Carlo study that the M-estimates are very sensitive to additive outliers. In order to improve robustness they propose to use RA-estimates with η, ψ and π bounded functions. In particular, they propose two types of η functions:

(i) Mallows type: $\eta(u,v) = \psi_1(u)\psi_2(v)$ where ψ_1 and ψ_2 are bounded and odd functions.

(ii) Hampel-Krasker type: $\eta(u,v) = \psi_3(u,v)$ where ψ_3 is a bounded and odd function.

If η, ψ, and π are bounded, the corresponding RA-estimates are qualitatively robust, (see Papantoni-Kazakos and Gray (1979), Bustos (1981), Boente, Fraiman and Yohai (1982) for a definition of qualitative robustness for dependent observations), for the $AR(p)$ model, but they are not qualitatively robust for an $ARMA(p,q)$ model with $q > 0$. However, in their Monte Carlo study, Bustos and Yohai (1983) show that even in this case, they are much less sensitive to additive outliers than the M-estimates.

3. Consistency.

In this section we show that it is possible to choose a root of (2.12) so that the resulting sequence of estimates be consistent (Theorem 3.1). We also give a criterium for choosing a consistent solution of (2.12): take the solution of (2.12) with minimum $\hat{\sigma}$. (Theorem 3.2).

We will need the following assumptions:

H1: ψ, η and π are bounded and continuously differentiable functions.

H2: ψ is odd and η is odd in each variable.

H3: $\pi(u)$ is even, non decreasing in $|u|$ and strictly increasing in a neighborhood of 0. Moreover, there exist $a > 0$, $a^* > 0$, $c > 0$ such that $\pi(0) = -a$ and $\pi(u) = a^*$ $\forall\, u \geq c$.

H4: Either (i) or (ii) is satisfied.

(i) $\dfrac{\partial\eta(u_1,u_2)}{\partial u_i}$, $i=1,2$ are bounded and $E(|U_t|) < \infty$,

(ii) $\left|\dfrac{\partial\eta(u_1,u_2)}{\partial u_1}\right| \leq Ku_2$, $\left|\dfrac{\partial\eta(u_1,u_2)}{\partial u_2}\right| \leq Ku_1$, and $E(U_t^2) < \infty$.

Remark 3.1. (i) is satisfied if η is of the Mallows type with ψ_1, ψ_2 bounded and $E(|U_t|) < \infty$; (ii) is satisfied if η is of the Hampel-Krasker type with ψ_3 bounded and $E(U_t^2) < \infty$.

H5: F is symmetric and continuous.

H6: $F(u)$ has a density $f(u)$ which is a non increasing function of $|u|$ and strictly decreasing for small u.

In Lemma 3.1 it is shown that under H3, H5, H6, and π continuous there exists a unique $\sigma_o > 0$ such that $E_F(\pi(\frac{U}{\sigma_o})) = 0$.

H7: (i) $E_F(U_{t-1}\,\eta_1(\frac{U_t}{\sigma_o}, \frac{U_{t-1}}{\sigma_o})) \neq 0$, where

$$\eta_i(u_1,u_2) = \partial\eta\,\frac{(u_1,u_2)}{\partial u_i} \qquad i = 1,2.$$

(ii) $E_F(\psi'(\frac{U}{\sigma_o})) \neq 0$.

H8: $\underset{\sim}{\varphi}_o \in R^{*p}$, $\underset{\sim}{\theta}_o \in R^{*q}$.

H9: $\underset{\sim}{\varphi}_o(B)$ and $\underset{\sim}{\theta}_o(B)$ have not common roots.

H10: $\ell_j(p,\underset{\sim}{\varphi})$ and $\ell_j(q,\underset{\sim}{\theta})$ $0 \leq j < \infty$ are continuously differentiable for $\underset{\sim}{\varphi} \in R^{*p}$ and $\underset{\sim}{\theta} \in R^{*q}$. Moreover, given $\underset{\sim}{\varphi}_o \in R^{*p}$, there exist $\epsilon > 0$, $A > 0$, $0 < b < 1$ such that

$$\sup\{|\ell_j(p,\underset{\sim}{\varphi})| : |\underset{\sim}{\varphi}-\underset{\sim}{\varphi}_o| \leq \epsilon\} \leq Ab^j \quad \forall\, j.$$

A similar condition holds for the $\ell_j(q,\underset{\sim}{\theta})$'s.

H11: Let $D(\underset{\sim}{\varphi},\underset{\sim}{\theta})$ be the $(p+q)\times(p+q)$ matrix defined by

$$D_{i,j}(\underset{\sim}{\varphi},\underset{\sim}{\theta}) = \sum_{h=0}^{\infty} \ell_h(p,\underset{\sim}{\varphi}) c_{h+j-i}(p,\underset{\sim}{\varphi}) \qquad 1 \le i \le j \le p$$

$$D_{i,j}(\underset{\sim}{\varphi},\underset{\sim}{\theta}) = \sum_{h=0}^{\infty} \ell_{h+i-j}(p,\underset{\sim}{\varphi}) c_h(p,\underset{\sim}{\varphi}) \qquad 1 \le j < i \le p$$

$$D_{p+i,p+j}(\underset{\sim}{\varphi},\underset{\sim}{\theta}) = \sum_{h=0}^{\infty} \ell_h(q,\underset{\sim}{\theta}) c_{h+j-i}(q,\underset{\sim}{\theta}) \qquad 1 \le i \le j \le q$$

$$D_{p+i,p+j}(\underset{\sim}{\varphi},\underset{\sim}{\theta}) = \sum_{h=0}^{\infty} \ell_{h+i-j}(q,\underset{\sim}{\theta}) c_h(q,\underset{\sim}{\theta}) \qquad 1 \le j < i \le q$$

$$D_{p+i,j}(\underset{\sim}{\varphi},\underset{\sim}{\theta}) = \sum_{h=0}^{\infty} \ell_h(q,\underset{\sim}{\theta}) c_{h+j-i}(p,\underset{\sim}{\varphi}) \qquad 1\le i\le q,\ 1\le j\le p,\ i\le j$$

$$D_{p+i,j}(\underset{\sim}{\varphi},\underset{\sim}{\theta}) = \sum_{h=0}^{\infty} \ell_{h+i-j}(q,\underset{\sim}{\theta}) c_h(p,\underset{\sim}{\varphi}) \qquad 1\le i\le q,\ 1\le j\le p,\ j<i$$

$$D_{i,p+j}(\underset{\sim}{\varphi},\underset{\sim}{\theta}) = \sum_{h=0}^{\infty} \ell_h(p,\underset{\sim}{\varphi}) c_{h+j-i}(p,\underset{\sim}{\theta}) \qquad 1\le i\le p,\ 1\le j\le q,\ i\le j$$

$$D_{i,p+j}(\underset{\sim}{\varphi},\underset{\sim}{\theta}) = \sum_{h=0}^{\infty} \ell_{h+i-j}(p,\underset{\sim}{\varphi}) c_h(q,\underset{\sim}{\theta}) \qquad 1\le i\le p,\ 1\le j\le q,\ j<i.$$

Then $D(\underset{\sim}{\varphi}_o,\underset{\sim}{\theta}_o)$ is non singular.

<u>Remark 3.2</u>. Suppose that H9 and H10 hold, and (as in the case of the RA-estimates) $\ell_j(p,\underset{\sim}{\varphi}) = c_j(p,\underset{\sim}{\varphi})$ and $\ell_j(q,\underset{\sim}{\theta}) = c_j(q,\underset{\sim}{\theta})$, then H11 holds. Effectively, in this case the matrix $D(\underset{\sim}{\varphi}_o,\underset{\sim}{\theta}_o)$ is the convariance matrix of $z_t, z_{t-1},\ldots,z_{t-p+1}, z_t^*, z_{t-1}^*,\ldots,z_{t-q+1}^*$ where $\underset{\sim}{\varphi}_o(B)z_t = U_t$, $\underset{\sim}{\theta}_o(B)z_t^* = U_t$ where the U_t's are i.i.d. $N(0,1)$ variables. Then, if we assume that $D(\underset{\sim}{\varphi}_o,\underset{\sim}{\theta}_o)$ is singular, there exists polinomials $\underset{\sim}{s}(B) = s_o + s_1B +\ldots+ s_{p-1}B^{p-1}$, $\underset{\sim}{r}(B) = r_o + r_1B +\ldots+ r_{q-1}B^{q-1}$ at least one of them non null such that

$$\underset{\sim}{s}(B)z_t = \underset{\sim}{r}(B)z_t^*.$$

Multiplying both sides by $\underset{\sim}{\varphi}_o(B)\underset{\sim}{\theta}_o(B)$ we get

$$\underset{\sim}{\theta}_o(B)\underset{\sim}{s}(B)U_t = \underset{\sim}{\varphi}_o(B)\underset{\sim}{r}(B)U_t.$$

This implies $\underset{\sim}{\theta}_o(B)\underset{\sim}{s}(B) = \underset{\sim}{\varphi}_o(B)\underset{\sim}{r}(B)$, and therefore both polinomials $\underset{\sim}{s}(B)$ and $\underset{\sim}{r}(B)$ should be different from 0. Suppose $p \ge q$. By H9 all the roots of $\underset{\sim}{\varphi}_o$ should also be roots of $\underset{\sim}{s}(B)$. But this is impossible since the degree of $\underset{\sim}{s}(B)$ is $q-1 < p$. Therefore $D(\underset{\sim}{\varphi}_o, \underset{\sim}{\theta}_o)$ is non singular.

Define $\underset{\sim}{\lambda}_o = (\underset{\sim}{\nu}_o, \sigma_o)$, then the following theorem guarantees the existence of a sequence of solutions of (2.12) which converges to $\underset{\sim}{\lambda}_o$.

<u>THEOREM 3.1</u>. Suppose that H1 to H11 hold, then almost surely there exists T_o such that for $T > T_o$ it is possible to choose a sequence $\hat{\underset{\sim}{\lambda}}_T$ of solutions of (2.12) which converges to $\underset{\sim}{\lambda}_o$.

The following theorem shows how to choose a consistent solution of (2.12).

<u>THEOREM 3.2</u>. Suppose that H1 to H11 hold and let C^q be a compact set included in R^{*q}. If $\tilde{\underset{\sim}{\lambda}}_T = (\tilde{\underset{\sim}{\varphi}}_T, \tilde{\underset{\sim}{\theta}}_T, \tilde{\mu}_T, \tilde{\sigma}_T)$ is the solution of (2.12) with $\tilde{\underset{\sim}{\varphi}}_T \in R^{*p}$ and $\tilde{\underset{\sim}{\theta}}_T \in C^q$ which has minimum σ, then $\tilde{\underset{\sim}{\lambda}}_T$ converges almost surely to $\underset{\sim}{\lambda}_o$.

<u>Remark 3.3</u>. For $q > 0$, the requirement that $\tilde{\underset{\sim}{\theta}}_T$ belongs to a compact is a serious shortcoming of this theorem. We conjecture that the theorem holds even without this restriction, but we have not been able to prove it.

<u>Remark 3.4</u>. We do not have a numerical procedure for finding the solution of (2.12) with the smallest σ. An heuristic possible approach will be to use an iterative algorithm, like Newton-Raphson or the one presented in Bustos and Yohai (1983), starting from different initial points and choosing among all the solutions found in this way the one with minimum σ.

Before proving Theorem 3.1 and Theorem 3.2 we need to prove some lemmas.

<u>LEMMA 3.1</u>. Assume H3, H5, H6, and π continuous. Define for $c \in R$ and $\sigma \in R^+ = \{\sigma \in R, \sigma > 0\}$

$$h(c,\sigma) = E_F\left(\pi\left(\frac{U+c}{\sigma}\right)\right). \tag{3.1}$$

Then we have:

(i) $h(c,\sigma)$ is continuous and non increasing in σ.

(ii) Given $\sigma_1 > 0$, $\sigma_2 > \sigma_1$, there exists $\varepsilon(\sigma_1,\sigma_2) > 0$

such that if $|c| < \varepsilon(\sigma_1,\sigma_2)$, $h(c,\sigma)$ is strictly decreasing in σ $\forall \sigma \in [\sigma_1,\sigma_2]$.

(iii) There exists $\varepsilon > 0$ such that for $|c| \leq \varepsilon$, there exists a unique $\sigma(c) > 0$ such that $h(c,\sigma(c)) = 0$.

(iv) For any $\sigma > 0$ $h(c,\sigma)$ takes a unique minimum at $c=0$.

<u>Proof</u>. (i). Follows from the fact that $\pi(u)$ is non decreasing in $|u|$ and bounded.

(ii). Follows from the fact that $\pi(u)$ is strictly increasing in $|u|$ for $|u| \leq \varepsilon$ and the fact that $F(u)-F(-u) > 0$ $\forall u > 0$ (from H6).

(iii). By the Lebesgue dominated convergence theorem and H3 we have

$$\lim_{\sigma\to 0} h(c,\sigma) = a^* \qquad \lim_{\sigma\to\infty} h(c,\sigma) = -a, \tag{3.2}$$

uniformly in a neighborhood of $c = 0$. Since h is continuous, there exists σ such that $h(c,\sigma) = 0$. We will show that there exists $\varepsilon > 0$ such that if $|c| \leq \varepsilon$, this σ is unique. Suppose that this is not true, then there exists $c_i \to 0$ and $\sigma_i < \sigma_i'$ such that

$$h(c_i,\sigma_i) = h(c_i,\sigma_i') = 0.$$

Because of (3.2) σ_i and σ_i' can neither converge to 0 nor to ∞. Therefore we can assume that $\sigma_i \to \sigma_o > 0$ and $\sigma_i' \to \sigma_o' \geq \sigma_o$. Then for any $\delta > 0$, if we take as $\sigma_1 = \sigma_o - \delta$ and $\sigma_2 = \sigma_o + \delta$, (ii) is contradicted.

(iv). Take U with distribution F, and put $V_c = \frac{|U-c|}{\sigma}$. Then H6 implies that the distribution function $F_c(u)$ of V_c satisfies $F_c(u) < F_o(u)$ $\forall u \geq 0$. Then $V_c \overset{s}{>} V_o$ and therefore $E_{F_c}(\pi(V)) > E_{F_O}(\pi(V))$. ■

<u>LEMMA 3.2</u>. Assume $E([\ln |U_t|]^+) < \infty$ where $[\ ]^+$ denotes positive part, then:

(i) $z_t = \mu_o + \sum_{i=0}^{\infty} g_i(p,q,\underset{\sim}{\varphi}_o,\underset{\sim}{\theta}_o)U_{t-i}$ a.s.

(ii) $U_t^{(\infty)}(\underset{\sim}{\nu}_o) = U_t$ a.s.

(iii) Let $C \subset R^p \times R^{*q} \times R$ be a compact set, then

$$\lim_{k\to\infty} \sup\{|U_t^{(k)}(\underset{\sim}{\nu})-U_t^{(\infty)}(\underset{\sim}{\nu})|: \underset{\sim}{\nu} \in C\} = 0 \quad \text{a.s.}$$

(iv) Let $V_t(\underset{\sim}{\nu},\epsilon) = \sup\{|U_t^{(\infty)}(\underset{\sim}{\nu}^*)-U_t^{(\infty)}(\underset{\sim}{\nu})|: |\underset{\sim}{\nu}^*-\underset{\sim}{\nu}| \le \epsilon\}$ and $x_t(\epsilon) = \sup\{V_t(\underset{\sim}{\nu},\epsilon): \underset{\sim}{\nu} \in C\}$, then

$$\lim_{\epsilon\to 0} x_t(\epsilon) = 0 \quad \text{a.s.}$$

<u>Proof</u>. The proof of (i) is similar to Lemma 1 of Yohai and Maronna (1977), and (ii) is proved similarly.

(iii). According to (2.6) we have

$$\sup\{|U_t^{(k)}(\underset{\sim}{\nu})-U_t^{(\infty)}(\underset{\sim}{\nu})|: \underset{\sim}{\nu} \in C\} \le A \sum_{i=k}^{\infty} b^i|z_{t-i}| + \frac{Ab^k}{1-b}|\mu|,$$

therefore by (i) and (2.6) there exist $A^* > 0$, $0 < b^* < 1$, $D^* > 0$ such that

$$\sup\{|U_t^{(k)}(\underset{\sim}{\nu})-U_t^{(\infty)}(\underset{\sim}{\nu})|: \underset{\sim}{\nu} \in C\} \le A^* \sum_{i=k}^{\infty} b^{*i}|U_{t-i}| + \frac{D^*b^k}{1-b}$$

Yohai and Maronna (1977), while proving Lemma 1, show

$$(3.3) \qquad \sum_{i=0}^{\infty} b^{*i}|U_{t-i}| < \infty \quad \text{a.s.}$$

Then (iii) follows.

(iv). We have

$$|U_t^{(\infty)}(\underset{\sim}{\nu}^*)-U_t^{(\infty)}(\underset{\sim}{\nu})| \le \sum_{j=0}^{\infty} |g_j(p,q,\underset{\sim}{\varphi}^*,\underset{\sim}{\theta}^*)-g_j(p,q,\underset{\sim}{\varphi},\underset{\sim}{\theta})||z_{t-j}| +$$

$$+ \sum_{j=0}^{\infty} |g_j(p,q,\underset{\sim}{\varphi}^*,\underset{\sim}{\theta}^*)-g_j(p,q,\underset{\sim}{\varphi},\underset{\sim}{\theta})||\mu^*|+|\sum_{j=0}^{\infty} g_j(p,q,\varphi,\theta)||\mu^*-\mu|.$$

Given $i \ge 1$ and $\delta > 0$, there exists $\epsilon(i,\delta)$ such that

$$\sup\{|g_j(p,q,\underset{\sim}{\varphi}^*,\underset{\sim}{\theta}^*)-g_j(p,q,\underset{\sim}{\varphi},\underset{\sim}{\theta})|: (\underset{\sim}{\varphi}^*,\underset{\sim}{\theta}^*) \in C^*, \ (\underset{\sim}{\varphi},\underset{\sim}{\theta}) \in C^*,$$
$$|\underset{\sim}{\varphi}^*-\underset{\sim}{\varphi}| \le \epsilon(i,\delta), \ |\underset{\sim}{\theta}^*-\underset{\sim}{\theta}| \le \epsilon(i,\delta)\} \le \delta b^j \quad \forall\ j \le i,$$

where C^* is the projection of C on $R^p \times R^{*q}$. Therefore we have

$$x_t(\epsilon(i,\delta)) \le \delta \sum_{j=0}^{i} b^j|z_{t-j}| + 2A \sum_{j=i+1}^{\infty} b^j|z_{t-j}| +$$
$$+ (K+1)\left(\frac{2Ab^{i+1}+\delta}{1-b}\right) + A\,\frac{b}{1-b}\,\epsilon(i,\delta).$$

where $K = \sup\{\mu: (\underset{\sim}{\varphi},\underset{\sim}{\theta},\mu) \in C\}$. Since as in (iii) $\sum_{j=0}^{\infty} b^j|z_{t-j}| < \infty$ a.s. we have

$$\limsup_{\epsilon\to 0} x_t(\epsilon) \le \lim_{\delta\to 0} \sup_{i\to\infty} x_t(\epsilon(i,\delta)) = 0,$$

therefore (iv) follows. ■

<u>LEMMA 3.3</u>. Let Z_t, $1 \le t < \infty$ be an stationary and ergodic process and let $\underset{\sim}{Z}_t = (Z_t, Z_{t-1}, \ldots)$. Suppose that for any $\lambda \in \Lambda$, where (Λ, d) is a metric space, $1 \le k \le \infty$, $t > T_o$ it is defined $U_t^{(k)}(\lambda) = h(\lambda, k, \underset{\sim}{Z}_t) \in R^h$. Let $W: R^h \to R$ and $r: R \to R$. Put $Y_t^{(k)}(\lambda) = W(U_t^{(k)}(\lambda)) + r(E(W(U_t^{(\infty)}(\lambda))))$, for $1 \le k \le \infty$. Assume

(i) W and r are continuous.

(ii) Let $s: N \to N$ be such that $\lim_{k\to\infty} s(k) = \infty$.

(iii) $\sup\{|Y_t^{(k)}(\lambda)| : \lambda \in \Lambda \ \ 1 \le k \le \infty\} \le Y_t^*$ a.s., where $E(Y_t^*) < \infty$.

(iv) $\lim_{\lambda^* \to \lambda} U_t^{(\infty)}(\lambda^*) = U_t^{(\infty)}(\lambda)$ a.s.

(v) $\forall \lambda \in \Lambda \ \ \exists \delta > 0$ such that

$$\lim_{k\to\infty} \sup\{|U_t^{(k)}(\lambda^*) - U_t^{(\infty)}(\lambda^*)| : d(\lambda, \lambda^*) \le \delta\} = 0 \quad \text{a.s.}$$

(vi) (a) $E(Y_t^{(\infty)}(\lambda)) > (\ge)\ A \quad \forall \lambda \in \Lambda$, or

(b) $E(Y_t^{(\infty)}(\lambda)) < (\le)\ A \quad \forall \lambda \in \Lambda$.

(vii) There exists an increasing sequence of compact sets $C_n \subset \Lambda$ such that $\bigcup_{n=1}^{\infty} C_n = \Lambda$ and such that

(a) $\underline{\lim}_{n\to\infty, k\to\infty} \inf\{Y_t^{(k^*)}(\lambda) : \lambda \in C_n', k^* \ge k\} > (\ge)\ A$, or

(b) $\overline{\lim}_{n\to\infty, k\to\infty} \sup\{Y_r^{(k^*)}(\lambda) : \lambda \in C_n', k^* \ge k\} < (\le)\ A$

where "′" denotes complement.

Put $Z_T(\lambda) = \frac{1}{T-T_o} \sum_{t=T_o+1}^{T} Y_t^{s(t)}(\lambda)$.

Then under (i)-(v), (vi) (a) and (vii) (a) we have

$$\underline{\lim}_{T\to\infty} \inf\{Z_T(\lambda) : \lambda \in \Lambda\} > (\ge)\ A \quad \text{a.s.,} \tag{3.4}$$

and under (i)-(v), (vi) (b) and (vii) (b) we have

$$\overline{\lim}_{T\to\infty} \sup\{Z_T(\lambda) : \lambda \in \Lambda\} < (\le)\ A \quad \text{a.s.} . \tag{3.5}$$

<u>Proof</u>. Suppose that (vi) (a) and (vii) (a) hold with $>$. Then by (i), (iii), (vii) (a) and the Lebesgue dominated convergence theorem, we can find a compact C, k_o and

$\varepsilon_o > 0$ such that

$$E(\inf\{Y_t^{(k)}(\lambda) : \lambda \in C', k \geq k_o\}) \geq A + \varepsilon_o.$$

Using (iii), (v), (vi) (a), the Lebesgue dominated convergence theorem and a standard compacity argument we can find sets $C_1,\ldots,C_n$, $\varepsilon_1 > 0,\ldots,\varepsilon_n > 0$, $k_1,\ldots,k_n$ such that $\bigcup_{i=1}^{n} C_i \supset C$ and

$$E(\inf\{Y_t^{(k)}(\lambda) : \lambda \in C_i, k \geq k_i\}) \geq A + \varepsilon_i.$$

Put $Y_{t,i}^* = \inf\{Y_t^{(k)}(\lambda) : \lambda \in C_i, k \geq k_i\}$ and

$$Z_{T,i}^* = \frac{1}{T-T_o} \sum_{t=T_o+1}^{T} Y_{t,i}^*,$$

then by (ii) it is easy to check that

$$\underline{\lim}_{T\to\infty} \inf\{Z_T(\lambda) : \lambda \in \Lambda\} \geq \min_{1\leq i\leq n} Z_{T,i}^*.$$

But by the ergodic theorem $\underline{\lim}_{T\to\infty} Z_{T,i}^* \geq A + \varepsilon_i$ a.s., and therefore (3.4) holds.

If we have the $\geq$ sign in (vi) (a) and (vii) (a), (3.4) should hold with the sign $>$ and $A = A-\varepsilon$ $\forall\, \varepsilon > 0$, then (3.4) holds with the $\geq$ sign. Finally, we get (3.5) in a similar way using (vi) (b) and (vii) (b). ■

<u>LEMMA 3.4</u>. Let $w: R^4 \to R$ be a continuous function satisfying

$$w(u,v,x,y) \leq Kx^m y^m, \tag{3.6}$$

where (m,n) may be $(0,0)$, $(1,0)$ or $(1,1)$. Put $s = m+n$ and assume $E(|U_t|^s) < \infty$ if $s > 0$ and $E([\ln |U_t|]^+) < \infty$ if $s = 0$. Assume $\underset{\sim}{\lambda}_o = (\underset{\sim}{\varphi}_o,\underset{\sim}{\theta}_o,\mu_o,\sigma_o)$ with $\underset{\sim}{\theta}_o \in R^{*q}$, $\sigma_o > 0$. Then there exists ε_o such that

(i) (3.7) $\lim_{T\to\infty} \sup\{|\frac{1}{T-T_o} \sum_{t=T_o+1}^{T} w(\frac{\hat{U}_t(\underset{\sim}{\nu})}{\sigma}, \frac{\hat{U}_{t-i}(\underset{\sim}{\nu})}{\sigma}, \frac{\hat{U}_{t-j}(\underset{\sim}{\nu})}{\sigma}, \frac{\hat{U}_{t-h}(\underset{\sim}{\nu})}{\sigma}) - E(w(\frac{U_t^{(\infty)}(\underset{\sim}{\nu})}{\sigma}, \frac{U_{t-i}^{(\infty)}(\underset{\sim}{\nu})}{\sigma}, \frac{U_{t-j}^{(\infty)}(\underset{\sim}{\nu})}{\sigma}, \frac{U_{t-h}^{(\infty)}(\underset{\sim}{\nu})}{\sigma}))| : |\underset{\sim}{\lambda}-\underset{\sim}{\lambda}_o| \leq \varepsilon_o\} = 0$ a.s., where $T_o = 1 + \max(i,j,h)$. (3.7) also holds if $\frac{\hat{U}_{t-j}}{\sigma}$ and $\frac{U_{t-j}^{(\infty)}}{\sigma}$ are replaced by z_{t-j}.

(ii) Assume $b_i(\underset{\sim}{\varphi},\underset{\sim}{\theta})$, $0 \leq i < \infty$ satisfying

$$(3.8)\quad \sup\{|b_i(\underset{\sim}{\varphi},\underset{\sim}{\theta})| : |\underset{\sim}{\varphi}-\underset{\sim}{\varphi}_o| \le \epsilon,\ |\underset{\sim}{\theta}-\underset{\sim}{\theta}_o| \le \epsilon\} \le A^* b^{*i}$$

for some $\epsilon > 0$, $A^* > 0$ and $b^* < 1$, then for any $j \ge 0$, $h \ge 0$, $s \ge 0$ there exists ϵ_o such that

$$(3.9)\quad \lim_{T\to\infty} \sup\{|\frac{1}{T-T_o}\sum_{t=T_o+1}^{T}\sum_{j=s}^{t-1} b_j(\underset{\sim}{\varphi},\underset{\sim}{\theta}) w(\frac{\hat{U}_t(\underset{\sim}{\nu})}{\sigma},\frac{\hat{U}_{t-i}(\underset{\sim}{\nu})}{\sigma},\frac{\hat{U}_{t-j}(\nu)}{\sigma},\frac{\hat{U}_{t-h}(\underset{\sim}{\nu})}{\sigma}) - \sum_{j=s}^{\infty} b_j(\underset{\sim}{\varphi},\underset{\sim}{\theta}) E(w(\frac{U_t^{(\infty)}(\underset{\sim}{\nu})}{\sigma},\frac{U_{t-i}^{(\infty)}(\underset{\sim}{\nu})}{\sigma},\frac{U_{t-j}^{(\infty)}(\underset{\sim}{\nu})}{\sigma},\frac{U_{t-h}^{(\infty)}(\underset{\sim}{\nu})}{\sigma}))| : |\underset{\sim}{\lambda}-\underset{\sim}{\lambda}_o| \le \epsilon_o\} = 0 \text{ a.s.},$$

where $T_o = 1 + \max(i,h,s)$. (3.9) also holds if $\frac{\hat{U}_{t-j}(\underset{\sim}{\nu})}{\sigma}$ and $\frac{U_{t-j}^{(\infty)}(\underset{\sim}{\nu})}{\sigma}$ are replaced by z_{t-j}.

(iii) Assume $b_i^{(1)}$ and $b_i^{(2)}$ satisfy (3.8). Then for any $s, s', j \ge 0$ there exists $\epsilon_o > 0$ such that

$$(3.10)\quad \lim_{T\to\infty} \sup\{|\frac{1}{T-T_o}\sum_{t=T_o+1}^{T}\sum_{j=s}^{t-1}\sum_{h=s'}^{t-1} b_j^{(1)}(\underset{\sim}{\varphi},\underset{\sim}{\theta}) b_h^{(2)}(\underset{\sim}{\varphi},\underset{\sim}{\theta}) w(\frac{\hat{U}_t(\underset{\sim}{\nu})}{\sigma},\frac{\hat{U}_{t-i}(\underset{\sim}{\nu})}{\sigma},\frac{\hat{U}_{t-j}(\underset{\sim}{\nu})}{\sigma},\frac{\hat{U}_{t-h}(\underset{\sim}{\nu})}{\sigma}) - \sum_{j=s}^{\infty}\sum_{h=s'}^{\infty} b_j^{(1)}(\underset{\sim}{\varphi},\underset{\sim}{\theta}) b_h^{(2)}(\underset{\sim}{\varphi},\underset{\sim}{\theta}) E(w(\frac{U_t^{(\infty)}(\underset{\sim}{\nu})}{\sigma},\frac{U_{t-i}^{(\infty)}(\underset{\sim}{\nu})}{\sigma},\frac{U_{t-j}^{(\infty)}(\underset{\sim}{\nu})}{\sigma},\frac{U_{t-h}^{(\infty)}(\underset{\sim}{\nu})}{\sigma}))| : |\underset{\sim}{\lambda}-\underset{\sim}{\lambda}_o| < \epsilon\} = 0 \text{ a.s.},$$

where $T_o = 1+\max(i,s,s')$. (3.10) also holds if we replace $\frac{\hat{U}_{t-j}}{\sigma}$ and $\frac{U_{t-j}^{(\infty)}}{\sigma}$ by z_{t-j}.

<u>Proof</u>. We will assume $m = n = 1$, and therefore $s = 2$. The proofs in the other cases are similar. In this case we have $E(Z_t^2) < \infty$. We can find ϵ_o such that if $\underset{\sim}{\lambda} = (\underset{\sim}{\varphi},\underset{\sim}{\theta},\mu,\sigma)$ is such that $|\underset{\sim}{\lambda}-\underset{\sim}{\lambda}_o| \le \epsilon_o$, then $\underset{\sim}{\theta} \in R^{*q}$ and $\sigma > 0$. Then by (2.6) there exist $A > 0$, $0 < b < 1$, $D > 0$ such that $\sup\{|\frac{U_t^{(k)}(\underset{\sim}{\nu})}{\sigma}| : |\underset{\sim}{\lambda}-\underset{\sim}{\lambda}_o| \le \epsilon_o,\ 1 \le k \le \infty\} \le$ $\le U_t^*$, where $U_t^* = A\sum_{i=0}^{\infty} b^i|z_{t-i}| + D$. Since $E(|z_t|^2) < \infty$ we also have $E(|U_t^*|^2) < \infty$ and therefore by (3.6)

$$(3.11)\quad \sup\{|w(\frac{U_t^{(k)}(\underset{\sim}{\nu})}{\sigma},\frac{U_{t-i}^{(k)}(\underset{\sim}{\nu})}{\sigma},\frac{U_{t-j}^{(k)}(\underset{\sim}{\nu})}{\sigma},\frac{U_{t-h}^{(k)}(\underset{\sim}{\nu})}{\sigma})| : |\underset{\sim}{\lambda}-\underset{\sim}{\lambda}_o|\le\epsilon_o,\ 1 \le k \le \infty\} \le K\, U_{t-j}^* U_{t-h}^* \text{ a.s.}\ .$$

Using the Cauchy-Schwartz inequality we have $E(|U_{t-j}^* U_{t-h}^*|) < \infty$

therefore we can apply the Lemma 3.3 putting $Z_t = z_t$,

$$U_t^{(k)}(\underset{\sim}{\lambda}) = \left(\frac{U_t^{(k)}(\underset{\sim}{\nu})}{\sigma}, \frac{U_{t-i}^{(k-i)}(\underset{\sim}{\nu})}{\sigma}, \frac{U_{t-j}^{(k-j)}(\underset{\sim}{\nu})}{\sigma}, \frac{U_{t-h}^{(k-h)}(\underset{\sim}{\nu})}{\sigma}\right),$$

$\Lambda = \{\underset{\sim}{\lambda}: |\underset{\sim}{\lambda}-\underset{\sim}{\lambda}_o| \leq \varepsilon_o\}$, which is itself compact, $h = 4$, $W = w$, $s(k) = k-1$, $r(u) = -u$. Since (vi)(a) and (vi) (b) are satisfied with $A = 0$, we get (3.4) and (3.5) which imply (i). The case with $\frac{\hat{U}_{t-j}}{\sigma}$ and $\frac{U_{t-j}^{(\infty)}}{\sigma}$ replaced by z_{t-j} is proved similarly putting $U_t^{(k)}(\lambda) = \left(\frac{U_t^{(k)}(\underset{\sim}{\nu})}{\sigma}, \frac{U_{t-i}^{(k-i)}(\underset{\sim}{\nu})}{\sigma}, z_{t-j}, \frac{U^{(k-h)}(\underset{\sim}{\nu})}{\sigma}\right)$.

According to (i), (3.8) and (3.11) in order to prove (ii) it is enough to prove

$$(3.12) \qquad \lim_{T\to\infty} \lim_{\ell\to\infty} \frac{1}{T-T_o} \sum_{t=T_o+1}^{T} U_{t-h}^* \sum_{j=\ell}^{\infty} b^{*j} U_{t-j}^* = 0 \quad \text{a.s.},$$

where $T_o = 1 + \max(j,h,\ell)$, and

$$(3.13) \qquad \lim_{\ell\to\infty} \sum_{j=\ell}^{\infty} b^{*j} E(U_{t-h}^* U_{t-j}^*) = 0.$$

Since $E(U_{t-h}^* U_{t-j}^*) \leq E(U_t^{*2}) < \infty$ (3.13) is true.

Put $x_{t,\ell} = U_{t-h}^* \sum_{j=\ell}^{\infty} b^{*j} U_{t-j}^*$ and $y_{T,\ell} = \frac{1}{T-T_o} \sum_{t=T_o+1}^{T} x_{t,\ell}$. Then, since $y_{T,\ell}$ is non increasing in ℓ, in order to prove (3.12) it is enough to show that $\lim_{T\to\infty} y_{T,\ell} = c_\ell$ a.s.,, where c_ℓ is a sequence of constants which converges to 0. But this is true, since by the ergodic theorem $\lim_{T\to\infty} y_{T,\ell} = E(x_{t,\ell})$ a.s., and $\lim_{\ell\to\infty} E(x_{t,\ell}) = 0$.

(iii) is proved similarly to (ii). ■

We need to introduce the following notation. Define for $1 \leq k,h \leq \infty$

$$(3.14)\begin{cases} r_{t,i}^{(h,k)}(\underset{\sim}{\lambda}) = \sum_{j=0}^{h} \ell_j(p,\underset{\sim}{\varphi})\eta\left(\frac{U_t^{(k)}(\underset{\sim}{\nu})}{\sigma}, \frac{U_{t-i-j}^{(k-i-j)}(\underset{\sim}{\nu})}{\sigma}\right), & i=1,\ldots,p \\ r_{t,p+i}^{(h,k)}(\underset{\sim}{\lambda}) = \sum_{j=0}^{h} \ell_j(q,\underset{\sim}{\theta})\eta\left(\frac{U_t^{(k)}(\underset{\sim}{\nu})}{\sigma}, \frac{U_{t-i-j}^{(k-i-j)}(\underset{\sim}{\nu})}{\sigma}\right), & i=1,\ldots,q \\ r_{t,p+q+1}^{(k)}(\underset{\sim}{\lambda}) = \psi\left(\frac{U_t^{(k)}(\underset{\sim}{\nu})}{\sigma}\right) \\ r_{t,p+q+2}^{(k)}(\underset{\sim}{\lambda}) = \pi\left(\frac{U_t^{(k)}(\underset{\sim}{\nu})}{\sigma}\right) \end{cases}$$

Observe that

$$(3.15) \qquad R_{T,i}(\underset{\sim}{\lambda}) = \frac{1}{T-T_{o,i}} \sum_{t=T_{o,i+1}}^{T} r_{t,i}^{(t-i-1,t-1)}(\underset{\sim}{\lambda})$$

where $T_{o,i} = i$ $1 \le i \le p$, $T_{o,p+i} = p+i$ $1 \le i \le q$, $T_{o,p+q+1} = T_{o,p+q+2} = 0$. Define also

$$(3.16)\begin{cases} r_i^*(\underset{\sim}{\lambda}) = E(r_{t,i}^{(\infty,\infty)}(\underset{\sim}{\lambda})), & 1 \le i \le p+q \\ r_i^*(\underset{\sim}{\lambda}) = E(r_{t,i}^{(\infty)}(\underset{\sim}{\lambda})), & p+q+1 \le i \le p+q+2. \end{cases}$$

The following lemma shows that $\underset{\sim}{R}_T(\underset{\sim}{\lambda})$ can be approximated by $\underset{\sim}{r}^*(\underset{\sim}{\lambda}) = (r_1^*(\underset{\sim}{\lambda}),\ldots,r_{p+q+2}^*(\underset{\sim}{\lambda}))$ in a neighborhood of $\underset{\sim}{\lambda}_o$.

<u>LEMMA 3.5</u>. Assume H1, H4, H8 and H10, then there exists $\epsilon_o > 0$ such that

(i) $\lim_{T\to\infty} \sup\{|R_{T,i}(\underset{\sim}{\lambda})-r_i^*(\underset{\sim}{\lambda})| : |\underset{\sim}{\lambda}-\underset{\sim}{\lambda}_o| \le \epsilon_o\} = 0$ a.s. $1 \le i \le p+q+2$.

(ii) $\lim_{T\to\infty} \sup\{|\frac{\partial R_{T,i}}{\partial\lambda_j}(\underset{\sim}{\lambda}) - \frac{\partial r_i^*(\underset{\sim}{\lambda})}{\partial\lambda_j}| : |\underset{\sim}{\lambda}-\underset{\sim}{\lambda}_o| \le \epsilon_o\} = 0$ a.s. $1 \le i \le p+q+2$, $1 \le j \le p+q+2$.

<u>Proof</u>. For $i = 1,\ldots,p+q$ (i) follows from Lemma 3.4 (ii) and for $i = p+q+1$, $p+q+2$ from Lemma 3.4 (i).

Since

$$\frac{\partial U_t^{(k)}(\underset{\sim}{\nu})}{\partial\nu_i} = \sum_{j=0}^{k} \frac{\partial g_j(p,q,\underset{\sim}{\varphi},\underset{\sim}{\theta})}{\partial\nu_i}(z_{t-j}-\mu) \qquad 1 \le i \le p+q$$

and

$$\frac{\partial U_t^{(k)}}{\partial\mu} = -\sum_{i=0}^{k} g_i(p,q,\underset{\sim}{\varphi},\underset{\sim}{\theta}) \to \frac{1-\sum_{i=1}^{p}\varphi_i}{1-\sum_{i=1}^{q}\theta_i}$$

therefore using assumptions H1, H4 and H8 and (2.7) we get (ii) by applying Lemma 3.4 (i) (ii) and (iii). ■

LEMMA 3.6. Assume H1 to H11. Let $Dr^*(\underset{\sim}{\lambda})$ be the differential matrix of $\underset{\sim}{r}^*(\underset{\sim}{\lambda})$, then $D^* = D\underset{\sim}{r}^*(\underset{\sim}{\lambda}_o)$ is non-singular and given by

$$
(3.17)\begin{cases}
D^*_{i,j} = E_F\left(\eta_1\left(\frac{U_t}{\sigma_o}, \frac{U_{t-1}}{\sigma_o}\right)\frac{U_{t-1}}{\sigma_o}\right) D_{i,j}(\underset{\sim}{\varphi}_o, \underset{\sim}{\theta}_o), & 1 \le i,j \le p+q \\
D^*_{p+q+1,j} = D^*_{j,p+q+1} = 0 & \text{if} \quad j \ne p+q+1 \\
D^*_{p+q+2,j} = D^*_{j,p+q+2} = 0 & \text{if} \quad j \ne p+q+2 \\
D^*_{p+q+1,p+q+1} = \left(1-\sum_{i=1}^{p}\varphi_i\right)\left(1-\sum_{i=1}^{q}\theta_i\right)^{-1} E_F(\psi'(U_t/\sigma_o)) \\
D^*_{p+q+2,p+q+2} = -E_F(\pi'(U_t/\sigma_o)U_t/\sigma_o^2)
\end{cases}
$$

where $D_{h,j}(\underset{\sim}{\varphi}, \underset{\sim}{\theta})$ is defined in H11.

Proof. It is easy to show that $\dfrac{\partial U_t^{(\infty)}(\underset{\sim}{\nu})}{\partial \varphi_i} = -\underset{\sim}{\varphi}^{-1}(B)U_{t-i}^{(\infty)}(\underset{\sim}{\nu})$,

$$\frac{\partial U_t^{(\infty)}(\underset{\sim}{\nu})}{\partial \theta_i} = \underset{\sim}{\theta}^{-1}(B)U_{t-i}^{(\infty)}(\underset{\sim}{\nu}), \quad \frac{\partial U_t^{(\infty)}(\underset{\sim}{\nu})}{\partial \mu} = \left(1-\sum_{i=1}^{p}\varphi_i\right)\Big/\left(1-\sum_{i=1}^{q}\theta_i\right),$$

then, since $U_t^{(\infty)}(\underset{\sim}{\nu}_o) = U_t$, (3.17) follows easily using assumptions H2, H3 and H5. Using H8, H11 and the fact that H3 and H6 imply $E(\pi'(U_t)U_t) > 0$, we get that $D^* = D\underset{\sim}{r}^*(\underset{\sim}{\lambda}_o)$ is non-singular. ■

We will use the following lemma from Functional Analysis.

LEMMA 3.7 (Ruskin (1978), The Zeros Lemma). Let $f_n, f\colon R^d \to R^d$ $n = 1,2,\ldots$ be continuously differentiable, and $\underset{\sim}{\lambda}_o \in R^j$ such that $f(\underset{\sim}{\lambda}_o) = 0$, and $Df(\underset{\sim}{\lambda}_o)$ be non-singular. Let $\gamma > 0$ so that $f_n \to f$ and $Df_n \to Df$ uniformly for $|\underset{\sim}{\lambda}-\underset{\sim}{\lambda}_o| < \gamma$. Then, there is $(\underset{\sim}{\lambda}_n)_n$ in R^d such that: $\underset{\sim}{\lambda}_n \to \underset{\sim}{\lambda}_o$, $f_n(\underset{\sim}{\lambda}_n) = 0$ for all n sufficiently large.

Proof of Theorem 3.1. It is an immediate consequence of Lemmas 3.5, 3.6 and 3.7 and the fact that $\underset{\sim}{r}^*(\underset{\sim}{\lambda}_o) = 0$. ■

Proof of Theorem 3.2. We will make the following change of

parameter. Define $\mu^+ = (\mu-\mu_o)(1-\sum_{i=1}^{p}\varphi_i)/(1-\sum_{i=1}^{q}\theta_i)$. Put $\underset{\sim}{\nu}^+ = (\underset{\sim}{\varphi},\underset{\sim}{\theta},\mu^+)$ and $\underset{\sim}{\nu}_o^+ = (\underset{\sim}{\varphi}_o,\underset{\sim}{\theta}_o,\mu_o^+)$. Since $\underset{\sim}{\varphi}_o \in R^{*p}$, $(1-\sum_{i=1}^{p}\varphi_{oi}) \neq 0$. Therefore, since Theorem 3.1 guarantees the existence of a solution of (2.12) $\hat{\underset{\sim}{\lambda}}_T^+ = (\hat{\underset{\sim}{\nu}}_T^+,\hat{\sigma}_T)$ with $\lim_{T\to\infty}\hat{\sigma}_T = \sigma_o$ a.s. and since by H3, π is non decreasing, it is enough to show that given $\delta > 0$, there exists $\sigma_1 > \sigma_o$, and $d > 0$ such that

$$\text{(3.17)}\qquad \underline{\lim}_{T\to\infty} \inf\{\frac{1}{T}\sum_{t=1}^{\infty}\pi(\frac{\hat{U}_t(\underset{\sim}{\nu}^+)}{\sigma_1}) : \underset{\sim}{\nu}^+ \in \Lambda\} \geq d \quad \text{a.s.}$$

where

$$\Lambda = \{\underset{\sim}{\nu}^+ = (\underset{\sim}{\varphi},\underset{\sim}{\theta},\mu^+) : (\underset{\sim}{\varphi},\underset{\sim}{\theta}) \in \bar{R}^{*p} \times C^q,\ |\underset{\sim}{\nu}^+-\underset{\sim}{\nu}_o^+| \geq \delta\}.$$

From Lemma 3.2 (i) we have

$$U_t^{(k)}(\underset{\sim}{\nu}) = [\underset{\sim}{\theta}^{-1}(B)\underset{\sim}{\varphi}(B)]_k\ \underset{\sim}{\varphi}_o^{-1}(B)\underset{\sim}{\theta}_o(B)U_t + [\underset{\sim}{\theta}^{-1}(B)\underset{\sim}{\varphi}(B)]_k\ (\mu-\mu_o)$$

where $[\underset{\sim}{\theta}^{-1}(B)\underset{\sim}{\varphi}(B)]_k = \sum_{j=0}^{k} g_j(p,q,\underset{\sim}{\varphi},\underset{\sim}{\theta})B^j$. Therefore since $\sum_{j=0}^{\infty} g_j(p,q,\underset{\sim}{\varphi},\underset{\sim}{\theta}) = (1-\sum_{j=1}^{p}\varphi_j)/(1-\sum_{j=1}^{q}\theta_j)$, we can write

$$\text{(3.18)}\qquad U_t^{(k)}(\underset{\sim}{\nu}^+) = U_t + \sum_{j=1}^{\infty} f_j^{(k)}(\underset{\sim}{\varphi},\underset{\sim}{\theta})U_{t-j} + d^{(k)}(\underset{\sim}{\varphi},\underset{\sim}{\theta})\mu^+$$

where $f_j^{(k)}(\underset{\sim}{\varphi},\underset{\sim}{\theta})$ are the coefficients of $[\underset{\sim}{\theta}^{-1}(B)\underset{\sim}{\varphi}(B)]_k\ \underset{\sim}{\varphi}_o^{-1}(B)\underset{\sim}{\theta}_o(B)$ and therefore by (2.6) it is easy to show that there exist $A > 0$, $0 < b < 1$ such that

$$\text{(3.19)}\qquad \sup\{|f_j^{(k)}(\underset{\sim}{\varphi},\underset{\sim}{\theta})| : (\underset{\sim}{\varphi},\underset{\sim}{\theta}) \in \bar{R}^{*p}\times C^q,\ 1 \leq k \leq \infty\} \leq Ab^j.$$

Moreover, (2.6) implies

$$\text{(3.20)}\qquad \lim_{k\to\infty} \sup\{|d^{(k)}(\underset{\sim}{\varphi},\underset{\sim}{\theta})-1| : (\underset{\sim}{\varphi},\underset{\sim}{\theta}) \in \bar{R}^{*p} \times C^q\} = 0.$$

Clearly we also have

$$f_j^{(\infty)}(\underset{\sim}{\varphi},\underset{\sim}{\theta}) = g_j(p+q,p+q,\underset{\sim}{\beta}(\underset{\sim}{\varphi},\underset{\sim}{\theta}),\underset{\sim}{\beta}^*(\underset{\sim}{\varphi},\underset{\sim}{\theta})),$$

where $\underset{\sim}{\beta}(\underset{\sim}{\varphi},\underset{\sim}{\theta})$ and $\underset{\sim}{\beta}^*(\underset{\sim}{\varphi},\underset{\sim}{\theta})$ are defined by

$$\underset{\sim}{\beta}(B) = \underset{\sim}{\varphi}(B)\underset{\sim}{\theta}_o(B) \qquad \underset{\sim}{\beta}^*(B) = \underset{\sim}{\varphi}_o(B)\underset{\sim}{\theta}(B).$$

H9 implies that if $\underset{\sim}{\varphi} \neq \underset{\sim}{\varphi}_o$ or $\underset{\sim}{\theta} \neq \underset{\sim}{\theta}$, then $\underset{\sim}{\beta}(\underset{\sim}{\varphi},\underset{\sim}{\theta}) \neq$ $\neq \underset{\sim}{\beta}^*(\underset{\sim}{\varphi},\underset{\sim}{\theta})$. Therefore, using the recursive relation (2.5), it is easy to show that there exists j, $1 \leq j \leq p+q$ such that $f_j^{(\infty)}(\underset{\sim}{\varphi},\underset{\sim}{\theta}) \neq 0$. Therefore we can write for $\underset{\sim}{\nu}^+ \in \Lambda$

(3.21) $$U_t^{(\infty)}(\underset{\sim}{\nu}^+) = U_t + \sum_{j=1}^{\infty} f_j^{(\infty)}(\underset{\sim}{\varphi},\underset{\sim}{\theta}) + \mu^+ ,$$

and

(3.22) $$\sum_{j=1}^{\infty} |f_j^{(\infty)}(\underset{\sim}{\varphi},\underset{\sim}{\theta})| > 0.$$

Let V be such that $P(V\neq 0) > 0$ and let F_V be its distribution function. Then, since

$$E(\pi(\frac{U_t+V}{\sigma})) = \int_{-\infty}^{+\infty} E_F(\pi(\frac{U_t+v}{\sigma}) dF_V(v),$$

according to Lemma 3.1 (iv), we have

$$E(\pi(\frac{U_t+V}{\sigma})) > E(\pi(\frac{U_t}{\sigma})),$$

therefore, according to (3.18) and (3.19) we have $\forall\ \underset{\sim}{\nu}^+ \in \Lambda$

(3.23) $$E(\pi(\frac{U_t(\underset{\sim}{\nu}^+)}{\sigma_o})) > 0.$$

We also have for $\underset{\sim}{\nu}^+ = (\underset{\sim}{\varphi},\underset{\sim}{\theta},\mu^+) \in \Lambda$

$$E_F(\pi(\frac{U_t^{(\infty)}(\underset{\sim}{\nu}^+)}{\sigma_o})) > E_F(\pi(\frac{U_t+\mu^+}{\sigma_o}))$$

and therefore if $\underset{\sim}{\nu}_i^+ = (\underset{\sim}{\varphi}_i,\underset{\sim}{\theta}_i,\mu_i^+) \in \Lambda$ with $|\mu_i| \to \infty$, by H3 and the Lebesgue dominated convergence theorem, we have

(3.24) $$\lim_{i\to\infty} E(\pi(\frac{U_t^{(\infty)}(\underset{\sim}{\nu}_i^+)}{\sigma_o}))) = a^* > 0.$$

We are going to prove now that there exists $\sigma_1 > \sigma_o$ and $d > 0$ such that

(3.25) $$\inf\{E(\pi(\frac{U_t^{(\infty)}(\underset{\sim}{\nu}^+)}{\sigma_1})) : \underset{\sim}{\nu}^+ \in \Lambda\} \geq d.$$

Suppose that (3.25) is not true, then there exists a sequence $(\underset{\sim}{\nu}_i^+,\sigma_i,b_i)$ such that $\underset{\sim}{\nu}_i^+ \in \Lambda$, and

(3.26) $$E_F(\pi(\frac{U_t^{(\infty)}(\underset{\sim}{\nu}_i^+)}{\sigma_i})) = b_i$$

and

(3.27) $$\lim_{i\to\infty} \sigma_i = \sigma_o , \qquad \lim_{i\to\infty} b_i = 0.$$

We can assume also that $\underset{\sim}{\nu}_i^+ = (\underset{\sim}{\varphi}_i,\underset{\sim}{\theta}_i,\mu_i^+) \to \underset{\sim}{\nu}^{+*} = (\underset{\sim}{\varphi}^*,\underset{\sim}{\nu}^*,\mu^{+*})$ where μ^{+*} may be eventually $\pm\infty$. If $\mu^{+*} \neq \infty$, by (3.26),

(3.27) and the Lebesgue dominated convergence theorem, we have $E(\pi(\frac{U_t^{(\infty)}(\underset{\sim}{\nu}^{++*})}{\sigma_o})) = 0$ contradicting (3.23). If $\mu^{++*} = \pm\infty$ we will contradict (3.24). Then, there exists $\sigma_1 > \sigma_o$ such that (3.25) holds for all $\underset{\sim}{\nu}^+ \in \Lambda$.

Let $C_m = \{\underset{\sim}{\nu}^+ = (\underset{\sim}{\varphi},\underset{\sim}{\theta},\mu^+) \in \Lambda,\ \mu^+ \le n\}$. Then by (3.18), (3.19) and (3.20) we have

$$\lim_{n\to\infty,k\to\infty} \inf\{|U_t^{(k^*)}(\underset{\sim}{\nu}^+)| : \underset{\sim}{\nu}^+ \in C_n',\ k^* \ge k\} \ge$$
$$\ge \lim_{n\to\infty} |n - \sum_{i=0}^{\infty} b^i|U_{t-i}|| = \infty$$

since, as we have already shown while proving Lemma 3.4, we have $\sum_{i=0}^{\infty} b^i|U_{t-i}| < \infty$ a.s. Then, by Assumption H3 we have

$$(3.28)\quad \lim_{n\to\infty,k\to\infty} \inf\{\pi(U_t^{(k^*)}(\underset{\sim}{\nu}^+)) : \underset{\sim}{\nu}^+ \in C_n',\ k^* > k\} = a^* \text{ a.s.,} \quad a^* > 0.$$

Then, according to (3.25), (3.28) and Lemma 3.2 (iii) and (iv), the assumptions of Lemma 3.3 hold and therefore we get (3.17). ■

4. Asymptotic normality.

In this section we show the asymptotic normality of the GRA-estimates and give their asymptotic covariance matrix.

Let $C(\underset{\sim}{\varphi},\underset{\sim}{\theta})$ be the $(p+q)\times(p+q)$ matrix defined as $D(\underset{\sim}{\varphi},\underset{\sim}{\theta})$ in H11 but replacing $c(p,\underset{\sim}{\varphi})$, $c(q,\underset{\sim}{\theta})$ by $\ell(p,\underset{\sim}{\varphi})$ and $\ell(q,\underset{\sim}{\theta})$ respectively, and let C^* be the symmetric $(p+q+2)\times(p+q+2)$ matrix defined by

$$(4.1)\quad \begin{cases} c^*_{ij} = E_F(\eta^2(\frac{U_t}{\sigma_o},\frac{U_{t-1}}{\sigma_o}))c_{ij}(\underset{\sim}{\varphi}_o,\underset{\sim}{\theta}_o) & 1 \le i,j \le p+q \\ c^*_{p+q+1,i} = c^*_{p+q+2,i} = c^*_{i,p+q+1} = c^*_{i,p+q+2} = 0 & 1\le i\le p+q \\ c^*_{p+q+1,p+q+2} = c^*_{p+q+2,p+q+1} = 0 & \\ c^*_{p+q+1,p+q+1} = E_F(\psi^2(\frac{U_t}{\sigma_o})) & \\ c^*_{p+q+2,p+q+2} = E_F(\pi^2(\frac{U_t}{\sigma_o})) & \end{cases}$$

THEOREM 4.1. Assume H1-H11 and assume that $\underset{\sim}{\hat{\lambda}}_T$ is a sequence of estimates satisfying

$$(4.2) \qquad \lim_{T\to 0} \sqrt{T}\, R_T(\hat{\underset{\sim}{\lambda}}_T) \overset{(P)}{=} 0$$

and such that $\hat{\underset{\sim}{\lambda}}_T \to \underset{\sim}{\lambda}_o$ a.s. Then

$$\sqrt{T}(\hat{\underset{\sim}{\lambda}}_T - \lambda_o) \xrightarrow{d} N(\underset{\sim}{0}, G)$$

where $G = D^{*-1}\, C^*\, D^{*-1'}$, D^* is defined in (3.17) and C^* is defined by (4.1).

Remark 4.1. In the case of RA-estimates, i.e., $\ell(p,\underset{\sim}{\varphi}) = c(p,\underset{\sim}{\varphi})$ and $\ell(q,\underset{\sim}{\theta}) = c(q,\underset{\sim}{\theta})$ we have for $1 \le i,j \le p+q$

$$G_{ij} = \sigma^2 \frac{E_F\left(\eta^2\left(\frac{U_t}{\sigma_o}, \frac{U_{t-1}}{\sigma_o}\right)\right)}{E_F^2\left(\eta_1\left(\frac{U_t}{\sigma_o}, \frac{U_{t-1}}{\sigma_o}\right) U_{t-1}\right)} D^{ij}(\underset{\sim}{\varphi}_o, \underset{\sim}{\theta}_o)$$

where $D^{ij}(\underset{\sim}{\varphi}_o, \underset{\sim}{\theta}_o)$ is the (i,j) element of $D^{-1}(\underset{\sim}{\varphi}_o, \underset{\sim}{\theta}_o)$. Observe that $D^{-1}(\underset{\sim}{\varphi}_o, \underset{\sim}{\theta}_o)$ is the asymptotic covariance matrix of the LS-estimates of $(\underset{\sim}{\varphi}, \underset{\sim}{\theta})$.

Before proving theorem 4.1, we need to prove the following lemmas.

LEMMA 4.1. Let $w\colon R^2 \to R$ be a continuously differentiable function and assume that either (a) or (b) holds where,

(a) $\left|\frac{\partial w(v_1,v_2)}{\partial v_i}\right| \le k \quad i=1,2$ and $E(|U_t|) < \infty$

and

(b) $\left|\frac{\partial w(v_1,v_2)}{\partial v_1}\right| \le k|v_2|$, $\left|\frac{\partial w(v_1,v_2)}{\partial v_2}\right| \le k|v_1|$ and $E(U_t^2) < \infty$.

Let C be a compact set included in $R^p \times R^{*q} \times R \times (0,\infty)$, then

(i) $$\sup\left\{\sum_{t=j+1}^{\infty} \left|w\left(\frac{\hat{U}_t(\underset{\sim}{\nu})}{\sigma}, \frac{\hat{U}_{t-j}(\underset{\sim}{\nu})}{\sigma}\right) - w\left(\frac{U_t^{(\infty)}(\underset{\sim}{\nu})}{\sigma}, \frac{U_{t-j}^{(\infty)}(\underset{\sim}{\nu})}{\sigma}\right)\right| : \lambda \in C\right\} < \infty \text{ a.s.}$$

(ii) Let $0 \le b < 1$, and $s \ge 1$ then

$$\sup\left\{\sum_{t=s+1}^{\infty} \sum_{j=s}^{t-1} b^j \left|w\left(\frac{\hat{U}_t(\underset{\sim}{\nu})}{\sigma}, \frac{\hat{U}_{t-j}(\underset{\sim}{\nu})}{\sigma}\right) - w\left(\frac{U_t^{(\infty)}(\underset{\sim}{\nu})}{\sigma}, \frac{U_{t-j}^{(\infty)}(\underset{\sim}{\nu})}{\sigma}\right)\right| : \lambda \in C\right\} < \infty \text{ a.s.}$$

Proof. We will assume (b), the proof when (a) holds is similar. By (2.16) there exist $A^* \ge 0$, $0 < b^* < 1$ such that

$$\sup\{|U_t^{(k)}(\underset{\sim}{\nu}) - U_t^{(\infty)}(\underset{\sim}{\nu})| : \underset{\sim}{\nu} \in C_1 \quad 1 \le k \le \infty\} \le b^{*k+1}\, U_{t-k-1}^* \text{ a.s. and}$$

$$\sup\{|U_t^{(k)}(\underset{\sim}{\nu})|\ \underset{\sim}{\nu} \in C_1, \quad 1 \le k \le \infty\} \le U_t^* \quad \text{a.s.,}$$

where $U_t^* = A \sum_{i=0}^{\infty} b^{*i}|U_{t-i}|$, and C_1 is the projection of C on the first p+q+1 coordinates. Clearly since $E(U_t^2) < \infty$ we also have $E(U_t^{*2}) < \infty$. Then using (b) and the Mean Value Theorem we have

$$(4.3)\quad \sup\{|w(\frac{\hat{U}_t(\underset{\sim}{\nu})}{\sigma},\frac{\hat{U}_{t-j}(\underset{\sim}{\nu})}{\sigma}) - w(\frac{U_t^{(\infty)}(\underset{\sim}{\nu})}{\sigma},\frac{U_{t-j}^{(\infty)}(\underset{\sim}{\nu})}{\sigma})| : \underset{\sim}{\lambda} \in C\} \le$$

$$\le kA(b^{*j}\, U_{t-j}^* U_t^* + b^{*t-j}\, U_t^* U_{t-j}^*) \le 2kA\, b^{*t-j}\, U_t^* U_{t-j}^* .$$

Therefore,

$$\sup\{\sum_{t=j+1}^{\infty} |w(\frac{\hat{U}_t(\underset{\sim}{\nu})}{\sigma},\frac{\hat{U}_{t-j}(\underset{\sim}{\nu})}{\sigma}) - w(\frac{U_t^{(\infty)}(\underset{\sim}{\nu})}{\sigma},\frac{U_{t-j}^{(\infty)}(\underset{\sim}{\nu})}{\sigma})| : \lambda \in C\} \le$$

$$\le 2kA \sum_{t=1}^{\infty} b^{*t-j}\, U_t^* U_{t-j}^* .$$

Since $E(U_t^{*2}) < \infty$, we have $E(U_t^* U_{t-j}^*) < \infty$ and therefore $\sum_{t=1}^{\infty} b^{*t-j}\, U_t^* U_{t-j}^* < \infty$ a.s. and (i) follows.

Let $\bar{b} = \max(b,b^*)$, $w_t = (1/t)\sum_{i=1}^{t} U_i^*$, then by the ergodic theorem $W = \sup w_t$ is bounded a.s.. Then by (4.3) we have

$$\sup\{\sum_{t=s+1}^{\infty} \sum_{j=s}^{t-1} b^j |w(\frac{\hat{U}_t(\underset{\sim}{\nu})}{\sigma},\frac{\hat{U}_{t-j}(\underset{\sim}{\nu})}{\sigma}) - w(\frac{U_t^{(\infty)}(\underset{\sim}{\nu})}{\sigma},\frac{U_{t-j}^{(\infty)}(\underset{\sim}{\nu})}{\sigma})| : \lambda \in C\} \le$$

$$\le 2k \sum_{t=s+1}^{\infty} \sum_{j=s}^{t-1} b^j b^{*t-j}\, U_t^* U_{t-j}^* \le 2kW \sum_{t=1}^{\infty} t\bar{b}^{t}\, U_t^* \quad \text{a.s.} .$$

Since $E(U_t^*) < \infty$ and $\sum_{t=1}^{\infty} t\bar{b}^{t} < \infty$, (ii) follows. ■

Define now for $0 \le h \le \infty$

$$(4.4)\quad R_{T,i}^{(h,\infty)}(\underset{\sim}{\lambda}) = \frac{1}{T-T_o} \sum_{t=T_o+1}^{T} r_{t,i}^{(h,\infty)}(\underset{\sim}{\lambda}) \qquad 1 \le i \le p+q,$$

where the $r_{t,i}^{(h,\infty)}$'s are defined in (3.14), $T_o = \max(p,q)$ and

$$(4.5)\quad R_{T,i}^{(h,\infty)}(\underset{\sim}{\lambda}) = \frac{1}{T-T_o} \sum_{t=T_o+1}^{T} r_{t,i}^{(\infty)}(\underset{\sim}{\lambda}) \qquad p+q+1 \le i \le p+q+2,$$

and put $\underset{\sim}{R}_T^{(h,\infty)}(\underset{\sim}{\lambda}) = (R_{T,1}^{(h,\infty)}(\underset{\sim}{\lambda}),\ldots,R_{T,p+q+2}^{(h,\infty)}(\underset{\sim}{\lambda}))$.

<u>LEMMA 4.2</u>. Assume H1, H4 and H10 and suppose that $\hat{\lambda}_T$ satisfies (4.2), and $\lim_{T\to\infty} \hat{\lambda}_T = 0$ a.s., then

$$(4.6)\qquad \lim_{T\to\infty} \sqrt{T}\, \underset{\sim}{R}^{(\infty,\infty)}(\underset{\sim}{\hat{\lambda}}_T) \overset{(P)}{=} 0.$$

<u>Proof</u>. Since $\hat{\underset{\sim}{\lambda}}_T$ satisfies (4.2) in order to prove (4.6) it is enough to show that for $1 \le i \le p+q+2$

$$(4.7) \qquad \lim_{T\to\infty} \sqrt{T}(R_{T,i}(\hat{\underset{\sim}{\lambda}}_T) - R_{T,i}^{(\infty,\infty)}(\hat{\underset{\sim}{\lambda}}_T)) \overset{(P)}{=} 0.$$

We will show (4.7) for $1 \le i \le p$, the proofs for the other cases are similar. Using (3.15) we get

$$(4.8) \quad \sqrt{T}|R_{T,i}(\hat{\underset{\sim}{\lambda}}_T)-R_{T,i}^{(\infty,\infty)}(\underset{\sim}{\lambda}_o)| \le \frac{(T_o-T_{i,o})\sqrt{T}}{(T-T_o)(T-T_{i,o})} |R_T(\hat{\underset{\sim}{\lambda}}_T)| +$$

$$+ \frac{\sqrt{T}}{T-T_o} | \sum_{i=T_o}^{T_{o,i}} r_{t,i}^{(t-i-1,t-1)}(\hat{\underset{\sim}{\lambda}}_T)| +$$

$$+ \frac{\sqrt{T}}{T-T_o} \sum_{t=T_o}^{\infty} |r_{t,i}^{(t-i-1,t-1)}(\hat{\underset{\sim}{\lambda}}_T)-r_{t,i}^{(t-i-1,\infty)}(\hat{\underset{\sim}{\lambda}}_T)| +$$

$$+ \frac{\sqrt{T}}{T-T_o} \sum_{t=T_o}^{\infty} |r_{t,i}^{(t-i-1,\infty)}(\hat{\underset{\sim}{\lambda}}_T)-r_{t,i}^{(\infty,\infty)}(\hat{\underset{\sim}{\lambda}}_T)|$$

$$= \frac{\sqrt{T}}{T-T_o}(A_T(\hat{\underset{\sim}{\lambda}}_T) + B_T(\hat{\underset{\sim}{\lambda}}_T) + C_T(\hat{\underset{\sim}{\lambda}}_T) + D_T(\hat{\underset{\sim}{\lambda}}_T)).$$

By (4.2) we have

$$(4.9) \qquad \lim_{T\to\infty} \frac{\sqrt{T}}{T-T_o} A_T(\hat{\underset{\sim}{\lambda}}_T) \overset{(P)}{=} 0.$$

Since η is assumed bounded we have $\lim_{T\to\infty} \frac{\sqrt{T}}{T-T_o} B_T(\hat{\underset{\sim}{\lambda}}_T) = 0$. Since $\hat{\underset{\sim}{\lambda}}_T$ converges to $\underset{\sim}{\lambda}_o$ a.s., by Lemma 4.2 (i) we have that $C_T(\hat{\underset{\sim}{\lambda}}_T)$ is bounded a.s. and therefore

$$(4.10) \qquad \lim_{T\to\infty} \frac{\sqrt{T}}{T-T_o} C_T(\hat{\underset{\sim}{\lambda}}_T) = 0 \quad \text{a.s.}$$

Let C be any compact included in $R^{*p} \times R^{*q} \times R \times (0,\infty)$, then since η is bounded, using H10 we get

$$(4.11) \qquad \sup\{ r_{t,i}^{(k,\infty)}(\underset{\sim}{\lambda}) - r_{t,i}^{(\infty,\infty)}(\underset{\sim}{\lambda})| : \underset{\sim}{\lambda} \in C\} \le \frac{Kb^{k+1}}{1-b},$$

where $0 < b < 1$ and therefore $D_T(\hat{\underset{\sim}{\lambda}}_T)$ is bounded, then we have

$$(4.12) \qquad \lim_{T\to\infty} \frac{\sqrt{T}}{T-T_o} D_T(\hat{\underset{\sim}{\lambda}}_T) = 0.$$

Using (4.8), (4.9), (4.10), (4.11) and (4.12) we get (4.7). ∎

<u>LEMMA 4.3</u>. Assume H1-H11, then

$$\sqrt{T-T_o}\, \underset{\sim}{R}_T^{(\infty,\infty)}(\underset{\sim}{\lambda}_o) \xrightarrow{d} N(\underset{\sim}{0}, C^*),$$

where C^* is given by (4.1).

<u>Proof</u>. It is enough to show that for any $\underset{\sim}{a} \in R^{p+q+2}$

$$\underset{\sim}{a}' \; R_T^{(\infty,\infty)}(\underset{\sim}{\lambda}_o) \xrightarrow{d} N(\underset{\sim}{0}, \underset{\sim}{a}' C^* \underset{\sim}{a}).$$

Using (4.11) we can show that for $1 \le i \le p+q$

$$\sup\{ |R_{T,i}^{(h,\infty)}(\underset{\sim}{\lambda}_o) - R_{T,i}^{(\infty,\infty)}(\underset{\sim}{\lambda}_o)| \; : \; 1 \le h \le \infty\} \le \frac{K}{(1-b)^2 T}$$

and therefore

$$\lim_{T\to\infty} \sup\{\sqrt{T-T_o} \; |\underset{\sim}{a}' R_T^{(h,\infty)}(\underset{\sim}{\lambda}_o) - \underset{\sim}{a}' R_T^{(\infty,\infty)}(\underset{\sim}{\lambda}_o)| \; : \; 1 \le h \le \infty\} = 0,$$

therefore by Theorem 7.71 of Anderson (1971) it is enough to prove that $\forall\, h$

$$(4.13) \qquad \sqrt{T-T_o} \; R_T^{(h,\infty)}(\underset{\sim}{\lambda}_o) \xrightarrow{d} N(0, \underset{\sim}{a}' C^{*h} \underset{\sim}{a}),$$

and $C^{*h} \to C^*$.

But since $U_t^{(\infty)}(\underset{\sim}{\lambda}_o) = U_t$, we have that $r_{t,i}^{(h,\infty)}(\underset{\sim}{\lambda}_o)$ depends only on $U_t, \ldots, U_{t-h}$, and therefore is independent of $r_{t-h-j,i}(\underset{\sim}{\lambda}_o)$ for $j \ge 1$. We also have $E_F(r_t^{(h,\infty)}(\underset{\sim}{\lambda}_o)) = 0$. Then using Theorem 7.75 of Anderson (1971) we get (4.11) with

$$(4.14) \qquad C_{i,j}^{*h} = \mathrm{cov}(r_{t,i}^{(h,\infty)}(\underset{\sim}{\lambda}_o), \; r_{t,j}^{(h,\infty)}(\underset{\sim}{\lambda}_o))$$

where $r_{t,i}^{(h,\infty)}(\underset{\sim}{\lambda}_o) = r_{t,i}^{(\infty)}(\underset{\sim}{\lambda}_o)$ for $p+q+1 \le i \le p+q+2$.

Using Assumptions H1, H2, H3 and H5, it is easy to check that if C^{*h} is given by (4.14), then $C^{*h} \to C^*$ where C^* is given by (4.1). ■

<u>Proof of Theorem 4.1</u>. By the Mean Value Theorem, we have

$$(4.15) \qquad \underset{\sim}{R}_T^{(\infty,\infty)}(\hat{\underset{\sim}{\lambda}}_T) = D \; \underset{\sim}{R}_T^{(\infty,\infty)}(\hat{\underset{\sim}{\lambda}}_T^*)(\hat{\underset{\sim}{\lambda}}_T - \underset{\sim}{\lambda}_o).$$

As in Lemma (3.5) we can prove that there exists $\varepsilon > 0$ such that

$$\lim_{T\to\infty} \sup\{ |D \; \underset{\sim}{R}_T^{(\infty,\infty)}(\underset{\sim}{\lambda}) - D \; \underset{\sim}{r}^{*(\infty)}(\underset{\sim}{\lambda})| \; : \; |\underset{\sim}{\lambda} - \underset{\sim}{\lambda}_o| < \varepsilon\} = 0 \quad \text{a.s.}$$

Therefore, since $\hat{\underset{\sim}{\lambda}}_T^* \to \underset{\sim}{\lambda}_o$ a.s. we have

$$(4.16) \qquad \lim_{T\to\infty} D \; \underset{\sim}{R}_T^{(\infty,\infty)}(\hat{\underset{\sim}{\lambda}}_T^*) = D \; \underset{\sim}{r}^{(\infty)}(\underset{\sim}{\lambda}_o) = D^*,$$

where D^* is given in (3.17). Then Theorem 3.2 is derived from Lemma 4.3, (4.15), (4.16) and the Slutzky's theorem. ■

References

Anderson, T.W. (1971). "The Statistical Analysis of Time Series" Wiley, New York.

Boente, G., Fraiman, R., and Yohai, V.J. (1982). "Qualitative Robustness for General Stochastic Processes", Technical Report No. 26, Department of Statistics, University of Washington, Seattle.

Bustos, O.H. (1981). "Qualitative Robustness for General Processes". Informes de Matemática, Série B, nº 002/81, Instituto de Matemática Pura e Aplicada, Rio de Janeiro.

Bustos, O.H. (1982). "General M-estimates for Contaminated p-th Order Autoregressive Processes: Consistency and Asymptotic Normality", Zeitschrift für Wahrscheinlichkeitstheorie und Verwandte Gebiete, 59, 491-504.

Bustos, O.H. and Yohai, V.J. (1983). "Robust Estimates for ARMA Models", Informes de Matemática, Série B, nº 012/83, Instituto de Matemática Pura e Aplicada, Rio de Janeiro.

Denby, L. and Martin, D.R. (1979). "Robust Estimation on the First Order Autoregressive Parameter", Journal of the American Statistic Association, 74, 140-146.

Hampel, F.R. (1974). "The Influence Curve and its Role in Robust Estimation", Journal of the American Statistic Association, 69, 383-393.

Lee, C.H. and Martin, R.D. (1982). "M-Estimates for ARMA Processes", Technical Report No. 23, Department of Statistics, University of Washington, Seattle.

Martin, R.D. (1981). "Robust Methods for Time Series", in Applied Time Series II, edited by D.F. Findley, Academic Press, New York.

Martin, R.D., and Yohai, V.J. (1984). "Influence Curve for Time Series", unpublished manuscript, Department of Statistics, University of Washington, Seattle.

Papantoni-Kazakos, P. and Gray, R.M. (1979). "Robustness of Estimators on Stationary Observations", Annals of Probability, 7, 989-1002.

Ruskin, D.M. (1978). "M-Estimates of Nonlinear Regression Parameters and their Jacknife Constructed Confidence Intervals", Ph.D. dissertation, University of California, Los Angeles.

Yohai, V.J. and Maronna, R.A. (1977). "Asymptotic Behaviour of Least Squares Estimates for Autoregressive Processes with Infinite Variances", Annals of Statistics, 5, 554-560.

PARAMETER ESTIMATION OF STATIONARY PROCESSES WITH SPECTRA CONTAINING STRONG PEAKS

Rainer Dahlhaus
Universität Essen
Fachbereich Mathematik
Postfach 103764
4300 Essen 1
West-Germany

Abstract. The advantage of using data tapers in parameter estimation of stationary processes are investigated. Consistency and a central limit theorem for quasi maximum likelihood parameter estimates with tapered data are proved, and data tapers leading to asymptotically efficient parameter estimates are determined. Finally the results of a Monte Carlo study on Yule-Walker estimates with tapered data are presented.

1. INTRODUCTION

Consider a real valued stationary time series $\{X(t)\}, t \in \mathbb{Z}$ with mean 0 and spectral density $f_\theta(\lambda)$, depending on a vector of unknown parameters $\theta=(\theta_1,\ldots,\theta_p) \in \Theta$. When θ is to be estimated from a sample $\underline{X}_N=(X_0,\ldots,X_{N-1})$ a natural approach is to maximize the log likelihood function which in the Gaussian case takes the form

$$L_N(\theta) = -\frac{N}{2}\log 2\pi - \frac{1}{2}\log \det T_N(f_\theta) - \frac{1}{2}\underline{X}_N T_N(f_\theta)^{-1}\underline{X}_N \quad (1)$$

where $T_N(f)$ is the Toeplitz matrix of f. Unfortunately the above function is difficult to handle and no expression for the estimate θ_N is known (even in the case of simple processes). One approach is to approximate $\log \det T_N(f_\theta)$ by $\frac{N}{2\pi}\int_\Pi \log 2\pi f_\theta(\lambda)\, d\lambda$ (with $\Pi=(-\pi,\pi]$) and the inverse $T_N(f_\theta)^{-1}$ by $T_N(\frac{1}{4\pi^2 f_\theta})$ (cp. Coursol and Dacunha-Castelle, 1982). This leads to the following function introduced by Whittle (1953)

$$L_N^W(\theta) = -\frac{N}{4\pi}\int_\Pi \left[\log(4\pi^2 f_\theta(\lambda)) + \frac{I_N(\lambda)}{f_\theta(\lambda)}\right] d\lambda \quad (2)$$

where $I_N(\lambda)$ is the ordinary periodogram. The maximization of $L_N^W(\theta)$ with respect to $\theta \in \Theta$ may now be interpreted as the search for that function f_θ with $\theta \in \Theta$ that approximates the calculated periodogram best (cp. Hosoya and Taniguchi, 1982, Lemma 3.1). A brief discussion of the Whittle function and the estimate $\hat{\theta}_N$ of θ obtained by maximizing $L_N^W(\theta)$ may be found in Dzhaparidze and Yaglom (1982). Let us just remark that $\hat{\theta}_N$ is asymptotically consistent and efficient in Fishers sense. If the integral in (2) is approximated by a sum over the Fourier frequencies $\frac{2\pi s}{N}, s=0,\dots,N-1$ the same results hold (cp. Robinson, 1978).

On the other hand it is well known that the periodogram $I_N(\lambda)$ is a very poor estimate of the spectral density f_θ and the question arises whether $\hat{\theta}_N$ can be improved by choosing a better estimate of the spectral density instead of $I_N(\lambda)$ in (2). Two bad properties of $I_N(\lambda)$ are known. The first is that the periodogram values at different frequencies are nearly uncorrelated which makes a smoothing necessary when the spectrum is estimated directly by $I_N(\lambda)$. However, smoothing $I_N(\lambda)$ in (2) would obviously yield no advantage since the choice of the (smooth) parametric f_θ does precisely this.

The second bad effect is the leakage effect. When the spectrum contains high peaks the spectrum at the other frequencies may be overestimated markedly. The result is that other lower peaks may possibly not be discovered (for an example see the analysis of the variable star data in

Bloomfield, 1976, Chapter 5.3). In contrast to the first effect this leakage effect seems to transfer to the parameter $\hat{\theta}_N$ and the parametric spectral density $f_{\hat{\theta}_N}$. This effect is demonstrated for autoregressive processes in a Monte Carlo study in section 3.

To avoid the leakage effect we suggest the following estimation procedure adapted from nonparametric spectral density estimation. Instead of the ordinary periodogram we use a tapered version of the periodogram, i.e. we multiply the process X(t) with a data taper $h_{t,N}$, t=0,...,N-1. We therefore define

$$I_N(\alpha) := \{2\pi H_2^{(N)}(0)\}^{-1}\, d_N(\alpha)\, d_N(-\alpha)$$

where

$$d_N(\alpha) := \sum_{t=0}^{N-1} h_{t,N}\, X(t)\, e^{-i\alpha t}$$

and

$$H_k^{(N)}(\alpha) := \sum_{t=0}^{N-1} h_{t,N}^k\, e^{-i\alpha t}$$

is the spectral window. The data taper normally has a maximum at t=N/2 and then decreases as t tends to 0 or N-1. We further assume the data taper to be of bounded variation, i.e.

$$BV(h_N) := \sum_{t=0}^{N} |h_{t,N} - h_{t-1,N}| \leq K \quad \text{independent of } N \ (h_{-1,N} = h_{N,N} = 0),$$

and define

$$H_k := \lim_{N\to\infty} \frac{1}{N} \sum_{t=0}^{N-1} h_{t,N}^k$$

and assume the existence of the above limit and $H_2 \neq 0$. Note that the tapered periodogram has asymptotically the same properties as the nontapered periodogram. It is asymptotically unbiased, inconsistent and the periodogram values at different frequencies are asymptotically uncorrelated (cp. Brillinger, 1975). Instead of the ordinary periodogram we now take the tapered periodogram in (2) and consider estimates $\hat{\theta}_N$ and $f_{\hat{\theta}_N}$ with $\hat{\theta}_N$ obtained by maximizing $L_N^W(\theta)$.

As an example consider the widely used cosine bell taper

$h_{t,N}=h_N(\frac{t}{N})$ with

$$h_N(x)=\begin{cases}\frac{1}{2}\,[1-\cos\{2\pi\frac{x}{\rho_N}+\frac{\pi}{N}\}]\;, & x\in[0,\frac{\rho_N}{2}]\\ 1 & ,\; x\in(\frac{\rho_N}{2},1-\frac{\rho_N}{2})\\ \frac{1}{2}\,[1-\cos\{2\pi\frac{1-x}{\rho_N}-\frac{\pi}{N}\}], & x\in[1-\frac{\rho_N}{2},1]\end{cases} \tag{3}$$

where $\rho_N\in[0,1]$ (cp. Tukey, 1967 and Bloomfield, 1976, Chapter 5.2). This taper leaves $100(1-\rho_N)$% of the data unaltered, the remaining $100\rho_N$% are multiplied by a cosine function. We therefore call ρ_N the 'degree' of tapering.

In section 2 we prove the consistency and asymptotic normality of $\hat{\theta}_N$ and investigate what kind of tapers lead to asymptotically efficient estimates $\hat{\theta}_N$. We further discuss the validity of a central limit theorem for the empirical covariances calculated with tapered data, because this validity is assumed in the central limit theorem for $\hat{\theta}_N$. In section 3 we present the results of a Monte Carlo study on Yule-Walker parameter estimates with tapered data. The proofs of the theorems are contained in section 4.

2. ASYMPTOTIC PROPERTIES OF TAPERED PARAMETER ESTIMATES

We first set down some notation. Let $k\in\mathbb{N}$ and $E|X(t)|^k<\infty$. If the cumulant $\text{cum}(X(u_1+t),\ldots,X(u_{l-1}+t),X(t))$ is independent of $t\in\mathbb{Z}$ for all $u_1,\ldots,u_{l-1}\in\mathbb{Z}$ and all $l\leq k$ we will call the process $X(t)$ k-th order stationary. In this case we define $c_k(u_1,\ldots,u_{k-1}):=\text{cum}(X(u_1+t),\ldots,X(u_{k-1}+t),X(t))$. As usual let further the k'th order cumulant spectrum be a function $f_k:\Pi^{k-1}\to\mathbb{C}$ with the property

$$c_k(u_1,\ldots,u_{k-1})=\int_{\Pi^{k-1}} f_k(\alpha_1,\ldots,\alpha_{k-1})\exp(i\sum_{j=1}^{k-1}u_j\alpha_j)\,\lambda^{k-1}(d\alpha)$$

for all $u_1,\ldots,u_{k-1}\in\mathbb{Z}$ with the Lebesgue measure λ^{k-1} on the Borel σ-algebra $\mathcal{B}_{\Pi}^{k-1}$ on Π^{k-1}. When no confusion arises we will write f_θ or just f instead of $f_{2,\theta}$ for the second order spectrum.

THEOREM 2.1. *Let $X(t)$, $t \in \mathbb{Z}$ be a fourth order stationary process with $EX(t) \equiv 0$ and the spectral density f_{θ_0}. Suppose that the parameter set Θ is a compact subset of $\mathbb{R}^p$, that $\theta_1 \neq \theta_2$ implies $f_{\theta_1} \neq f_{\theta_2}$ on a set of positive Lebesgue measure that $f_\theta(\lambda) > 0$ for all $\lambda \in \Pi$ and all $\theta \in \Theta$ and that $f_\theta(\lambda)$ is continuous on $\Theta \times \Pi$. Suppose further that $\|f_4\|_3 := (\int_{\Pi^3} f_4 d\lambda^3)^{1/3} < \infty$. If $\hat{\theta}_N$ exists uniquely and is an interior point of Θ we have*

$$\hat{\theta}_N \xrightarrow{P} \theta_0.$$

Thus also in the case of non Gaussian processes the maximization of the Whittle function (2) leads to consistent parameter estimates. Note that under some additional conditions on the spectra up to the 8-th order we can also obtain strong consistency. The integrability condition may be weakened if the process is stationary up to a higher order.

We now state a central limit theorem for $\hat{\theta}_N$. To specify the limit distribution we define the matrices Γ and B by

$$\Gamma_{ij} = \frac{1}{4\pi} \int_\Pi \frac{\partial}{\partial\theta_i} \log f_{\theta_0}(\lambda) \frac{\partial}{\partial\theta_j} \log f_{\theta_0}(\lambda)\, d\lambda$$

$$B_{ij} = \frac{1}{8\pi} \int_{\Pi^2} \frac{f_4(\lambda_1, -\lambda_1, \lambda_2)}{f_{\theta_0}(\lambda_1) f_{\theta_0}(\lambda_2)} \frac{\partial}{\partial\theta_i} \log f_{\theta_0}(\lambda_1) \frac{\partial}{\partial\theta_j} \log f_{\theta_0}(\lambda_2)\, \lambda^2(d\lambda)$$

$(i,j=1,\ldots,p)$. In the Gaussian case we have under suitable regularity conditions the relation $\Gamma = \lim_{N\to\infty} \frac{1}{N} I_N$ where I_N is the Fisher information matrix (cp. Dzhaparidze and Yaglom, 1982, section 3). We further define the empirical covariances calculated with tapered data by

$$c_N(u) = \{H_2^{(N)}(0)\}^{-1} \sum_{0 \le t, t+u \le N-1} h_{t,N} X_t\, h_{t+u,N} X_{t+u}. \qquad (4)$$

THEOREM 2.2. *Let the following conditions hold:*

(i) $X(t)$, $t \in \mathbb{Z}$ is fourth order stationary with $EX(t) \equiv 0$ and continuous second and fourth order spectra.

(ii) Θ is a compact subset of $\mathbb{R}^p$, $\theta_1 \neq \theta_2$ implies $f_{\theta_1} \neq f_{\theta_2}$ on a set of positive Lebesgue measure, f_θ is positive, $f_\theta(\lambda)$ is a twice continuously differentiable function of $\theta \in \Theta$, and these second derivatives are continuous in $\lambda \in \Pi$.

(iii) For all $k \in \mathbb{N}$, $u_1,\ldots,u_k \in \mathbb{Z}$ $\sqrt{N}\{c_N(u_j)-Ec_N(u_j)\}_{j=1,\ldots,k}$ *converges weakly to a Gaussian random vector* $\{C(u_j)\}_{j=1,\ldots,k}$ *with*

$cov\{C(u_1),C(u_2)\} = \lim_{N\to\infty} N\, cov\{c_N(u_1),c_N(u_2)\}$ *for all* $u_1,u_2 \in \mathbb{Z}$.

(iv) $\frac{\delta}{\partial\theta_j} f_{\theta_0}^{-1} \in Lip_\alpha$ $(j=1,\ldots,l)$, $f_{\theta_0} \in Lip_\beta$ *with* $\alpha+\beta>1/2$ *and* $\alpha,\beta \geq 0$ *(where* $\phi \in Lip_0$ *means that* ϕ *is continuous).*

(v) Γ *is nonsingular.*

Then we have

$$\sqrt{N}\,(\hat{\theta}_N - \theta_0) \xrightarrow{\mathcal{D}} \mathcal{N}\left(0,\ \frac{H_4}{H_2^2}\,\Gamma^{-1}(\Gamma+B)\Gamma^{-1}\right). \tag{5}$$

Using the Cauchy-Schwarz inequality it follows $H_4/H_2^2 \geq 1$ which means that the asymptotic variance of the tapered parameter estimate is never less than the variance of the nontapered one where $H_4/H_2^2 = 1$ (which is at least in the case of a Gaussian process obvious, since $\Gamma = \lim_{N\to\infty} \frac{1}{N} I_N$ which means that the nontapered estimate is asymptotically efficient). In fact for the cosine bell taper (3) with $\rho_N = 1.0$ we have $H_4/H_2^2 = 2.5$ which implies that we have asymptotically a considerable loss of information. To clarify this we consider in the following corollary an important subclass of tapers.

<u>COROLLARY 2.3.</u> *Suppose in addition to the assumptions of Theorem 2.2 that* $X(t), t \in \mathbb{Z}$ *is Gaussian and that* $h_{t,N} = h_N(\frac{t}{N})$ *with continuous functions* $h_N: [0,1] \to [0,\infty]$ *of uniformly bounded variation* $BV(h_N)$ *that converge almost everywhere to a function* $h(x)$*. Then* $\hat{\theta}_N$ *is asymptotically efficient if and only if* $h(x)=1$ *almost everywhere.*

Dzhaparidze (1982) (cp. also Dzhaparidze and Yaglom, 1982) has shown in the nontapered case that the covariance structure $\Gamma^{-1}(\Gamma+B)\Gamma^{-1}$ is the smallest obtainable for a great class of estimates even in the non Gaussian case. Using the result of Dzhaparidze the above corollary may also be extended to non Gaussian processes.

Note that nearly all tapers used in practice belong to the above class. Corollary 2.3 means that with increasing N more data should be left nearly unmodified. This is plausible since the leakage effect in the periodogram also takes off when N increases. As an example consider the cosine bell taper (3). If $\rho_N \to 0$ we get $\lim_{N\to\infty} h_N(x) = I_{(0,1)}(x)$ which means that the estimate $\hat{\theta}_N$ is asymptotically efficient.

At this point the problem of determination of the 'optimal' taper for fixed N arises ('optimal' e.g. could mean in this context to minimize $E\int_{\Pi} [\{f_{\hat{\theta}_N}(\lambda) - f_{\theta_0}(\lambda)\}/f_{\theta_0}(\lambda)]^2 d\lambda$ for fixed N with respect to ρ_N). This problem seems to be unsolvable at the present state since the optimal value will not only depend on N but also on the special structure of f_{θ_0}, especially on the magnitude of the high peaks. Instead we suggest the following heuristic procedure for practice. Increase ρ_N as long as a substantial effect in the periodogram is seen and take that last value for parameter estimation.

Let us remark on a surprising special case of Theorem 2.2. If we choose $h_{t,N} \in \{-1,1\}$ with a finite number of jumps, e.g. $h_{t,N} = h(\frac{t}{N})$ with $h(x) = 1 - 2\cdot\chi_{(1/2,1]}$, the estimate $\hat{\theta}_N$ is asymptotically efficient. However there seems to be no sense in using such a taper which is also confirmed by simulations the author carried out.

At the end of this section we will discuss the assumptions made in Theorem 2.2. The most stringent assumption is the convergence of the empirial covariances (assumption (iii)), especially since the used covariances are also calculated with tapered data which makes the assumption difficult to verify. Instead we will now give a condition which is independent of the taper used. We denote by $D^k[0,1]$ the cartesian product of $D[0,1]$, the space of functions on $[0,1]$ that are right continuous and have left hand limits, with the Skorohod topology (cp. Billingsley, 1968, p.111), by $C^k[0,1]$ the cartesian product of the space of continuous functions on $[0,1]$, and by $V_a^b(h)$ the variation of h on $[a,b]$.

<u>THEOREM 2.4.</u> *Let $X(t), t \in \mathbb{Z}$ be a fourth order stationary process with $EX(t)=0$ and continuous second and fourth order spectra and let further for all $k \in \mathbb{N}$ and all $u_1,\ldots,u_k \in \mathbb{Z}$*

$$\underline{B}_N(\underline{u},s) = N^{-1/2} \sum_{t=0}^{[Ns]} \{X(t)X(t+u_j) - c(u_j)\}_{j=1,\ldots,k}, \quad s \in [0,1]$$

converge weakly in $D^k[0,1]$ *to a Gaussian process* $\underline{B}(\underline{u},s) = \{B(u_j,s)\}_{j=1,..,k}$ *with*

$$cov\{B(u_1,s_1),B(u_2 s_2)\} = \lim_{N\to\infty} cov\{B_N(u_1,s_1),B_N(u_2,s_2)\}.$$

Then condition (iii) of Theorem 2.2 holds for all tapers of the type $h_{t,N}=h_N(\frac{t}{N})$ *with functions of bounded variation* h_N *that fulfill* $\lim_{N\to\infty} h_N(x)=h(x)$ *and* $\lim_{N\to\infty} V^x_{-1}(h_N)=V^x_{-1}(h)$ *for all continuity points* $x\geq 0$ *of the function* $h(x)$ *with* $H_2\neq 0$ *where* $h_N(x)=h(x)=0$ *for all* $x\notin[0,1]$.

The advantage of Theorem 2.4 is that the functional convergence of $\underline{B}_N(\underline{u},\cdot)$ is much more easier to verify than the convergence of the tapered covariances. To do this, note that with X(t) also X(t)X(t+u) is a stationary process with mean c(u) for all $u\in\mathbb{Z}$. Thus the functional convergence of $\underline{B}_N(\underline{u},\cdot)$ may in various situations be deduced from the numerous articles on the ordinary functional limit theorem for stationary sequences (e.g. Heyde, 1975; Ibragimov, 1975; McLeish, 1975). Below we will state as examples two important situations in which the functional convergence holds.

(i) The functional convergence holds if the process X(t) is a linear transformation of martingale differences.

(ii) The functional convergence holds if the process X(t) is ϕ-mixing with $\sum_{m=1}^{\infty} \phi_X(m)^{2-2/q}<\infty$ $(q\geq 2)$ or strong mixing with $\sum_{m=1}^{\infty} \alpha_X(m)^{1-2/q}<\infty$ $(q>2)$ and the moments of X(t) exist up to the order 2q.

Both results may be obtained by using a result of McLeish (1975) on mixingales. They are stated in full in Dahlhaus (1983b, Corollary 3.1 and Corollary 3.2).

Note that in the case of ϕ-mixing or strong-mixing processes (case ii)) Theorem 2.2 contains a result that was unknown even in the nontapered case.

Tapered parameter estimates have not been considered so far (note that Dunsmuir and Robinson (1981) proved the consistency and a central limit theorem for parameter estimates for amplitude modulated processes Y(t)=a(t)X(t) with view on missing data).

<u>Remark.</u> All results extend to vector valued processes. In that case the estimate $\hat{\theta}_N$ is obtained by maximizing the function

$$L_N^W(\theta) = -\frac{N}{4\pi}\int_{\Pi}\left[\log(4\pi^2 \det f_\theta(\lambda)) + \operatorname{tr}(f_\theta^{-1}(\lambda) I^{(N)}(\lambda))\right] d\lambda$$

(for the definition of the matrix $I^{(N)}(\lambda)$ see Brillinger, 1975). All results of this section still hold. The covariance of the limit distribution in Theorem 2.2 becomes $\Gamma^{-1}C\Gamma^{-1}$ where Γ and C are defined by

$$\Gamma_{ij} = \frac{1}{4\pi}\int_{\Pi}\operatorname{tr}\left[f_{\theta_0}^{-1}(\lambda)\left\{\frac{\partial}{\partial\theta_i} f_{\theta_0}(\lambda)\right\} f_{\theta_0}^{-1}(\lambda)\left\{\frac{\partial}{\partial\theta_j} f_{\theta_0}(\lambda)\right\}\right] d\lambda$$

$$C_{ij} = \frac{1}{8\pi}\sum_{a,b,c,d=1}^{r}\frac{H_{abcd}}{H_{ab}H_{cd}}\cdot$$

$$\cdot\Big[\int_{\Pi^2} f_{abcd}(\gamma_1,-\gamma_1,\gamma_2)\frac{\partial}{\partial\theta_i} f_{\theta_0}^{-1}(\gamma_1)_{ba}\frac{\partial}{\partial\theta_j} f_{\theta_0}^{-1}(\gamma_2)_{dc}\,\lambda^2(d\gamma)$$

$$+\int_{\Pi} f_{ac}(\gamma)\, f_{db}(\gamma)\frac{\partial}{\partial\theta_i} f_{\theta_0}^{-1}(\gamma)_{ba}\frac{\partial}{\partial\theta_j} f_{\theta_0}^{-1}(\gamma)_{cd}\, d\gamma$$

$$+\int_{\Pi} f_{ad}(\gamma)\, f_{cb}(\gamma)\frac{\partial}{\partial\theta_i} f_{\theta_0}^{-1}(\gamma)_{ba}\frac{\partial}{\partial\theta_j} f_{\theta_0}^{-1}(\gamma)_{dc}\, d\gamma\Big]$$

with

$$H_{a_1\ldots a_k} := \lim_{N\to\infty}\frac{1}{N}\sum_{t=0}^{N-1}\left\{\prod_{j=1}^{k} h_{a_j,t,N}\right\}.$$

If H_{abcd} and H_{ab} are independent of $a,b,c,d\in\{1,\ldots,r\}$ the covariance structure reduces to a simple form analogous to (5). This is for example fulfilled if the taper is the same for all components.

3. A SIMULATION STUDY

To maximize $L_N^W(\theta)$ the system of equations

$$\frac{\partial L_N(\theta)}{\partial\theta_j} = 0 \qquad (j=1,\ldots,p) \tag{6}$$

has to be solved. This normally is not possible directly and algorithms such as the Newton-Raphson algorithm are needed. A discussion of such procedures in the nontapered case may be found in Dzhaparidze and Yaglom (1982). Often the integral in (2) must be replaced by a sum over the Fourier frequencies (cp. Robinson, 1978).

An important case in which the system (6) can be solved directly is the case of autoregressive processes

$$\sum_{j=0}^{p} a_j X_{t-j} = \varepsilon_t$$

with $a_0=1$, $E\varepsilon_t=0$, and $\mathrm{cov}(\varepsilon_s,\varepsilon_t)=\delta_{st}\sigma^2$. a simple calculation shows that (6) reduces in this case to the Yule-Walker equations and therefore the estimates $a_1,\ldots,a_p,\sigma^2$ are obtained by solving the equations

$$\sum_{j=0}^{p} \hat{a}_j\, c_N(l-j) = \delta_{0l}\hat{\sigma}^2 \qquad l=0,\ldots,p$$

(with $\hat{a}_0=1$ and the Kronecker symbol δ_{st}) where the empirical covariances (4) are also calculated with tapered data.

The simulation reported was performed with FORTRAN programmes on a PRIME computer in double precision using pseudo normal random numbers. Uniformly distributed random

Table 1. Exact and estimated coefficients of an AR(14)-process.

	exact coeff.	estimated coefficients		
		$\rho=0.0$	$\rho=0.6$	$\rho=1.0$
a_0	1.00	1.00	1.00	1.00
a_1	1.27	-0.50	1.27	1.24
a_2	2.33	1.29	2.46	2.31
a_3	2.21	-0.73	2.24	2.05
a_4	2.58	0.67	2.58	2.34
a_5	1.70	-0.78	1.41	1.25
a_6	2.35	0.90	2.01	1.95
a_7	1.62	-0.91	1.15	1.18
a_8	2.13	0.84	2.01	2.03
a_9	1.40	-0.67	1.37	1.31
a_{10}	1.90	0.47	2.01	1.91
a_{11}	1.41	-0.27	1.38	1.31
a_{12}	1.45	0.10	1.42	1.40
a_{13}	0.84	-0.03	0.70	0.71
a_{14}	0.59	-0.02	0.54	0.60
σ^2	1.00	130.35	1.68	1.42

numbers were taken from a FORTRAN routine and transformed to normal random numbers by a rational approximation to the inverse distributuion function (see Abramowitz and Stegun, 26.2.23). However the same results were obtained when using non normal random numbers. As the data taper we took the cosine bell taper (3) with $\rho=0.0$, $\rho=0.6$, and $\rho=1.0$. We further set N=256 and considered a simulated AR(14)-process (of which the coefficients are listed in the first column of Table 1). Fitting also an AR(14)-model to the data, we estimated the coefficients with the tapered Whittle function (2) (the tapered Yule-Walker equations). The estimated coefficients are listed in the second column (without taper, i.e. $\rho=0.0$) and the third and fourth column (with taper) of Table 1. To interpret and discuss the results we have plotted

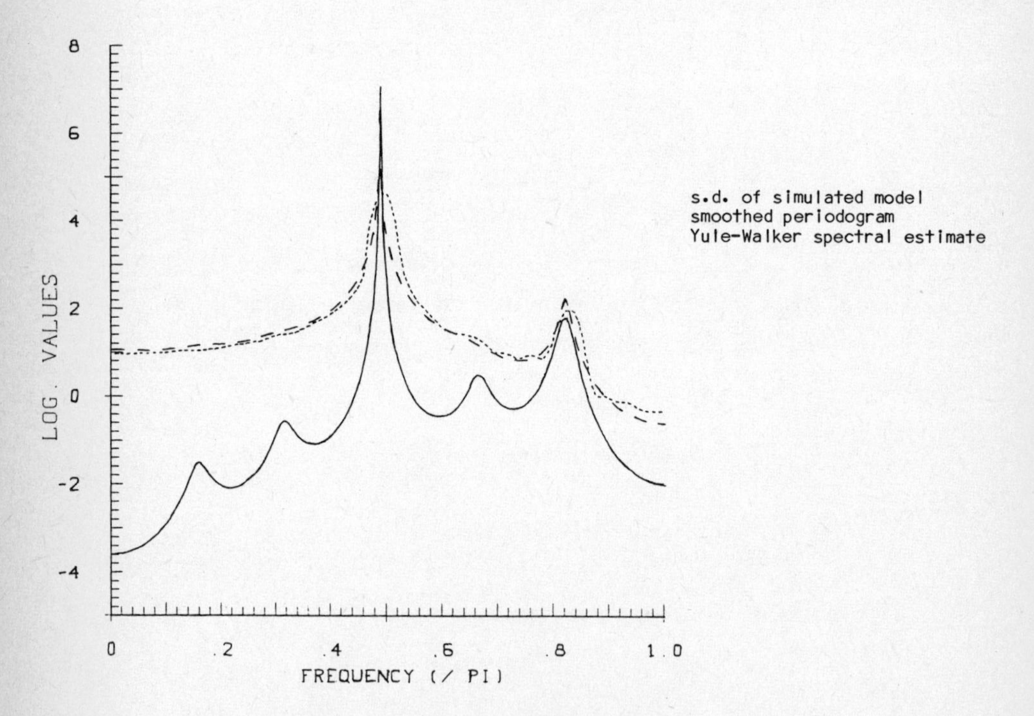

Fig.1. Spectral density and Yule-Walker spectral estimate with nontapered data

the theoretical spectral density together with $f_{\hat{\theta}_N}$ in the nontapered case and the nontapered periodogram values in Figure 1 (for reasons of clearness the periodogram was smoothed by a weighted moving average over 7 values before plotting). The leakage effect in the periodogram caused by the high peak at $\lambda=0.49\pi$ is so marked that it overlaps the lower peaks. Only the lower peak at $\lambda=0.82\pi$ is strong enough to be discovered. Since the Whittle estimate $\hat{\theta}_N$ may be interpreted as the value such that $f_{\hat{\theta}_N}$ approximates $I_N(\lambda)$ best it is not surprising that the same effects transfer to $f_{\hat{\theta}_N}$. In Figure 2 we have plotted the tapered periodogram together with the tapered version of $f_{\hat{\theta}_N}$. The leakage effect disappears and the estimated spectral density is very close

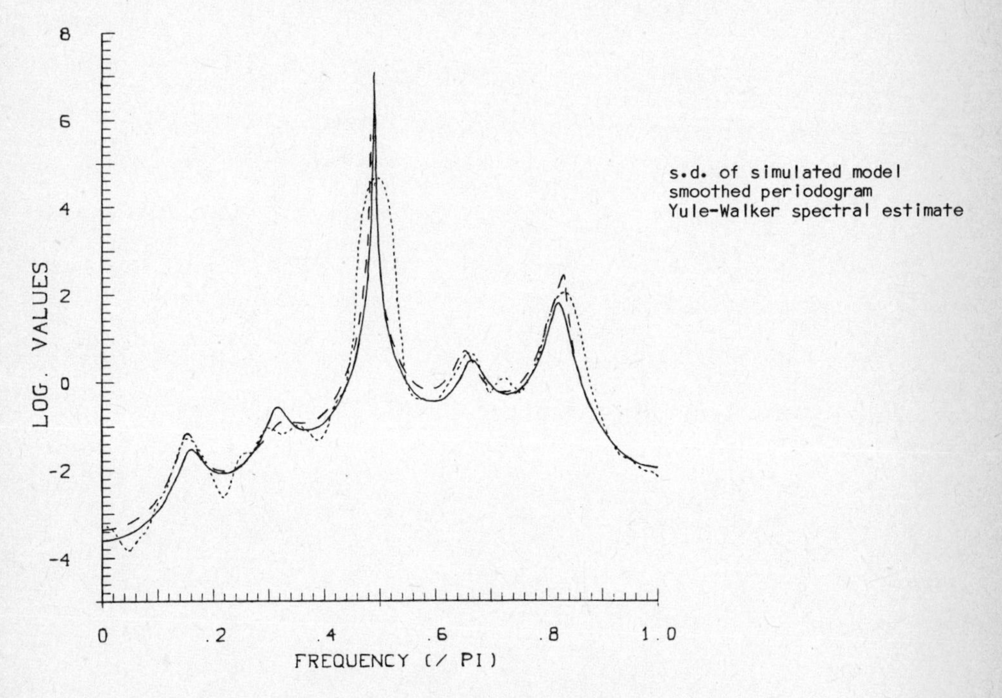

Fig.2. Spectral density and Yule-Walker spectral estimate with tapered data

to the true function. The same holds for the estimated parameters (Table 1). Remarkable is the improvement in the predicted variance $\hat{\sigma}^2$ from 130.35 to 1.68 or to 1.42 since nearly all criteria for the estimation of the order of an autoregressive process are based on this value (cp. Priestley, 1981, section 5.4.5). Indeed the order of the underlying process was estimated in the nontapered case with the BIC criterion suggested by Akaike (1978) to be 11 while it was estimated with $\rho=1.0$ correctly to be 14.

The loss of information with $\rho=0.6$ and $\rho=1.0$ is considerable (cp. the discussion below Theorem 2.2). For nonparametric spectral analysis Tukey (1967) has suggested that values up to $\rho=0.25$ are suitable. However the choice of ρ clearly depends on the magnitude of the strongest peak which, in the above example, is very strong. The heuristic procedure suggested below Corollary 2.3 yielded a value of $\rho=0.6$.

4. PROOFS OF THE THEOREMS

Proof of Theorem 2.1. By means of Lemma 3.1 of Hosoya and Taniguchi (1982) it is sufficient to prove that $\int_{\Pi}\psi(\alpha)I_N(\alpha)d\alpha$ converges to $\int_{\Pi}\psi(\alpha)f_{\theta_0}(\alpha)d\alpha$ in probability for all continuous functions ψ. Following the proof of Lemma 6 of Dahlhaus (1983a)where the behaviour of

$N \operatorname{cov}\{\int_{\Pi} \psi_1(\alpha)\, I^{(N)}(\alpha)\, d\alpha, \int_{\Pi} \psi_2(\beta)\, I^{(N)}(\beta)\, d\beta)\}$ is investigated we obtain (with $\operatorname{cum}_k(X):=\operatorname{cum}(\underbrace{X,\ldots,X}_{k\text{-times}})$)

$$|\operatorname{cum}_2(\int_{\Pi} \psi(\alpha)I_N(\alpha)d\alpha)| \leq KN^{-1+3/p}$$

with a constant K. Using this we get

$$P(|\int_{\Pi} \psi(\alpha)[I_N(\alpha)-EI_N(\alpha)]d\alpha| \geq \varepsilon) \to 0$$

from which the result follows since $\int_{\Pi}\psi(\alpha)EI_N(\alpha)d\alpha$ tends to $\int_{\Pi}\psi(\alpha)f_{\theta_0}(\alpha)d\alpha$.

To prove Theorem 2.2 we need the following lemma for tapered data.

LEMMA 4.1. *Let the following conditions hold:*
(i) $X(t), t\in\mathbb{Z}$ is fourth order stationary with $EX(t)\equiv 0$ and continuous second and fourth order spectra.
(ii) For all $k\in\mathbb{N}$, $u_1,\ldots,u_k\in\mathbb{Z}$ $\sqrt{N}\{c_N(u_j) - Ec_N(u_j)\}_{j=1,\ldots,k}$

converges weakly to a Gaussian random vector $\{C(u_j)\}_{j=1,\ldots,k}$ *with*

$$cov\{C(u_1),C(u_2)\} = \lim_{N\to\infty} N\, cov\{c_N(u_1),c_N(u_2)\} \text{ for all } u_1,u_2\in\mathbb{Z}.$$

(iii) Let $\underline{\phi}: \Pi^l \to \mathbb{R}$ *be a vector of continuous functions and* $\phi_j \in Lip_\alpha$ $(j=1,\ldots,l)$, $f_2\in Lip_\beta$ *with* $\alpha+\beta>1/2$ *and* $\alpha,\beta\geq 0$ *(where* $\phi\in Lip_0$ *means that* ϕ *is continuous).*

Then $\{\sqrt{N}\int_\Pi \phi_j(\alpha)[I_N(\alpha)-f_2(\alpha)]d\alpha\}_{j=1,\ldots,l}$ *converges weakly to a Gaussian random vector* $\{\xi(\phi_j)\}_{j=1,\ldots,l}$ *with*

$$E\xi(\phi_j) = 0 \qquad (j=1,\ldots,l),$$

$$cov\{\xi(\phi_i),\xi(\phi_j)\} = 2\pi\,\frac{H_4}{H_2^2}\,\Big[\int_{\Pi^2}\phi_i(\gamma_1)\phi_j(-\gamma_2)f_4(\gamma_1,-\gamma_1,\gamma_2)\lambda^2(d\gamma)$$

$$+\int_\Pi \phi_i(\gamma)\phi_j(\gamma)f(\gamma)^2\,d\gamma + \int_\Pi \phi_i(\gamma)\phi_j(-\gamma)f(\gamma)^2\,d\gamma\ \Big]. \qquad (7)$$

Proof. Let us denote $A_N(\phi):=\sqrt{N}\int_\Pi\phi(\alpha)[I_N(\alpha)-EI_N(\alpha)]d\alpha$. We prove the result by approximating ϕ by its Cesaro sums (cp. e.g. Dunsmuir, 1979). Let $\phi_j^{(M)}$ be the M-th Cesaro sum with $\sup_{\lambda\in\Pi}|\phi_j(\lambda)-\phi_j^{(M)}(\lambda)| \leq \varepsilon$ and $\delta_j=\phi_j-\phi_j^{(M)}$ $(j=1,\ldots,l)$. We have $A_N(\phi_j) = A_N(\phi_j^{(M)}) + A_N(\delta_j)$. Following the proof of Lemma 6 of Dahlhaus (1983a) we obtain $|var(A_N(\delta_j))| \leq K\varepsilon^2$ with a constant $K\in\mathbb{R}$ independent of N and M. We further have (with $\hat{\phi}(u)=\int_\Pi\phi(\alpha)\exp(-i\alpha u)d\alpha$)

$$A_N(\phi_j^{(M)}) = \frac{\sqrt{N}}{2\pi}\sum_{u=-(M-1)}^{M-1}[c_N(u)-Ec_N(u)]\,(1-\frac{|u|}{M})\,\hat{\phi}_j(u).$$

Thus using assumption (ii) $\{A_N(\phi_j^{(M)})\}_{j=1,\ldots,l}$ tends weakly to the Gaussian random vector $\{S_M(\phi_j)\}_{j=1,\ldots,l}$ with

$$S_M(\phi) = \frac{1}{2\pi}\sum_{u=-(M-1)}^{M-1} C(u)\,(1-\frac{|u|}{M})\,\hat{\phi}(u).$$

From Dahlhaus (1983a), Lemma 6 we get

$$cov\{C(u_1),C(u_2)\} = 2\pi\,\frac{H_4}{H_2^2}\,\cdot$$

$$\cdot\ \Big[\int_{\Pi^2} f_4(\gamma_1,-\gamma_1,\gamma_2)\exp(i\gamma_1u_1+i\gamma_2u_2)\,\lambda^2(d\gamma) +$$

$$+ \int_{\Pi} f(\gamma)^2 \{\exp(i\gamma(u_1-u_2)) + \exp(i\gamma(u_1+u_2))\} d\gamma].$$

With the Fejer kernel $K_M(\alpha) = \frac{1}{2\pi M} \sin^2(M\alpha/2)/\sin^2(\alpha/2)$ we now obtain

$$\operatorname{cov}\{S_M(\phi_i),S_M(\phi_j)\} = 2\pi \frac{H_4}{H_2^2} \cdot$$

$$\cdot[\int_{\Pi^2} f_4(\gamma_1,-\gamma_1,\gamma_2)\int_{\Pi}\phi_i(\alpha_1)K_M(\gamma_1-\alpha_1)d\alpha_1\int_{\Pi}\phi_j(\alpha_2)K_M(\gamma_2+\alpha_2)d\alpha_2\lambda^2(d\gamma)$$

$$+ \int_{\Pi} f(\gamma)^2 \int_{\Pi} \phi_i(\alpha_1)K_M(\gamma-\alpha_1)d\alpha_1 \int_{\Pi} \phi_j(\alpha_2)K_M(-\gamma+\alpha_2)d\alpha_2 d\gamma$$

$$+ \int_{\Pi} f(\gamma)^2 \int_{\Pi} \phi_i(\alpha_1)K_M(\gamma-\alpha_1)d\alpha_1 \int_{\Pi} \phi_j(\alpha_2)K_M(\gamma+\alpha_2)d\alpha_2\, d\gamma].$$

Since all ϕ_j are continuous this expression converges to the covariance structure (7) and thus the Gaussian random vector $\{S_M(\phi_j)\}_{j=1,...,l}$ and therefore also the random vector $\{A_N(\phi_j)\}_{j=1,...,l}$ tend to the Gaussian random vector $\{\xi(\phi_j)\}_{j=1,...,l}$. The proof is complete if we show that

$$\sqrt{N} \int_{\Pi} \phi_j(\alpha) \,[EI_N(\alpha) - f(\alpha)]\, d\alpha$$

$$= \sqrt{N} \int_{\Pi} \phi_j(\alpha) \int_{\Pi} [f(\alpha-\beta) - f(\alpha)] \frac{|H_1^{(N)}(\beta)|^2}{2\pi H_2^{(N)}(0)} d\beta\, d\alpha$$

tends to zero $(j=1,...,l)$. Using the Parseval equality and the definition

$$D_N(u) := H_2^{(N)}(0)^{-1} \sum_{0\leq t,t+u\leq N-1} h_{t,N}h_{t+u,N}$$

the last expression is equal to

$$- \frac{\sqrt{N}}{2\pi} \sum_{u=-\infty}^{\infty} \hat{f}(u)\hat{\phi}_j(u)[1-D_N(u)].$$

Since $h_{t,N}$ is of bounded variation we have $|1-D_N(u)| \leq K\frac{|u|}{N}$ for $|u| \leq N$. We further have

$4 \sum_{u=-\infty}^{\infty} \hat{f}(u)^2\sin^2 hu = 2\pi \int_{\Pi} [f(x+h)-f(x-h)]^2\, dx \leq Kh^{2\beta}$ for all $h \in \mathbb{R}$ which implies $\sum_{u=n}^{2n-1} \hat{f}(u)^2 \leq Kn^{-2\beta}$ for all $n \in \mathbb{N}$. This gives

$\sum_{u=1}^{N} u\hat{f}(u)^2 \leq KN^{1-2\beta}$ and $\sum_{u>N} \hat{f}(u)^2 \leq KN^{-2\beta}$ and the same inequalities with $\hat{\phi}_j(u)$ and α instead of $\hat{f}(u)$ and β from which the required result follows.

A similar theorem for tapered data but for noncontinuous ϕ_j was proved under additional assumptions with a completely different method of proof in Dahlhaus (1983b, Theorem 2.1). Note that the condition of the continuity of the fourth order spectrum may be weakened.

Proof of Theorem 2.2. Since $\int_{\Pi} \psi(\alpha)\, I_N(\alpha)\, d\alpha$ tends in probability to $\int_{\Pi} \psi(\alpha)\, f_{\theta_0}(\alpha)\, d\alpha$ for every continuous function ψ (Theorem 2.1) the assertion follows from Lemma A3.2 of Hosoya and Taniguchi (1982) if we prove that

$$\left\{\frac{\sqrt{N}}{4\pi} \int_{\Pi} [I_N(\alpha)-f(\alpha)]\, \frac{\partial}{\partial\theta_j} f_{\theta_0}(\alpha)^{-1}\, d\alpha\right\}_{j=1,\ldots,p}$$

tends weakly to a Gaussian distribution with mean 0 and covariance matrix $\frac{H_4}{H_2^2}(\Gamma+B)$ (the matrix Γ coincides with $\frac{1}{4\pi} M_f$ in Lemma A3.2 of Hosoya and Taniguchi). This follows using Lemma 4.1.

Note that it is also possible to use the same method of proof employed by the most authors who have previously considered this problem (e.g. Dunsmuir, 1979; Dunsmuir and Hannan, 1976). In fact this is essentially the same as using Lemma A3.2 of Hosoya and Taniguchi.

Proof of Corollary 2.3. The corollary follows from Theorem 2.2 with the relation

$$\left|\frac{1}{N} \sum_{t=0}^{N-1} h_N\left(\frac{t}{N}\right)^k - \int_0^1 h_N(x)^k dx\right| \leq \frac{BV(h_N^k)}{N},$$

the dominated convergence theorem, and the fact that in the Cauchy-Schwarz inequality we have equality if and only if both factors are the same almost everywhere.

<u>Proof of Theorem 2.4.</u> For the special case when all functions h_N are the same the theorem was proved in Dahlhaus (1983b, Theorem 2.2). We therefore just give a short sketch. By calculating the covariance structure of $\underline{B}(\underline{u},\cdot)$ we obtain that $B(u_j,\cdot)$ is a Wiener process on $[0,1]$ $(j=1,\ldots,k)$ and therefore $P(\underline{B}(\underline{u},\cdot)\in C^k[0,1])=1$. Now let $h_N^*(x)=\lim_{y\downarrow x} h_N(y)^2$ $(N\in\mathbb{N})$ and $h^*(x)=\lim_{y\downarrow x} h(y)^2$. h_N^* $(N\in\mathbb{N})$ and h^* induce signed measures on $([0,1],\mathcal{B}_{[0,1]})$. We define the functions $f_N: D^k[0,1]\to\mathbb{R}^k$ by

$$\{x_j(\cdot)\}_{j=1,\ldots,k} \to \{H_2^{-1}\int_{[0,1]} x_j(s)\, h_N^*(ds)\}_{j=1,\ldots,k}$$

and the function $f: D^k[0,1]\to\mathbb{R}^k$ analogously with h^* instead of h_N^*. If $x(\cdot)\in C^k[0,1]$ and $x^{(N)}(\cdot)\in D^k[0,1]$ is a sequence with $d(x,x^{(N)})\to 0$ (where d is the Skorohod metric, cp. Billingsley, 1968, p. 111), we have $f_N(x^{(N)})\to f(x)$ from which we conclude (see Billingsley, 1968, Theorem 5.5) that $f_N(\underline{B}_N(\underline{u},\cdot))$ and therefore also $\sqrt{N}\{c_N(u_j)-Ec_N(u_j)\}_{j=1,\ldots,k}$ converges weakly to the Gaussian random vector $f(\underline{B}(\underline{u},\cdot))$. Using the formula (integration by parts)

$$\int_{[0,1)} s h^*(s) h^*(ds) = h^*(-1)^2 - \int_{[0,1)} h^*(s-) h^*(s)ds - \int_{[0,1)} s h^*(s-) h^*(ds)$$

we calculate the covariance structure of $f(\underline{B}(\underline{u},\cdot))$ and conclude to the convergence of the covariances in assumption (iii) of Theorem 2.2.

ACKNOWLEDGEMENT

The author is grateful to K.O. Dzhaparidze for some useful conversations and to the referee for his helpful comments.

REFERENCES

Abramowitz, M. and I.A. Stegun (1964). Handbook of Mathematical Functions. Dover Publications, New York.

Akaike, H. (1978). A bayesian analysis of the minimum AIC procedure. Ann. Inst. Statist. Math., 30A, 9-14.

Billingsley, P. (1968). Convergence of Probability Measures. Wiley, New York.

Bloomfield, P. (1976). Fourier Analysis of Time Series. Wiley. New York.

Brillinger, D.R. (1969). Time Series: Data Analysis and Theory. Holt, Rinehart and Winston, New York.

Coursol, J. and D. Dacunha-Castelle (1982). Remarks on the approximation of the likelihood function of a stationary Gaussian process. Theory Prob. Appl., 27, 162-166.

Dahlhaus, R. (1983a). Spectral analysis with tapered data. J. Time Ser. Anal., to appear.

Dahlhaus, R. (1983b). Asymptotic normality of spectral estimates. J. Multivariate Anal., to appear.

Dunsmuir, W. (1979). A central limit theorem for parameter estimation in stationary vector time series and its application to models for a signal observed with noise. Ann. Statist., 7, 490-506.

Dunsmuir, W. and E.J. Hannan (1976). Vector linear time series models. Adv. Appl. Prob., 8, 339-364.

Dunsmuir, W. and P.M. Robinson (1981). Parametric estimators for stationary time series with missing observations. Adv. Appl. Prob., 13, 129-146.

Dzhaparidze, K.O. (1982). On asymptotically efficient estimation of spectrum parameters. Mathematisch Centrum, Amsterdam, Holland.

Dzhaparidze, K.O. and A.M. Yaglom (1982). Spectrum parameter estimation in time series analysis, in: Developments in Statistics (ed. P.R. Krishnaiah), Vol.4, Ch.1, pp. 1-96, Academic Press, New York.

Heyde, C.C. (1975). On the central limit theorem and iterated logarithm law for stationary processes. Bull. Austral. Math. Soc., 12, 1-8.

Hosoya, Y. and M. Taniguchi (1982). A central limit theorem for stationary processes and the parameter estimation of linear processes. Ann. Statist., 10, 132-153.

Ibragimov, I.A. (1975). A note on the central limit theorem for dependent random variables. Theory Prob. Appl., 20, 135-141.

McLeish, D.L. (1975). Invariance principles for dependent variables. Z.Wahrscheinlichkeitsth.verw.Geb.,32,165-178.

Priestley, M.B. (1981). Spectral Analysis and Time Series, Vol.1. Academic Press, London.

Robinson, P.M. (1978). Alternative models for stationary stochastic processes. Stoch. Proc. Appl., 8, 141-152.

Tukey, J.W. (1967). An introduction to the calculations of numerical spectrum analysis, in: Spectral Analysis of Time Series (ed. B. Harris), pp. 25-46, Wiley, New York.

Whittle, P. (1953). Estimation and information in stationary time series. Ark. Mat., 2, 423-434.

Linear Error-in-Variables Models

Prof.M.Deistler

Inst.f.Oekonometrie,TU Wien
Argentinerstrasse 8
A-1040 Wien

Oestereich

1. Introduction

In this paper we are concerned with the statistical analysis of systems, where both, inputs and outputs, are contaminated by errors. Models of this kind are called error-in-variables (EV) models. Let x_t^* and y_t^* denote the "true" inputs and outputs respectively and let x_t and y_t denote the observed inputs and outputs, then the situation can be illustrated as follows:

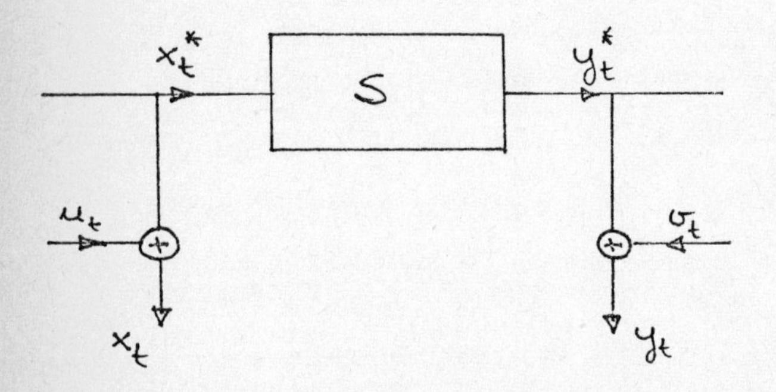

Fig 1: Schematic representation of an EV model

Thereby u_t and v_t are the errors of the inputs and the outputs respectively.

This setting is different from the conventional setting in the statistical analysis of systems where all errors are attributed to the outputs (in this case we use the term errors-in-equations model).

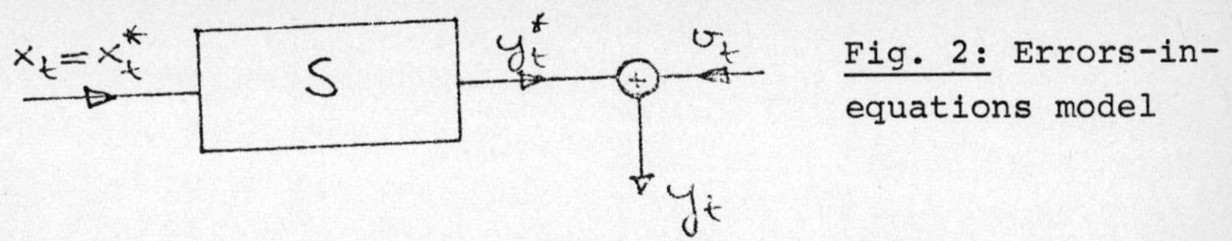

Fig. 2: Errors-in-equations model

Of course, if we are interested in prediction of the observed outputs from the observed inputs only, then the errors-in-equations setting is adequate. If, however, our main interest is in the analysis of the system S, then the EV-setting is more adequate in many cases, if we cannot be sure a-priori that the inputs are observed without errors. In addition EV models provide a symmetric treatment for all variables, so that we even need not to distinguish a-priori between inputs and outputs. For a detailed description of the advantages of EV modelling see Kalman (1983).

We are dealing with linear systems only here. Our primary interest is in the characteristics of the system. i.e. in the transfer function (or the parameters of the transfer function); but also the characteristics of the errors and of (x_t^*) are of interest. We deal only with the case where inputs and outputs are scalar processes.

Our main emphasis is on two problems: The first is the problem of identifiability, i.e. the problem whether the characteristics of interest mentioned above can be uniquely determined from certain characteristics of the observations as e.g. from their (ensemble) second moments or from their probability law (see Deistler and Seifert (1978)). If the answer is negative then the second problem is to describe the sets of observationally equivalent characteristics of interest, i.e. the sets of characteristics of interest which correspond to the same characteristics of the observations.

These questions are questions preceding estimation in the narrow sense and turn out to be the main difficulty in the process of estimation (or inference) in EV models. This difficulty is the reason why not very much attention has been paid to EV models for a long time. However, in the last decade there has been a resurging interest in EV models, especially in econometrics (see e.g. Aigner and Goldberger (1977), Aigner et.al. (1984), Maravall (1979), Schneeweiß und Mittag (1984), Wegge (1983)), but also in system theory (Kalman (1982), Kalman (1983), Söderström (1980)).

2. The Static Case

Here we consider the special case where the system is static, i.e. the transfer function is simply the slope parameter of a line and all processes are white noise. This case has been discussed in great detail in the literature and the reader is refered to the survey papers by Madansky (1959) and Moran (1971).

This static EV model is written as:

(2.1) $y_t^* = ax_t^*, \quad a \in \mathbb{R}$

(2.2) $x_t = x_t^* + u_t$

(2.3) $y_t = y_t^* + v_t$

where (x_t^*), (u_t) and (v_t) are white noise processes, i.e.

(2.4) $Ex_t^* = Eu_t = Ev_t = 0$

(this assumption could be easily generalized) and

(2.5) $Ex_s^*x_t^* = \delta_{st} \cdot c_{x^*x^*}; \; Eu_su_t = \delta_{st} \cdot c_{uu}; \; Ev_sv_t = \delta_{st} \cdot c_{vv} \qquad \forall s,t \in \mathbb{Z}$

In addition we assume

(2.6) $Ex_t^*u_s = 0; \; Ex_t^*v_s = 0; \; Eu_tv_s = 0; \qquad \forall s,t \in \mathbb{Z}$

The parameters of interest are $\theta = (a, c_{x^*x^*}, c_{uu}, c_{vv})$. The relation between these parameters and the second moments of the observations is given by

(2.7) $c_{xx} = Ex_t^2 = c_{x^*x^*} + c_{uu}$

(2.8) $c_{xy} = Ex_ty_t = Ex_t^*y_t^* = a \cdot c_{x^*x^*}$

(2.9) $c_{yy} = Ey_t^2 = a^2c_{x^*x^*} + c_{vv}$

Thus the problem of identifiability from second moments for this model is whether θ is uniquely determined from c_{xx}, c_{xy}, c_{yy}. A slightly more general model would be of the form

(2.10) $by_t^* = ax_t^*$ (where a and b are suitably normalized)

(2.2) - (2.6). Sloppy speaking here we allow for the case $a = \infty$ in (2.1). Then the problem of observational equivalence is equivalent to the following "Frisch" problem (see Kalman (1982)): Given the covariance matrix $K = \begin{pmatrix} c_{xx}, & c_{xy} \\ c_{yx}, & c_{yy} \end{pmatrix}$ find all decompositions

(2.11) $K = \hat{K} + \tilde{K}$

into covariance (i.e. symmetric, nonnegative definite) matrices $\hat{K}$ and $\tilde{K}$, such that $\hat{K}$ is singular and $\tilde{K}$ is diagonal. This equivalence is straightforward since, as here $\hat{K}$ is the covariance matrix of (x_t^*, y_t^*), a and b, after suitable normalization, are defined from the linear dependence relations in $\hat{K}$, and as here

$$\tilde{K} = \begin{pmatrix} c_{uu} & 0 \\ 0 & c_{vv} \end{pmatrix}$$

In the case (2.1) (which excludes the possibility $b=0$ in (2.10) and which is the only one we treat here, unless the contrary has been explicitely stated)

$$\hat{K} = c_{x^*x^*} \cdot \begin{pmatrix} 1 & a \\ a & a^2 \end{pmatrix}$$

holds. We do assume throughout that $c_{x^*x^*} > 0$ and that $\det K > 0$.

By the singularity of $\hat{K}$ we have

(2.12) $\det \hat{K} = c_{x^*x^*} \cdot c_{y^*y^*} - c_{xy}^2 = 0$,

where $c_{y^*y^*} = Ey_t^{*2}$ and furthermore

(2.13) $0 < c_{x^*x^*} \leq c_{xx}$

(2.14) $\quad 0 \leq c_{y^*y^*} \leq c_{yy}$

and these are the only restrictions on $c_{x^*x^*}$ and $c_{y^*y^*}$. Thus the range of pairs $(c_{x^*x^*}, c_{y^*y^*})$ compatible with the given second moments of the observations is a part of the hyperbola (2.12).

Then the range of compatible slope parameters $a = c_{xy}\, c_{x^*x^*}^{-1}$ is given by the intervals

$$\begin{aligned} &[c_{xy}\cdot c_{xx}^{-1},\ c_{yy}\cdot c_{xy}^{-1}] && \text{for } c_{xy} > 0 \\ (2.15)\quad &[c_{yy}\cdot c_{xy}^{-1},\ c_{xy}\cdot c_{xx}^{-1}] && \text{for } c_{xy} < 0 \\ &\{0\} && \text{for } c_{xy} = 0\ . \end{aligned}$$

From what was said above, it also follows that <u>every</u> covariance matrix K can be decomposed as in (2.11).
Let us summarize:

<u>Theorem 1:</u> For every covariance matrix K a corresponding EV system (2.10), (2.2) - (2.6) exists. Consider the (static) EV model (2.1) - (2.6). Then, under the assumptions $\det K > 0$ and $c_{x^*x^*} \neq 0$ the set of parameters $\theta = (a, c_{x^*x^*}, c_{uu}, c_{vv})$ compatible with the second moments of the observations c_{xx}, c_{xy}, c_{yy} for the case $c_{xy} > 0$ is given by

$$(2.16)\quad \begin{aligned} \{\theta = (a,\ a^{-1}c_{xy},\ c_{xx}-a^{-1}\cdot c_{xy},\ c_{yy}-ac_{xy}) \in \mathbb{R}^4\ |\ \\ a \in [c_{xy}\cdot c_{xx}^{-1},\ c_{yy}\cdot c_{xy}^{-1}]\} \end{aligned}$$

and analogous for the case $c_{xy} < 0$. If $c_{x^*x^*} \neq 0$ then $c_{xy} = 0$ is equivalent to $a = 0$ and thus a is uniquely determined. However $c_{x^*x^*}$ and c_{uu} are not unique in this case.

This result is due to Gini (1921) and has explicitely been stated by Frisch (1934).

In the case of non Gaussian observations we may utilize information coming <u>from moments (or cumulants) of order greater than two</u> (Geary (1942), Reiersøl (1950)). Here, for simplicity, we assume throughout that all moments exist. If $z_1 \ldots z_n$ are random variables, then their joint n-th order cumulant $c_{z_1 \ldots z_n}$ is given by the coefficient of $(i)^n t_1 \ldots t_n$ in the Taylor series expansion of $\ln E \exp\{i \sum_{j=1}^{n} z_j t_j\}$ about the origin.

If, in our model, in addition we assume:

(2.17) (x_t^*), (u_t) and (v_t) are i.i.d. (i.e. independent and identically distributed), the processes are also (mutually) independent, (and all moments exist)

then, by the properties of cumulants (see Brillinger (1981)), we have the following relations between the n-th order cumulants:

$$(2.18) \quad c_{x^n} = c_{x^{*n}} + c_{u^n}$$

$$(2.19) \quad c_{y^r x^{(n-r)}} = c_{y^{*r} x^{*(n-r)}} = a\, c_{y^{*(r-1)} x^{*(n+1-r)}}; \; r=1..n-1$$

$$(2.20) \quad c_{y^n} = c_{y^{*n}} + c_{v^n}$$

where we have used the notation $c_{z^r w^{n-r}} = c_{\underbrace{z \ldots z}_{r\ \text{times}} \underbrace{w \ldots w}_{n-r\ \text{times}}}$.

Let us assume that x_t^* is <u>non Gaussian</u> (and that $c_{x^*x^*} > 0$, i.e. $x_t^* \not\equiv 0$). Then there is a $n > 2$ such that $c_{x^{*n}} \neq 0$. If in addition we assume $a \neq 0$ then, by (2.19), $c_{y^{r-1} x^{(n+1-r)}} \neq 0$, and

$$(2.21) \quad a = \frac{c_{y^r x^{(n-r)}}}{c_{y^{(r-1)} x^{(n+1-r)}}}; \quad r-1 > 0, \; n-r > 0$$

and thus a is uniquely determined from the cumulants of the observed processes. Note that for $n > 2$ (as opposed to the case $n=2$) there are at least two cumulants of the form $c_{y^r x^{(n-r)}}$, $r > 0$, $n-r > 0$ and for these (2.19) holds. Once a is determined, $c_{x^*x^*}$, c_{uu} and c_{vv} can be uniquely determined from (2.7) - (2.9).

Thus we have shown:

<u>Theorem 2:</u> Consider the static EV model (2.1) - (2.6) and (2.17). Then, under the assumptions $c_{x^*x^*} \neq 0$, x_t^* is non Gaussian and $a \neq 0$, the model is identifiable.

3. Dynamic Models and Second Moments of the Observations: The General Case

We now consider linear dynamic systems; the analysis in this section is based on the second moments of the observations only or, equivalently (as the first moments are assumed to be zero) on the assumption that the observations are Gaussian.

The systems are of the form

(3.1) $\quad y_t^* = w(z)x_t^* \; ; \; t \in \mathbb{Z}$

where (x_t^*) and (y_t^*) are the true input and output processes respectively, where

(3.2) $\quad w(z) = \sum_{i=-\infty}^{\infty} W(i)z^i$

is the transfer function and where z is the backward-shift operator as well as a complex variable. The infinite sum on the r.h.s. of (3.1) is understood in the sense of mean squares convergence of random variables.

In order to provide a symmetric treatment of the variables we did not impose a causality assumption on the system, i.e. the summation in (3.2) is running over the negative integers too.

The observed inputs and outputs respectively are given by

(3.3) $\quad x_t = x_t^* + u_t$

(3.4) $\quad y_t = y_t^* + v_t$

where (u_t) are the input (measurement) errors and (v_t) are the output (measurement) errors.

We assume throughout:

(3.5) All processes considered are (wide sense) stationary

(3.6) $\quad Ex_t^* = Eu_t = Ev_t = 0$

(3.7) $\quad Ex_s^*u_t = Ex_s^*v_t = Eu_sv_t = 0; \quad \forall s,t \in \mathbb{Z}$

(3.8) $\quad (u_t)$ and (v_t) have finite spectral densities.

The assumption $Ex_t^* = 0$ is imposed for notational convenience only and may easily be relaxed. (3.7) is natural in our context; it is just a definition of the errors saying that they contain no common effect between the variables.

If the contrary is not stated explicitely we will also assume that (x_t^*) has a finite spectral density.

If we commence from the second moments of the observed processes (x_t) and (y_t) rather than from (3.1) - (3.4), again we obtain a slightly more general model from the following Frisch problem:

For the given (2×2) spectral density (matrix) $f = \begin{pmatrix} f_{xx} & f_{xy} \\ f_{yx} & f_{yy} \end{pmatrix}$ of $\begin{pmatrix} x_t \\ y_t \end{pmatrix}$ find all decompositions

$$(3.9) \qquad f = \hat{f} + \tilde{f}$$

where $\hat{f}$, $\tilde{f}$ are also spectral densities (corresponding to real processes) and where $\hat{f} = \begin{pmatrix} f_{x^*x^*}, & f_{x^*y^*} \\ f_{y^*x^*}, & f_{y^*y^*} \end{pmatrix}$ is singular (λ-a.e.) and where $\tilde{f} = \begin{pmatrix} f_{uu} & 0 \\ 0 & f_{vv} \end{pmatrix}$ is diagonal. Note that a spectral density f is nonnegative Hermitian (λ-a.e.) and integrable (and that $f(-\lambda) = f'(\lambda)$ for real processes) and that conversely every complex-matrix valued function f on $[-\pi, \pi]$ with these properties is a spectral density (of a real process) (see e.g. Hannan (1970)).

Since $\hat{f}$ is singular (λ-a.e.) there exist $w^{(x)}(e^{-i\lambda})$ and $w^{(y)}(e^{-i\lambda})$ such that

$$(3.10) \qquad w^{(x)}(e^{-i\lambda})(f_{x^*x^*}(\lambda), f_{x^*y^*}(\lambda)) = w^{(x)}(e^{-i\lambda})(f_{y^*x^*}(\lambda), f_{y^*y^*}(\lambda) \qquad \lambda\text{-a.e.}$$

Choosing a suitable normalization, the corresponding system then is of the form

$$w^{(y)}(z)\, y_t^* = w^{(x)}(z) x_t^*.$$

Unless the contrary is stated explicitely we restrict ourselves to the case where

(3.11) $\quad w^{(y)}(z) \neq 0 \quad \text{for } |z| = 1$

i.e. to the case (3.1) with $w(z) = w^{(y)}(z)^{-1} \cdot w^{(x)}(z)$.

Of course, $\tilde{f}$ in (3.9) is the spectral density of the errors $\begin{pmatrix} u_t \\ v_t \end{pmatrix}$.

(3.9) can be written as:

(3.12) $\quad f_{xx} = f_{x^*x^*} + f_{uu}$

(3.13) $\quad f_{yy} = f_{y^*y^*} + f_{vv}$

(3.14) $\quad f_{xy} = f_{x^*y^*}$

and since $\hat{f}$ is singular, we have

(3.15) $\quad |f_{xy}|^2 = f_{x^*x^*} \cdot f_{y^*y^*}$

In addition taking into account the nonnegativity of $\hat{f}$ and $\tilde{f}$ gives the following characterization of the set of all decompositions (3.9) (see Kalman (1981), Anderson and Deistler (1984)).

<u>Lemma 1:</u> For given f, a decomposition (3.9) always exists. The set of all decompositions (3.9) for given f is characterized

(3.16) $\quad 0 \leq f_{x^*x^*} \leq f_{xx} \;; \quad f_{x^*x^*}(\lambda) = f_{x^*x^*}(-\lambda)$

(3.17) $\quad 0 \leq f_{y^*y^*} \leq f_{yy} \;; \quad f_{y^*y^*}(\lambda) = f_{y^*y^*}(-\lambda)$

(3.14) and (3.15).

<u>Proof:</u> The second part is immediate, as every choice of $f_{x^*x^*}$, $f_{y^*y^*}$, f_{xy}, $f_{uu} = f_{xx} - f_{x^*x^*} \cdot f_{vv} = f_{vv} - f_{y^*y^*}$ corresponding to (3.14) - (3.17) gives nonnegative (and integrable) matrices $\hat{f}$ and $\tilde{f}$. The existence of a Frisch decomposition follows from the fact that by the nonnegativity of f we have

$$f_{xx} \cdot f_{yy} \geq |f_{xy}|^2$$

and thus $f_{x^*x^*}$ and $f_{y^*y^*}$ can always be found so to satisfy (3.15).

Thus for (λ-almost) every λ, the set of feasible $(f_{x^*x^*}(\lambda), f_{y^*y^*}(\lambda))$ is part of a hyperbola (depending on λ) (Anderson and Deistler (1984)).

Analogously, the set of feasible f_{uu}, f_{vv} can be described.

Note that once $f_{x^*x^*}$ (or $f_{y^*y^*}$) is fixed and $f_{x^*x^*} > 0$ (or $f_{y^*y^*} > 0$) the Frisch decomposition is unique.

The considerations above include the general case (3.10) where y_t does not necessarely have an explicite solution (3.1). This explicite solution is guaranteed by the assumption

$$(3.18) \qquad f_{x^*x^*}(\lambda) > 0 \qquad \forall\lambda \in [-\pi,\pi] .$$

Then we can define

$$(3.19) \qquad w(e^{-i\lambda}) = f_{yx}(\lambda)\cdot f^{-1}_{x^*x^*}(\lambda).$$

The transfer function $w(z)$ is the main internal characteristic of interest for the model; but we are also interested in $f_{x^*x^*}$, f_{uu} and f_{vv}.

From (3.19) and (3.14) - (3.17) it is clear that, whenever $f_{yx}(\lambda) \neq 0$, then

$$(3.20) \qquad f_{yx}(\lambda)\cdot f^{-1}_{xx}(\lambda) \leqq w(e^{-i\lambda}) \leqq f_{yy}(\lambda)\cdot f^{-1}_{xy}(\lambda)$$

where $a \geqq b$, $a,b \in C$ means that there exists a $,c \in [0,1]$ such that $ca = b$; in other words (3.20) means that the <u>phase</u> $\phi(\lambda)$ of $w(e^{-i\lambda}) = |w(e^{-i\lambda})|\cdot e^{i\phi(\lambda)}$ is uniquely determined and the <u>gain</u> $|w(e^{-i\lambda})|$ satisfies

$$|f_{yx}(\lambda)|\cdot f^{-1}_{xx}(\lambda) \leq |w(e^{-i\lambda})| \leq f_{yy}(\lambda)\cdot|f_{xy}(\lambda)|^{-1} .$$

The other internal characteristics of interest are "parametrized" by $|w(e^{-i\lambda})|$ in the sense that

$$(3.21) \qquad f_{x^*x^*}(\lambda) = |f_{xy}(\lambda)|\cdot|w(e^{-i\lambda})|^{-1}$$

$$(3.22) \qquad f_{uu}(\lambda) = f_{xx}(\lambda) - |f_{xy}(\lambda)||w(e^{-i\lambda})|^{-1}$$

$$(3.23) \qquad f_{vv}(\lambda) = f_{yy}(\lambda) - |f_{xy}(\lambda)||w(e^{-i\lambda})|$$

If $f_{yx}(\lambda) = 0$ then $w(e^{-i\lambda}) = f_{yx}(\lambda) \cdot f^{-1}_{x^*x^*}(\lambda) = 0$ (provided that (3.18) holds).

Now let us discuss some special cases: $\hat{f}$ may have either rank 1 on a set of λ-measure unequal to zero and rank zero (almost) elsewhere or may have rank equal to zero (λ-a.e.). Clearly the second case can occur if and only if f is diagonal. Of course, even if f is diagonal (and if e.g. $f > 0$), then there are also $\hat{f}$ of rank 1; e.g. $\hat{f} = \begin{pmatrix} f_{x^*x^*} & 0 \\ 0 & 0 \end{pmatrix} \neq 0$. This corresponds to the fact that the decompositions with the minimal rank of $\hat{f}$ give the maximal extraction of individual factors from the variables.

If f itself is singular, then $\hat{f} = f$ and $\tilde{f} = 0$ defines a decomposition corresponding to an error-free system. This decomposition is unique whenever $f_{xy}(\lambda) \neq 0$. For λ's where $f_{xy}(\lambda) = 0$ we may have e.g. $f_{yy}(\lambda) = 0$, ($f_{xx}(\lambda) > 0$ and $f_{x^*x^*}(\lambda) > 0$), and $f_{uu}(\lambda) > 0$ gives rise to another decomposition. Of course in this case we have for the corresponding transfer function $w(e^{-i\lambda}) = 0$. For the rest of the paper we always assume that $f(\lambda)$ is nonsingular on a set of Lebesque measure greater than zero.

Let us summarize:

Theorem 3: Consider the (dynamic) EV model (3.1) - (3.8). Then, under the assumption $f_{x^*x^*}(\lambda) \neq 0 \ \forall \lambda$, the set of internal characteristics $(w, f_{x^*x^*}, f_{uu}, f_{vv})$ compatible with the second moments f_{xx}, f_{xy}, f_{yy} of the observations, for $f_{xy}(\lambda) \neq 0 \ \forall\lambda$, is given by (3.23) - (3.26). If $f_{xy}(\lambda) = 0$ then $w(e^{-i\lambda}) = 0$, $f_{vv}(\lambda) = f_{yy}(\lambda)$ and $f_{uu}(\lambda)$ satisfies

$$0 \leq f_{uu}(\lambda) < f_{xx}(\lambda).$$

4. Dynamic Models and Second Moments of the Observations: Conditions for Identifiability

Now let us investigate additional a priori assumptions which guarantee identifiability from the second moments of a model; i.e. which guarantee that for a class of systems (3.1) - (3.4) the internal characteristics of interest, i.e. w and the second moments of (x_t^*), (u_t) and (v_t) are uniquely determined from the (true) second moments of the observed processes (x_t) and (y_t).

From now on we assume that w is causal, i.e. that the sum in (3.2) ranges from zero to infinity only and unless the contrary has been stated, that all transfer functions considered are rational, i.e.:

$$(4.1) \qquad w = a^{-1}\cdot b;\ w_3 = d^{-1}\cdot e;\ w_2 = c^{-1}\cdot h;\ w_4 = f^{-1}\cdot g$$

(where we have omitted the complex argument z and) where

$$a(z) = \sum_{i=0}^{na} A(i)z^i;\quad b(z) = \sum_{i=0}^{nb} B(i)z^i;\quad d(z) = \sum_{i=0}^{nd} D(i)z^i$$

$$(4.2) \qquad e(z) = \sum_{i=0}^{ne} E(i)z^i;\quad c(z) = \sum_{i=0}^{nc} C(i)z^i;\quad h(z) = \sum_{i=0}^{nh} H(i)z^i$$

$$f(z) = \sum_{f=0}^{nf} F(i)z^i;\quad g(z) = \sum_{i=0}^{ng} G(i)z^i$$

and where

$$f_{x^*x^*}(\lambda) = (2\pi)^{-1} w_3(e^{-i\lambda})\cdot\sigma_\varepsilon w_3^*(e^{-i\lambda}),$$

$$f_{uu}(\lambda) = (2\pi)^{-1} w_2(e^{-i\lambda})\cdot\sigma_\mu w_2^*(e^{-i\lambda}),$$

$$f_{vv}(\lambda) = (2\pi)^{-1} w_4(e^{-i\lambda})\cdot\sigma_\nu w_4^*(e^{-i\lambda}).$$

Here

$$w_i^*(z) = w_i(z^{-1}).$$

In the rational case, we also assume:

(4.3) a,b are relatively prime and so are d,e and c,h and f,g.

$$(4.4) \qquad d(z) \neq 0 \quad |z| \leq 1;\qquad e(z) \neq 0 \quad |z| < 1$$

$$c(z) \neq 0 \quad |z| \leq 1;\qquad h(z) \neq 0 \quad |z| < 1$$

$$f(z) \neq 0 \quad |z| \leq 1;\qquad g(z) \neq 0 \quad |z| < 1$$

$$(4.5) \qquad d(0) = e(0) = c(0) = h(0) = f(0) = g(0) = 1$$

All these assumptions are costless in the sense they do not restrict the class of transfer functions w and spectra $f_{x^*x^*}$, f_{uu}, f_{vv} considered, but serve only as norming conditions to obtain unique parameters. We in addition assume

$$(4.6) \qquad a(z) \neq 0 \qquad |z| \leq 1; \quad a(0) = 1 .$$

The first part of (4.6) is an a priori causality assumption. Note that we do not impose the miniphase assumption

$$\det b(z) \neq 0 \quad |z| < 1 \quad .$$

This would not be justified in our context as our interest is in the true transfer function w of the system and not only in a spectral factor. Analogously we do not assume $b(0) = 1$.

In addition we assume that the degrees $na,\dots,nh$ of the polynominals $a(z),\dots,h(z)$ in (4.2) have been prescribed a priori. This means that we are considering a class of EV-systems where the actual degrees $\delta a,\dots,\delta h$ say, of the respective polynominals are bounded by the respective prescribed degrees $na,\dots,nh$.

Now, the <u>parameters of interest</u> are $\theta = (A(1),\dots,A(n0), B(0),\dots,B(nb), D(1),\dots,D(nd), E(1),\dots,E(ne), C(1),\dots,C(nc), H(1),\dots,H(nh), F(1),\dots,F(nf), G(1),\dots,G(ng), c_{\varepsilon\varepsilon}, c_{\mu\mu}, c_{\nu\nu}$. These parameters θ are in one-to-one relation with the corresponding $(w, f_{x^*x^*}, f_{uu}, f_{vv})$ (see e.g. Hannan (1970)) and our <u>parameterspace</u> Θ is a subset of the Euclidian $\mathbb{R}^n$; $n=na+\dots+ng+4$. Two parameters θ_1 and θ_2 are called <u>observationally equivalent</u> if they correspond to the same characteristics of the observations i.e. in the case considered here to the same f. Θ is called <u>identifiable</u> if it contains no different observationally equivalent parameters.

Now, equations (3.12), (3.13) and (3.22) can be written as:

$$(4.7) \qquad \tilde{f}_{xx}(z) = d^{-1}(z)e(z)\sigma_\varepsilon \cdot e^*(z) \cdot d^{-1*}(z) + c^{-1}(z)h(z)\sigma_\mu \cdot h^*(z) \cdot c^{-1*}(z)$$

$$(4.8) \qquad \tilde{f}_{xy}(z) = d^{-1}(z)\, e(z)\sigma_\varepsilon \cdot e^*(z) \cdot d^{-1*}(z) \cdot b(z)a^{-1*}(z)$$

$$(4.9)\quad \tilde{f}_{yy}(z) = a^{-1}(z)b(z)\cdot d^{-1}(z)e(z)\sigma_\varepsilon\cdot e^*(z)d^{-1}(z)b(z)a^{-1*}(z) + f^{-1}(z)g(z)\sigma_\nu\cdot g^*(z)f^{-1}(z)$$

where $\tilde{f}_{xx}(z)$ is the (unique) rational extension of $2\pi f_{xx}(e^{-i\lambda}) = 2\pi\hat{f}_{xx}(\lambda)$ from the unit circle to C (except for the poles).

In the following we discuss some special cases:

(i) Let the inputs (x_t^*) be stationary with spectral distribution function $F_{x^*x^*}$ given by:

$$(4.10)\quad F_{x^*x^*}(\lambda) = \int_{[-\pi,\lambda]} f_{x^*x^*}d\lambda + \sum_{j:\lambda_j\le\lambda} F_{x^*x^*,j}\ ,$$

$$F_{x^*x^*,j} > 0$$

i.e $F_{x^*x^*}$ consists of an absolutely continuous and a discrete part (see e.g. Hannan (1970)): (u_t) and (v_t) have spectral densities. Then, using an obvious notation we obtain:

$$(4.11)\quad F_{xx}(\lambda) = \int_{[-\pi,\lambda]} (f_{x^*x^*} + f_{uu})d\lambda + \sum_{j:\lambda_j\le\lambda} F_{x^*x^*,j}$$

and

$$(4.12)\quad F_{xy}(\lambda) = \int_{[-\pi,\lambda]} w^*(e^{-i\lambda})dF_{x^*x^*}(\lambda)$$

$$= \int_{[-\pi,\lambda]} w^*(e^{-i\lambda})(f_{x^*x^*} + f_{uu})d\lambda + \sum_{j:\lambda_j\le\lambda} w^*(e^{-i\lambda_j})F_{x^*x^*,j}\quad .$$

Thus, from the jumps $F_{xx,j}$ and $F_{xy,j}$ in F_{xx} and F_{xy} we obtain:

$$(4.13)\quad w^*(e^{-i\lambda_j}) = F_{xy,j}(\lambda_j)\cdot F^{-1}_{xx,j}(\lambda_j)\quad .$$

If $w = a^{-1}\cdot b$ is rational with prescribed degrees na and nb for a and b respectively then w is determined under our assumptions from $na + nb + 1$ (different) values (see e.g. Stöhr (1979)), $(\lambda_j, w(e^{-i\lambda_j}))$, $j=1\ldots na+nb+1$. Once w is determined, if $w(z)\neq 0$ $|z| = 1$, then also the rest is determined from (3.21) - (3.23).

(ii) If it is a priori known, that the input errors satisfy

$$f_{uu}(\lambda) = 0 \qquad \forall\, \lambda \in A \subset [0,\pi]$$

where A contains a nontrivial open set, then from (3.12) and (3.19) we have

$$f_{x^*x^*}(\lambda) = f_{xx}(\lambda) \qquad \forall \lambda \in A$$

and

$$(4.14) \qquad w(e^{-i\lambda}) = f_{yx}(\lambda) \cdot f_{xx}^{-1}(\lambda) \qquad \forall\, \lambda \in A$$

(if $f_{x^*x^*}(\lambda) > 0$). Thus, as A contains an open set, the rational function w is uniquely determined by (4.14), e.g. from the derivatives of $w(e^{-i\lambda})$ at an interior point of A.

(iii) If the inputs satisfy

$$f_{x^*x^*}(\lambda) = 0 \quad \forall\, \lambda \in A \quad [0,\pi]$$

where again A contains a nontrivial open set, then analogously as in (ii), $c^{-1} \cdot h$ and σ_ν are determined from $f_{xx}(\lambda) = f_{uu}(\lambda)$, $\lambda \in A$ and thus also $f_{x^*x^*}$, w and f_{vv} are unique.

Now we shall turn to a more general condition. Let us write b as

$$b = b^+ \cdot b^-$$

where b^+ has no zeros inside the unit circle and b^- has no zeros outside or on the unit circle and where $b^-(0) = 1$.

For the next theorem see Nowak (1983) and Anderson and Deistler (1984):

<u>Theorem 4:</u> A class of EV systems satisfying

(4.15) $\quad b^+$ and b^{-*} have no common zeros

(4.16) $\quad a$ and e have no common zeros

(4.17) $\quad d$ and b^{-*} have no common zeros

(4.18) $\quad d$ and c have no common zeros

(4.19) $e(z) \neq 0 \quad |z| \leq 1$

(4.20) $\delta d > 0$

is identifiable.

The proof of this theorem is given in Anderson and Deistler (1984).

Condition (4.20) is a requirement of minimal dynamics. A more general condition is given in Nowak (1983). Such a condition of minimal dynamics is nessesary in some form, as by Theorem 1 the static model is nonidentifiable from second moments. Conditions (4.15) - (4.19) are generic, they are fulfilled "almost everywhere" in the parameter space. E.g. condition (4.16) is to exclude polezero constellation between the system and the input-generating mechanism.

6. Dynamic Models: Identifiability from Higher Order Moments

Analogously to the static case, we now investigate the problem of identifiability of dynamic systems utilizing information coming from moments of order greater than two (see Deistler (1984)). We here assume: (3.1) - (3.8) and in addition

(6.1) (x_t^*), (u_t) and (v_t) are strictly stationary processes; the processes are (mutually) independent and all moments exist and the cumulants of (x_t^*) and of (y_t^*) satisfy conditions of the form

$$\sum_{t_1..t_{n-1}=-\infty}^{\infty} |c_{y^*_{t_1} y^*_{t_2} .. y^*_{t_r} x^*_{t_{r+1}} ... x^*_{t_{n-1}} x_o}| < \infty$$

and the same holds for (u_t) and (v_t).

Then (see e.g. Brillinger (1981)) the corresponding n-th order cumulant spectrum exists and is given by

(6.2)
$$f_{y^{*r}x^{*(n-r)}}(\lambda_1..\lambda_{n-1})$$
$$= (2\pi)^{-n+1} \sum_{t_1..t_{n-1}=-\infty}^{\infty} c_{y^*_{t_1}..y^*_{t_r} x^*_{t_{r+1}}..x^*_{t_{n-1}} x^*_o} \exp\{-i \sum_{j=1}^{n-1} \lambda_j t_j\}$$

and analogously for (u_t) and (v_t). As easily seen, due to linearity and continuity of cumulants with respect to one variable (when the others are kept constant), we obtain from (6.2) and (3.1):

(6.3) $$f_{y^{*r}x^{*(n-r)}}(\lambda_1\ldots\lambda_{n-1})$$

$$= (2\pi)^{-n+1} \cdot \sum_{t_1..t_{n-1}=-\infty}^{\infty} \Big(\sum_{i=-\infty}^{\infty} W(i)\, c_{x^*_{t_1-i}\, y^*_{t_2}..y^*_{t_r}..x^*_0} \Big) \exp\Big\{-i \sum_{j=1}^{n-1} \lambda_j t_j\Big\}$$

$$= w(e^{-i\lambda_1}) \cdot f_{y^{*(r-1)}x^{*(n-r+1)}}(\lambda_1\ldots\lambda_{n-1}).$$

Furthermore from the properties of the cumulants we obtain

(6.4) $$f_{y^r x^{n-r}}(\lambda_1\ldots\lambda_{n-1})$$

$$= f_{y^{*r}x^{*(n-r)}}(\lambda_1\ldots\lambda_{n-1}) + f_{v^r u^{(n-r)}}(\lambda_1\ldots\lambda_{n-r})$$

and

(6.5) $$f_{v^r u^{(n-r)}}(\lambda_1\ldots\lambda_{n-r}) = 0 \quad \text{for } r > 0,\ n-r > 0$$

Thus we have:

<u>Theorem 5:</u> Consider the (dynamic) EV-model (3.1) - (3.8) and (6.1).

If

(6.6) $$f_{y^{(r-1)}x^{(n-r+1)}}(\lambda_1\lambda_2\ldots\lambda_{n-1}) \neq 0 \qquad \forall\, \lambda_1,$$

for suitable $\lambda_2\ldots\lambda_{n-1}$ and suitable $n > 2;\ r-1 > 0,\ n-r > 0$ then w is uniquely determined from

(6.7) $$w(e^{-i\lambda_1}) = f_{y^r x^{(n-r)}}(\lambda_1\ldots\lambda_{n-1}) \cdot f^{-1}_{y^{(r-1)}x^{(n-r+1)}}(\lambda_1\ldots\lambda_{n-1}).$$

If (x^*_t) has a Wold decomposition

$$x^*_t = w_3(z)\,\varepsilon_t$$

where (ε_t) is i.i.d., then (see e.g. Brillinger (1981)).

$$(6.8)\quad f_{x*^n}(\lambda_1\ldots\lambda_{n-1}) = (2\pi)^{-n+1}\cdot w_3(e^{-i\lambda_1})\ldots w_3(e^{-i\lambda_{n-1}}) \cdot w_3(\exp i \sum_{i=1}^{n-1}\lambda_j)\cdot c_{\varepsilon^n}$$

Thus if all moments of ε_t exist, if ε_t is non Gaussian (provided that $\varepsilon_t \neq 0$) there is a $n>2$ such that $c_{\varepsilon^n} \neq 0$. If in addition $w_3(e^{-i\lambda}) \neq 0 \quad \forall\lambda$, then $f_{x*^n}(\lambda_1\ldots\lambda_{n-1}) = 0$ $\forall\ \lambda_1\ldots\lambda_{n-1}$ and then due to (6.3) condition (6.6) is fulfilled.

References

Aigner,D.J. and A.S.Goldberger (eds.), (1977): Latent Variables in Socio-Economic Models. North Holland P.C., Amsterdam

Aigner,D.J., C.Hsiao, A.Kapteyn and T.Wansbeek (1984): Latent Variable Models in Econometrics. In: Griliches, Z. and M.D. Intrilitator (Eds.) Handbook of Econometrics. North Holland P.C., Amsterdam

Anderson, B.D.O. and M.Deistler (1984): Identifiability in Dynamic Errors-in-Variables Models. Journal of Time Series Analysis 5, 1-13

Brillinger, D.R. (1981): Time Series: Data Analysis and Theory. Expanded Edition. Holden Day, San Francisco

Deistler,M. (1984): Linear Dynamic Errors-in-Variables Models. Paper for: Gani,J. and M.B.Priestly (Eds.) Essays in Time Series and Allied Processes. Festschrift in honour of Professor E.J.Hannan

Deistler,M. and H.G.Seifert (1978): Identifiability and Consistent Estimability in Dynamic Econometric Models. Econometrica 46, 969 - 980

Frisch,R. (1934): Statistical Confluence Analysis by Means of Complete Regression Systems. Publication No. 5, University of Oslo, Economic Institute

Geary,R.C. (1942): Inherent Relations between Random Variables. Proceedings of the Royal Irish Academy. Sec. A, 47, 63-76

Geary,R.C. (1943): Relations between Statistics: The General and the Sampling Problem When the Samples are Large. Proceedings of the Royal Irish Academy. Se. A, 49, 177 - 196

Gini,C. (1921): Sull'interpolazione di una retta quando i valori della variable independente sono affetti da errori accidentali. Metron 1, 63 - 83

Hannan,E.J. (1970): Multiple Time Series. Wiley, New York

Kalman,R.E. (1982): System Identification from Noisy Data. In: A.Bednarek and L.Cesari (Eds.) Dynamical Systems II, a University of Florida International Symposium, Academic Press, New York

Madansky,A. (1959): The fitting of Straight Lines when Both Variables are Subject to Error. Journal of the American Statistical Association 54, 173 - 205

Maravall,A. (1979): Identification in Dynamic Shock-Errors Modells. Springer-Verlag, Berlin

Moran,P.A.P. (1971): Estimating Structural and Functional Relationships. Journal of Multivariable Analysis 1, 232 - 255

Nowak,E. (1983): Identification of the Dynamic Shock-Error Model with Autocorrelated Errors. Journal of Econometrics 23, 211 - 221

Reiersøl,O. (1941): Confluence Analysis by Means of Lag Moments and other Methods of Confluence Analysis. Econometrica 9,,1 - 24

Reiersøl,O. (1950): Identifiability of a Linear Relation Between Variables which are Subject to Error. Econometrica 18, 375 - 389

Schneeweiß,H. und H.J.Mittag (1984): Lineare Modelle mit fehlerbehafteten Daten. Physica Verlag, Würzburg

Söderström, T.(1980): Spectral Decomposition with Application to Identification. In: Archetti,F. and M. Cugiani (Eds.) Numerical Techniques for Stochastic Systems. North Holland P.C., Amsterdam

Stoer,J. (1979): Einführung in die Numerische Mathematik I. Springer-Verlag, Berlin

Wegge,L. (1983): ARMAX-Models Parameter Identification without and with Latent Variables. Working Paper. Dept. of Economics, Univ. of California, Davis.

MINIMAX-ROBUST FILTERING AND FINITE-LENGTH ROBUST PREDICTORS

Jürgen Franke
Fachbereich Mathematik
Johann Wolfgang Goethe-
Universität Frankfurt/Main
D-6000 Frankfurt/Main 1

H. Vincent Poor
Coordinated Science Laboratory
University of Illinois
1101 W. Springfield Avenue
Urbana, Illinois 61801 USA

1. Introduction

This paper consists basically of two parts. In the first part (Sections 2 and 3), we consider some aspects of linear filtering under uncertainty about the spectral features of the relevant time series. The second part (Sections 4 and 5), considers a time domain approach to the general problem of filtering under probabilistic uncertainty and applies this general formulation to aspects of the linear filtering problem. Each of these approaches represents a different means for solving the minimax robust filtering problem, and both are applied to the specific problem of robust q-step prediction to illustrate the similarities of the two approaches.

To introduce the problem of interest suppose we have an observation process $\{X_k\}_{k=-\infty}^{\infty}$ which is a noisy version of a random signal $\{S_k\}_{k=-\infty}^{\infty}$:

$$X_k = S_k + N_k \quad , \qquad -\infty < k < \infty.$$

We assume that the signal $\{S_k\}$ and noise $\{N_k\}$ are uncorrelated, zero-mean, real-valued weakly stationary stochastic processes with spectral densities f_S and f_N, respectively. Our goal is to form a "good" linear estimate of a linear transformation of $\{S_k\}$, the so-called desired signal:

$$S_k^d = \sum_{j=-\infty}^{\infty} d_j S_{k-j} \quad . \tag{1.1}$$

The estimate depends only on a certain given part of the observation process (e.g., $\{X_\ell\}_{\ell=-\infty}^{k}$). As a performance criterion we consider exclusively the mean-square difference of the desired signal and its estimate.

This general linear filtering problem encompasses a whole range of more familiar filtering problems. For example, suppose d_j vanishes for all $j \neq -q \leq 0$ and $d_{-q} = 1$. Then the desired signal at time k is the original signal at time k+q. If the whole observation process is given, then we have a noncausal filtering problem. If only observations up to time k+q are available, then we want to estimate S_{k+q} from the past and present of the observation process - we call this the causal filtering problem. If there is no noise, i.e. $f_N \equiv 0$, then signal and observation process coincide. If in this case the observations up to time k are given, then we have to predict S_{k+q} from the S_j, $j \leq k$ - we call this the q-step prediction problem.

The traditional theory of linear filtering, due mainly to Wiener and Kolmogorov, requires exact knowledge of signal and noise spectral densities f_S and f_N. However, in practice one usually has only some partial spectral information available. The usual way of proceeding is to consider parametric or nonparametric spectral estimates for f_S, f_N, or simple theoretically or computationally motivated nominal spectral densities. Then, one applies the traditional filter theory and chooses the optimal filter pretending that the estimated or nominal spectral densities are the true ones. This procedure can result in dangerous losses in performance as Vastola and Poor (1983) have demonstrated with some examples. Therefore, it makes sense to search for linear filters which are not optimal for one nominal pair f_S, f_N but which behave uniformly well over a class of spectral density pairs that are compatible with our information on the spectral features of signal and noise.

The study of linear filters which perform uniformly well under spectral uncertainty have found some interest during the past decade. Breiman (1973), Kassam and Lim (1977), Hosoya (1978), and Taniguchi (1981) have discussed particular filtering problems under special types of spectral information. Poor (1980), Franke (1981a,b), and Kassam (1982) have given general formulations for the problems of noncausal filtering, one-step prediction and interpolation and described general methods for determining explicitly the appropriate filters for large classes of spectral information.

Recently, Vastola and Poor (1984) have investigated the general filtering problem described above, and have given theorems on the existence and characterization of linear filters which satisfy the requirement of uniformly good performance against spectral uncertainty. We follow their approach and introduce in Section 2 the notion of a minimax-robust linear filter with respect to given spectral information. Then we present the theorems of Vastola and Poor under weaker conditions making them handier for application to some important types of spectral information which are discussed in more detail in Section 3.

In Section 3 we show how the theorems of Vastola and Poor can be used directly to calculate minimax-robust filters. We illustrate this general method with some examples. In particular, we demonstrate that if we have no noise and if we consider q-step prediction of the signal processes, then the minimax-robust predictor will automatically be finite length (i.e. it depends only on a finite part of the observation process) if our spectral information consists only of information about a finite part of the autocovariance sequence of the process which we want to predict. For 1-step prediction this result was obtained previously by Franke (1981a). A similar result holds in the presence of noise, as well.

As noted above, Sections 4 and 5 consider a treatment of general minimax robust filtering in a time-domain (rather than a frequency-domain) setting. This is accomplished by first considering general nonlinear MMSE estimation with uncertainty in the probability structure on the underlying probability space. Section 4 considers the general structure of this problem while Section 5 illustrates the application of this approach with the minimax q-step prediction problem that is also treated in Section 3. Section 5 also considers the problems of minimax linear filtering and of minimax nonlinear prediction under information only about the statistics of finite blocks of data. Again the minimax robust solutions are seen to be finite-length structures in each of these cases.

As an additional consequence of the general approach in Section 4 we can give a precise formulation of the common belief that linear filtering is the best one can do if one has only information on the moments up to second order of the time series concerned. We show that linear filters are minimax under all filters based on a certain amount of information with respect to sets of measures on processes that are defined by requirements on second moments only. Of course, it is well known that one can do much better by nonlinear filtering for non-Gaussian time series. Our result has to be interpreted in the following sense: If one cannot exclude the Gaussian case then linear filters, which are optimal for the Gaussian case, give the least possible upper bound on the filtering error. This error can be rather large, but every nonlinear filter can result under these circumstances in an even greater mean-square error.

2. Minimax-Robust Filters- Existence and Characterization in the Frequency Domain

Due to stationarity, we may consider without loss of generality the problem of estimating the desired signal, given by (1.1), at a fixed time instant, say $k = 0$. Let

$$X_{-k},\ k \in \mathcal{T}$$

be that part of the observation process which is known to us. Here, the index set $\mathcal{T}$ is a given subset of the integers. We consider only linear estimators for S_0^d of the form

$$(2.1) \qquad \hat{S}_0 = \sum_{k \in \mathcal{T}} h_k X_{-k}, \quad \text{where} \quad \sum_{k \in \mathcal{T}} |h_k|^2 < \infty.$$

Let L^2 denote the Hilbert-space of measurable functions on $(-\pi,\pi]$ that are square-integrable with respect to Lebesgue measure. Let H denote the filter transfer function of the estimation filter given by (2.1), i.e.

$$H(\omega) = \sum_{k \in \mathcal{T}} h_k e^{-ik\omega}, \quad -\pi < \omega \leq \pi.$$

H is contained in the subset

$$(2.2) \qquad \mathcal{L} = \{G \in L^2;\ \frac{1}{2\pi} \int e^{ik\omega}\, G(\omega)d\omega = 0 \quad \text{for all } k \notin \mathcal{T}\}$$

of L^2 which consists of all functions in L^2 for which certain Fourier coefficients vanish. Here and in the following, the range of integration is always the interval $(-\pi,\pi]$. Functions in $\mathcal{L}$ and admissible linear estimation filters of the form (2.1) correspond to each other in a unique fashion. Therefore, with a slight abuse of language, we call functions in $\mathcal{L}$ simply filters as well.

For the sake of simplicity, we restrict our attention to desired signals for which the Fourier transform

$$D(\omega) = \sum_{k=-\infty}^{\infty} d_k e^{-ik\omega}$$

of the coefficient sequence d_k, $-\infty < k < \infty$, is a bounded function.

If we use an estimate $\hat{S}_0$, given by (2.1), then the mean-square error depends only on the transfer function H and on the spectral densities f_S, f_N of signal and noise, and it is given by the well-known formula

$$E\{(S_0^d - \hat{S}_0)^2\} = \frac{1}{2\pi} \int \{|D(\omega)-H(\omega)|^2 f_S(\omega) + |H(\omega)|^2 f_N(\omega)\} d\omega$$
$$=: e(f_S, f_N; H).$$

We call the function e, depending on the signal and noise spectral densities and on the filter H, the error function.

In the traditional Wiener-Kolmogorov filtering theory f_S, f_N are fixed. Under weak regularity assumptions, there exists a unique filter $H_W \in \mathcal{L}$, which we call the Wiener filter with respect to f_S, f_N, minimizing the mean-square error, i.e.

$$e(f_S, f_N; H_W) = \min_{H \in \mathcal{L}} e(f_S, f_N; H) \quad .$$

In minimax-robust filtering theory we do not start from precisely known signal and noise spectral densities. Rather, we follow Huber's (1964) fundamental approach to robust estimation. We assume that we have only partial information on f_S and f_N. This knowledge is summarized in the statement that (f_S, f_N) is contained in a given set $\mathcal{J}$ of pairs of spectral densities, which we call the spectral information set. Instead of searching for a filter that is optimal for one particular pair of signal and noise spectral densities we want to find a filter for which the error function is uniformly bounded over $\mathcal{J}$ and for which the uniform bound on the mean-square error is as small as possible. A filter $H_R \in \mathcal{L}$, satisfying these requirements, i.e.

$$\text{(2.3)} \quad \sup_{(f_S,f_N)\in\mathscr{S}} e(f_S,f_N;H_R) = \min_{H\in\mathfrak{L}} \sup_{(f_S,f_N)\in\mathscr{S}} e(f_S,f_N;H)$$

is called a minimax-robust filter with respect to $\mathscr{S}$.

The following theorem guarantees the existence of such a filter for almost any reasonable spectral information set. Except for some slight relaxation of the assumptions, this theorem coincides with Theorem 1 of Vastola and Poor (1984). In particular, we allow for unbounded spectral densities as this makes it more convenient to apply the theorem to certain spectral information sets which we want to consider below.

<u>Theorem 2.1</u>: *Let $\mathfrak{L} \subseteq L^2$ be of the form (2.2), and let the spectral information set $\mathscr{S}$ satisfy*

(i) $\frac{1}{2\pi} \int f_S(\omega)d\omega \leq c_S < \infty$ *for all* $(f_S,f_N) \in \mathscr{S}$,

i.e. the total power of admissible signals is uniformly bounded.

(ii) There exists a pair $(f_S^*,f_N^*) \in \mathfrak{L}$ *and* $\Delta > 0$ *such that* $f_S^*(\omega) + f_N^*(\omega) \geq \Delta > 0$ a.e.

Then, there exists a filter $H_R \in \mathfrak{L}$ which is minimax-robust with respect to $\mathscr{S}$.

<u>Proof</u>: (1) Let $H_0 \equiv 0$ be the zero filter, which is contained in $\mathfrak{L}$. By the boundedness of D and by assumption (i)

$$\inf_{\mathfrak{L}} \sup_{\mathscr{S}} e(f_S,f_N;H) \leq \sup_{\mathscr{S}} e(f_S,f_N;H_0)$$

$$= \sup_{\mathscr{S}} \frac{1}{2\pi} \int |D(\omega)|^2 f_S(\omega)d\omega$$

$$\leq c_S \sup_{\omega} |D(\omega)|^2 =: M < \infty.$$

In particular, we have to search for minimax-robust filters only among those filters satisfying

$$\text{(2.4)} \quad e(f_S,f_N;H) \leq M \qquad \text{for all } (f_S,f_N) \in \mathscr{S} .$$

(2) Let $\|.\|$ denote the Hilbert-space norm of L^2, let the constant μ be given by

$$\mu = \sqrt{2}\ (\|D\| + \sqrt{4\pi M/\Delta}) ,$$

and let $\mathfrak{L}(\mu)$ denote the subset of that $H \in \mathfrak{L}$ for which $\|H\| \leq \mu$. Below we prove that

(2.5) $e(f_S^*, f_N^*; H) > M \qquad \text{if } \|H\| > \mu$.

Therefore, by (2.4), all candidates for the minimax-robust filter are contained in $\mathcal{L}(\mu)$, i.e. we have

(2.6) $\inf_{\mathcal{L}} \sup_{\mathcal{S}} e(f_S, f_N; H) = \inf_{\mathcal{L}(\mu)} \sup_{\mathcal{S}} e(f_S, f_N; H)$.

To prove (2.5), let

$B = \{\omega; f_S^*(\omega) \geq \Delta/2\}$ and $C = \{\omega; f_S^*(\omega) < \Delta/2\} \subseteq \{\omega; f_N^*(\omega) \geq \Delta/2\}$ a.e., where the last relation follows from assumption (ii). Let 1_B, 1_C denote the indicators of the sets B,C, and suppose H satisfies

$$\mu^2 < \|H\|^2 = \|1_B H\|^2 + \|1_C H\|^2, \text{ implying } \|1_B H\|^2 > \mu^2/2 \text{ or}$$

(2.7a) $$\|1_C H\|^2 > \mu^2/2 \geq 4\pi M/\Delta \quad .$$

The condition $\|1_B H\|^2 > \mu^2/2$ implies

(2.7b) $$\begin{aligned} \|1_B(D-H)\|^2 &\geq (\|1_B D\| - \|1_B H\|)^2 \\ &> (\|D\| + \sqrt{4\pi M/\Delta} - \|1_B D\|)^2 \\ &\geq 4\pi M/\Delta \quad . \end{aligned}$$

By definition of the error functional and of the sets B, C, we finally have

$$\begin{aligned} e(f_S^*, f_N^*; H) &= \frac{1}{2\pi} \int |D(\omega) - H(\omega)|^2 f_S^*(\omega)\, d\omega + \frac{1}{2\pi} \int |H(\omega)|^2 f_N^*(\omega)\, d\omega \\ &\geq \frac{\Delta}{4\pi} \{\|1_B(D-H)\|^2 + \|1_C H\|^2\} \\ &> M \end{aligned}$$

as at least one of the inequalities (2.7a), (2.7b) is satisfied.

(3) Before showing that the infimum on the right-hand side of (2.6) is attained we have to derive some properties of $\sup_{\mathcal{S}} e(f_S, f_N; \cdot)$ as a functional on L^2. If f_S, f_N are bounded spectral densities then $e(f_S, f_N; \cdot)$ is a continuous functional on L^2, as follows from some elementary inequalities (Vastola and Poor, 1984, proof of Theorem 1). For arbitrary f_S, f_N let f_S^C, f_N^C denote the truncated densities

$$f_S^C(\omega) = \min\{f_S(\omega), C\} \quad \text{and} \quad f_N^C(\omega) = \min\{f_N(\omega), C\}.$$

By a monotone convergence argument

$$e(f_S,f_N;\cdot) = \sup_{C>0} e(f_S^C,f_N^C;\cdot) .$$

The supremum of lower semicontinuous functionals is lower semicontinuous again (Hewitt,Stromberg, 1965), and we conclude that, first $e(f_S,f_N;\cdot)$ for arbitrary spectral densities and, then, $\sup_{\mathscr{S}} e(f_S,f_N;\cdot)$ are lower semicontinuous functionals on L^2. Furthermore, the latter functional is convex as the supremum of convex functionals.

(4) $\mathfrak{L}(\mu)$ is a convex, closed and bounded subset of the Hilbert-space L^2, and $\sup_{\mathscr{S}} e(f_S,f_N;\cdot)$ is lower semicontinuous and convex. By Theorem 1.2 in Chapter 2 of Barbu and Precupanu (1978) the infimum on the right-hand side of (2.6) is achieved by some H_R which, by (2.6), is the desired minimax-robust filter with respect to $\mathscr{S}$.

QED

Before we can formulate the second theorem, which gives a characterization of minimax-robust filters for a wide range of spectral information sets $\mathscr{S}$, we have to introduce the notion of a least favorable pair of spectral densities. Least favorability corresponds to that situation compatible with our spectral information where traditional Wiener-Kolmogorov filtering gives rise to the largest error. Intuitively, achieving a small filtering error is most difficult if the least favorable pair happens to be the true pair of signal and noise spectral densities.

Definition: $(f_S^L,f_N^L) \in \mathscr{S}$ *is a least favorable pair for the spectral information set* $\mathscr{S}$ *iff*

$$(2.8) \quad \min_{H\in\mathfrak{L}} e(f_S^L,f_N^L;H) = \max_{(f_S,f_N)\in\mathscr{S}} \min_{H\in\mathfrak{L}} e(f_S,f_N;H) .$$

The following theorem generalizes Theorem 2 of Vastola and Poor (1984) to situations where unbounded signal and noise spectral densities cannot easily be excluded from considerations and where the assumption that all admissible f_S,f_N are bounded away from 0 uniformly over $\mathfrak{L}$ does not seem natural. The theorem characterizes minimax-robust filters as Wiener filters of least favorable pairs provided the latter exist, which, however, is true in a wide range of situations. Intuitively, the theorem tells us that we get a uniformly satisfactory performance if we choose that filter which is optimal in the worst possible circumstances.

<u>Theorem 2.2</u>: *Let* $\mathfrak{L} \subseteq L^2$ *be of the form (2.2), and let* $\mathcal{J}$ *be a convex spectral information set satisfying assumptions (i) and (ii) of Theorem 2.1. Let* $(f_S^L, f_N^L) \in \mathcal{J}$ *, and let* H_W^L *be the Wiener filter with respect to* f_S^L, f_N^L*. Then* (f_S^L, f_N^L) *is the least favorable pair for* $\mathcal{J}$ *iff*

$$(2.9) \quad e(f_S^L, f_N^L; H_W^L) = \min_{H \in \mathfrak{L}} \sup_{(f_S, f_N) \in \mathcal{J}} e(f_S, f_N; H) \quad .$$

In particular, the Wiener filter with respect to a least favorable pair for $\mathcal{J}$ *is minimax-robust with respect to* $\mathcal{J}$.

<u>Proof</u>: For $\delta > 0$ let

$$\mathcal{J}(\delta) = \{(f_S, f_N) \in \mathcal{J} \; ; \; f_S(\omega) + f_N(\omega) \geq \delta > 0 \quad \text{a.e.}\} \quad .$$

In exactly the same manner as in the proof of Theorem 2.1 can we modify the proof of Theorem 2 of Vastola and Poor (1984) and conclude that

$$(2.10) \quad \min_{\mathfrak{L}} \sup_{\mathcal{J}(\delta)} e(f_S, f_N; H) = \sup_{\mathcal{J}(\delta)} \min_{\mathfrak{L}} e(f_S, f_N; H)$$

Applying the technical lemma 2.1 below to $\mathcal{X} \mathrel{\hat{=}} \mathcal{J}$, $\mathcal{X}(\delta) \mathrel{\hat{=}} \mathcal{J}(\delta)$, $\mathcal{Y} \mathrel{\hat{=}} \mathfrak{L}$, $F \mathrel{\hat{=}} e$, we conclude from (2.10) that

$$(2.11) \quad \min_{\mathfrak{L}} \sup_{\mathcal{J}} e(f_S, f_N; H) = \sup_{\mathcal{J}} \min_{\mathfrak{L}} e(f_S, f_N; H)$$

(As, by Theorem 2.1, there is a minimax-robust filter, and as for each pair (f_S, f_N) the Wiener filter exists we may write "$\min_{\mathfrak{L}}$" instead of "$\inf_{\mathfrak{L}}$" on both sides of (2.11).)

By the same argument which Vastola and Poor used at the end of the proof of their Theorem 2, (2.11) implies the assertion of the theorem. QED

<u>Lemma 2.1</u>: *Let* $\mathcal{X}$ *be a convex subset of a real vector space, and let* $\mathcal{Y}$ *be an arbitrary set. Let* F *be a function from* $\mathcal{X} \times \mathcal{Y}$ *into* $[0,\infty]$ *such that* $F(.,y)$ *is a concave function on* $\mathcal{X}$ *for all* $y \in \mathcal{Y}$.

Let $\mathcal{X}(\delta)$, $\delta > 0$, *be a decreasing sequence of subsets of* $\mathcal{X}$, *i.e.* $\mathcal{X}(\varepsilon) \subseteq \mathcal{X}(\delta) \subseteq \mathcal{X}$ *for all* $0 < \delta \leq \varepsilon$, *and for some* $\Delta > 0$ *let there exist an* $x^* \in \mathcal{X}(\Delta)$ *for which*

$$(1-\gamma)x + \gamma x^* \in \mathcal{X}(\gamma\Delta) \quad \textit{for all } x \in \mathcal{X}, \; 0 < \gamma \leq 1.$$

If

$$(2.12) \quad \inf_{y \in \mathcal{Y}} \sup_{x \in \mathcal{X}(\delta)} F(x,y) = \sup_{x \in \mathcal{X}(\delta)} \inf_{y \in \mathcal{Y}} F(x,y) \quad \textit{for all } 0 < \delta \leq \Delta$$

then we also have

(2.13) $\inf_{y \in \mathcal{Y}} \sup_{x \in \mathcal{X}} F(x,y) = \sup_{x \in \mathcal{X}} \inf_{y \in \mathcal{Y}} F(x,y)$.

<u>Proof</u>: Let $\mathcal{X}(0) = \cup\{\mathcal{X}(\delta);\ \delta > 0\}$. If $\mathcal{X} = \mathcal{X}(0)$ then (2.13) follows immediately from (2.12) by letting δ converge to 0. As we do not make this assumption we have to argue a bit more carefully.

For $\delta \in (0, \Delta]$ we set $\gamma = \delta/\Delta$. By assumption, we have

$$(1-\gamma)x + \gamma x^* \in \mathcal{X}(\delta) \quad \text{for all } x \in \mathcal{X}\ .$$

We conclude, using the concavity of F in its first argument, for any $y \in \mathcal{Y}$:

$$\begin{aligned} \sup_{x \in \mathcal{X}} F(x,y) &\geq \sup_{x \in \mathcal{X}(\delta)} F(x,y) \\ &\geq \sup_{x \in \mathcal{X}} F((1-\gamma)x + \gamma x^*, y) \\ &\geq (1-\gamma) \sup_{x \in \mathcal{X}} F(x,y) + \gamma F(x^*,y). \end{aligned}$$

Taking the infimum over $\mathcal{Y}$ and using $\gamma = \delta/\Delta$, we get

$$\begin{aligned} \inf_{y \in \mathcal{Y}} \sup_{x \in \mathcal{X}} F(x,y) &\geq \inf_{y \in \mathcal{Y}} \sup_{x \in \mathcal{X}(\delta)} F(x,y) \\ &\geq (1-\frac{\delta}{\Delta}) \inf_{y \in \mathcal{Y}} \sup_{x \in \mathcal{X}} F(x,y) \\ &+ \frac{\delta}{\Delta} \inf_{y \in \mathcal{Y}} F(x^*,y). \end{aligned}$$

If $\delta \to 0+$ the left-hand side and the right-hand side both converge to the same limit, and we get the first of the following equalities

$$\begin{aligned} \inf_{y \in \mathcal{Y}} \sup_{x \in \mathcal{X}} F(x,y) &= \lim_{\delta \to 0+} \inf_{y \in \mathcal{Y}} \sup_{x \in \mathcal{X}(\delta)} F(x,y) \\ &= \lim_{\delta \to 0+} \sup_{x \in \mathcal{X}(\delta)} \inf_{y \in \mathcal{Y}} F(x,y) \\ &= \sup_{x \in \mathcal{X}(0)} \inf_{y \in \mathcal{Y}} F(x,y) \\ &\leq \sup_{x \in \mathcal{X}} \inf_{y \in \mathcal{Y}} F(x,y) \end{aligned}$$

where we have used (2.12) for the second equality, and where the rest follows from the definition of $\mathcal{X}(0)$ and from $\mathcal{X}(0) \subseteq \mathcal{X}$. This proves (2.13) as the reverse inequality is always satisfied. QED

<u>Remark</u>: The two theorems of this section remain unchanged if one considers a more general class of linear filters,

analogously to Franke (1981a), instead of admitting only filters with transfer functions in L^2. For the sake of simplicity, we chose to discuss only the smaller class of filters which also suffices for practical purposes.

3. Finite-Length Minimax-Robust Predictors: An Example for Explicit Calculation of Minimax-Robust Filters

If we are able to find a least favorable pair of spectral densities in a given spectral information set then, by Theorem 2.2, we immediately have a minimax-robust filter. Up to now, there have been described two generally applicable methods determining least favorable spectral density pairs. One of them, due to Poor (1980), uses the analogy of least favorable spectral density pairs for filtering problems to least favorable probability density pairs for certain hypothesis testing problems. The other approach, discussed by Franke (1981a,b) in the prediction and interpolation context, is based on the interpretation of least favorable pairs as solutions of convex extremum problems and uses convex optimization method.

However, Theorem 2.2 itself contains a powerful method for determining least favorable pairs in a somewhat disguised form. By Theorem 2.2, (f_S^L, f_N^L) is least favorable in $\mathcal{J}$ if and only if $(f_S^L, f_N^L; H_W^L)$ is a saddlepoint of the concave-convex functional e on $\mathcal{J} \times \mathcal{L}$. By definition of a saddlepoint this is equivalent to

$$(3.1) \quad e(f_S, f_N; H_W^L) \leq e(f_S^L, f_N^L; H_W^L) \leq e(f_S^L, f_N^L; H) \quad \text{for all } (f_S, f_N) \in \mathcal{J},\ H \in \mathcal{L}.$$

The second inequality is always satisfied as H_W^L is the Wiener filter with respect to f_S^L, f_N^L, and we get

Corollary 3.1: *Let $\mathcal{L}$, $\mathcal{J}$ be as in Theorem 2.2. Suppose $(f_S^L, f_N^L) \in \mathcal{J}$, and let H_W^L be the Wiener filter with respect to f_S^L, f_N^L.*

a) (f_S^L, f_N^L) is a least favorable pair for $\mathcal{J}$ iff

$$(3.2) \quad e(f_S, f_N; H_W^L) \leq e(f_S^L, f_N^L; H_W^L) \quad \textit{for all } (f_S, f_N) \in \mathcal{J}.$$

b) If H_W^L *is a bounded function then (3.2) is equivalent to requiring that the functional* $-e(.,.;H_W^L)$ *be a support of the convex set* $\mathcal{J}$ *at the point* (f_S^L, f_N^L).

If $\mathcal{J}$ is a convex subset of a Banach space B then a continuous linear functional Φ on B is called a support of the set $\mathcal{J}$ at the point $\beta^* \in \mathcal{J}$ if

$$\Phi(\beta) \geq \Phi(\beta^*) \quad \text{for all } \beta \in \mathcal{J}.$$

Therefore, part b) is only a rephrasing of part a) of the corollary. The criterion of part b) has two main advantages. First, supports of convex sets in $L^1 \times L^1$ are often already known from work on convex extremum problems or, otherwise, easy to derive. Furthermore, if $\mathcal{J}$ is pieced together from different types of spectral information, i.e. if $\mathcal{J}$ is the intersection of spectral information sets $\mathcal{J}_k$, $k = 1,\ldots,K$, then frequently the supports of $\mathcal{J}$ are exactly the sums of supports of $\mathcal{J}_k$, $k = 1,\ldots,K$. Second, the particular filtering problem, i.e. the specification of the set $\mathcal{L}$, does not influence the form of supports. Once the supports for a certain spectral information set $\mathcal{J}$ have been derived, they can be used via Corollary 3.1 for finding minimax-robust filters for all kinds of filtering problems. The connection of Corollary 3.1 and the method used by Franke (1981a,b) is discussed in Appendix 2.

We illustrate the power and limitations of Theorem 2.2 and Corollary 3.1 with a discussion of the q-step prediction problem, i.e. we have no noise ($f_N \equiv 0$) and, for $q \geq 1$, we want to predict S_q from S_k, $k \leq 0$. This corresponds to the choice

$$D(\omega) = e^{iq\omega} \; ; \quad \mathcal{L} = L^2_-, \text{ where}$$

$$L^2_- = \{G \in L^2; \; \frac{1}{2\pi} \int e^{ik\omega} G(\omega) d\omega = 0 \text{ for all } k < 0\}$$

is the closed linear hull in L^2 of the trigonometric functions $e^{ik\omega}$, $k \leq 0$.

First, let us write down the well-known form of the Wiener filter for q-step prediction. Let f_S^L be the spectral density of a purely nondeterministic time series. It allows for the canonical factorization (see e.g. Hannan, 1970):

$$f_S^L(\omega) = \sigma^2|b(\omega)|^2 \quad , \text{ where } b(\omega) = \sum_{k=0}^{\infty} b_k e^{-ik\omega} \in L^2,\ b_o = 1,$$

(3.3)

$$\sigma^2 = \exp\{\frac{1}{2\pi} \int \log(f_S^L(\omega))d\omega \quad \text{and} \quad 1/b(\omega) \in L_-^2(f_S^L),$$

where $L_-^2(f_S^L)$ denotes the closed linear hull of the trigonometric functions $e^{ik\omega}$, $k \leq 0$, in the space of complex functions which are square-integrable with respect to the measure with density f_S^L. For the q-step prediction problem, the Wiener filter with respect to $(f_S^L,0)$ is given by (see e.g. Hannan,1970):

$$(3.4) \quad H_W^L(\omega) = e^{iq\omega}\{1 - (\sum_{k=0}^{q-1} b_k e^{-ik\omega})/b(\omega)\} .$$

In the q-step prediction problem a convex spectral information set $\mathcal{J}$ is of the form $\mathcal{J} = \mathcal{D} \times \{0\}$ where $\mathcal{D}$ is a convex set of spectral densities. As the only uncertainty relates to the signal spectrum we call $f_S^L \in \mathcal{D}$ a most indeterministic spectral density with respect to $\mathcal{D}$ if $(f_S^L,0)$ is a least favorable pair for $\mathcal{J}$. From the definition of supports, Corollary 3.1 and Theorem 2.2 we get immediately:

<u>Corollary 3.2</u>: *Let $\mathfrak{L} = L_-^2$, and suppose $\mathcal{D}$ be a convex set of spectral densities satisfying*

$$\frac{1}{2\pi} \int f_S(\omega)d\omega \leq c_S < \infty \qquad \textit{for all } f_S \in \mathcal{D} \quad .$$

Suppose $f_S^L \in \mathcal{D}$ is the spectral density of a purely nondeterministic time series, and let H_W^L, given by (3.4), be bounded. Then, f_S^L is most indeterministic with respect to $\mathcal{D}$ for q-step prediction iff $-e(.,0;H_W^L)$ is a support of $\mathcal{D}$ at f_S^L.

In particular, the q-step Wiener predictor with respect to a most indeterministic spectral density is minimax-robust with respect to $\mathcal{D}$.

We consider a particular form of spectral information sets for which the minimax-robust q-step predictor will automatically be finite. This feature is most desirable from the practical point of view. Let $r = (r_0,\ldots,r_p)'$, be the vector of the first p autocovariances of the signal process. Following Franke (1981a), we assume that the information on the signal process consists of some information on the autocovariance vector r only, which can be summarized in the requirement $r \in \mathcal{R}$. Here, $\mathcal{R}$ is a convex subset of $\mathbb{R}^{p+1}$ which, e.g., could be a confidence region for an estimate of the signal autocovariance vector. Assuming $r \in \mathfrak{R}$ to be the only information on f_S is equivalent to choosing

$$\mathcal{D} = \mathcal{D}_{ac} = \{f_S \in L^1;\ (r_0,\ldots,r_p)' \in \mathfrak{R}$$

$$\text{with } r_k = \frac{1}{2\pi}\int \cos(k\omega) f_S(\omega)\,d\omega,\ k = 0,\ldots,p\}.$$

We call spectral information sets of this form autocovariance information sets. As a subset of L^1, $\mathcal{D}_{ac}$ contains also functions which assume negative values. However, it is not necessary to impose the additional constraint of nonnegativity. If we can find a spectral density f_S^L for which $-e(.,0;H_W^L)$ is a support of $\mathcal{D}_{ac}$ at f_S^L then the functional is automatically a support for the smaller set $\mathcal{D}_{ac}^+$, which consists of the nonnegative functions in $\mathcal{D}_{ac}$, and, by Corollary 3.2, f_S^L is most indeterministic with respect to $\mathcal{D}_{ac}^+$.

The following theorem contains the basic results on q-step minimax-robust prediction with respect to autocovariance information sets. For $p < q$, $\mathcal{D}_{ac}$ does not provide enough information on the time series for predicting S_q from S_k $k \leq 0$ by anything else than its mean value 0. In other words, the zero predictor is minimax-robust with respect to $\mathcal{D}_{ac}$ for q-step prediction if $p < q$.

If $p \geq q$ and if there exists a most indeterministic spectral density with respect to $\mathcal{D}_{ac}$ for q-step prediction then the minimax-robust predictor for S_q depends only on $S_0,\ldots,S_{q-1}$ and not on the complete past S_k, $k \leq 0$, i.e. it is automatically finite-length. As will be discussed below, there are however, situations where an autocovariance information set does not contain a most indeterministic spectral density for q-step prediction.

<u>Theorem 3.1</u>: *Let either $\mathfrak{R}$ consist of one positive definite vector $r^m = (r_0^m,\ldots,r_p^m)'$ only, or let $\mathfrak{R}$ be a convex set in $\mathbb{R}^{p+1}$ with nonempty interior containing a positive definite vector $r^m = (r_0^m,\ldots,r_p^m)'$ for which*

$$r_0^m = \max_{r \in \mathfrak{R}} r_0.$$

Let $\mathcal{D}_{ac}$ be the autocovariance information set given by $\mathfrak{R}$.

a) Let $p < q$. Then, there exists $f_S^L \in \mathcal{D}_{ac}$ which is the spectral density of a MA(p)-process and which is most indeterministic with respect to $\mathcal{D}_{ac}$ for q-step prediction. In *particular, the zero predictor* $H_0 \equiv 0$ *is minimax-robust.*

b) *Let* $p \geq q$. *Let* f_S^L *be a spectral density in* $\mathcal{D}_{ac}$, *and let* $r_0^L, \ldots, r_p^L$ *be its first* p+1 *Fourier coefficients.* f_S^L *is most indeterministic with respect to* $\mathcal{D}_{ac}$ *for q-step prediction iff there are real parameters* $\sigma^2 > 0$, $b_1, \ldots, b_{q-1}, a_q, \ldots, a_p$ *such that with* $b_0 = 1$

$$(3.5) \quad f_S^L(\omega) = \sigma^2 \left| \sum_{k=0}^{q-1} b_k e^{-ik\omega} \right|^2 / \left| 1 + \sum_{k=q}^{p} a_k e^{-ik\omega} \right|^2$$

and

$$(3.6) \quad \sum_{k=0}^{p} \lambda_k r_k \leq \sum_{k=0}^{p} \lambda_k r_k^L \text{ for all } r \in \mathcal{R}$$

where $\lambda_0, \ldots, \lambda_p$ *are given by*

$$\sum_{k=0}^{p} \lambda_k \cos(k\omega) = \left| 1 + \sum_{k=q}^{p} a_k e^{-ik\omega} \right|^2 .$$

If f_S^L *is most indeterministic, then, in particular, the finite q-step predictor*

$$H_W^L(\omega) = - \sum_{k=q}^{p} a_k e^{-ik\omega}$$

is minimax-robust with respect to $\mathcal{D}_{ac}$.

<u>Proof</u>: 1) Let $f_S^L \in \mathcal{D}_{ac}$, and let $r_0^L, \ldots, r_p^L$ be its first p+1 Fourier coefficients. By Lemmas 3, 4 of Franke (1981a) Φ is a support functional of $\mathcal{D}_{ac}$ at f_S^L iff

$$\Phi(f_S) = \frac{1}{2\pi} \int \{ \sum_{k=0}^{p} \lambda_k \cos(k\omega) \} f_S(\omega) d\omega$$

where $\lambda_0, \ldots, \lambda_p$ are real Lagrange multipliers satisfying

$$(3.7) \quad \sum_{k=0}^{p} \lambda_k r_k \leq \sum_{k=0}^{p} \lambda_k r_k^L \quad \text{for all } r \in \mathcal{R}$$

2) If $f_S^L \in \mathcal{D}_{ac}$ is a spectral density of a purely nondeterministic time series, and if the corresponding q-step Wiener prediction filter H_W^L, given by (3.4), is bounded, then, by Corollary 3.2 and by 1) we have: f_S^L is most indeterministic with respect to $\mathcal{D}_{ac}$ iff there exist real $\lambda_0, \ldots, \lambda_p$ satisfying (3.7) such that

$$-e(f_S, 0; H_W^L) = \frac{-1}{2\pi} \int |e^{iq\omega} - H_W^L(\omega)|^2 f_S(\omega) d\omega$$

$$= \frac{-\sigma^2}{2\pi} \int \left| \sum_{k=0}^{q-1} b_k e^{-ik\omega} \right|^2 f_S(\omega) / f_S^L(\omega) d\omega$$

$$= \frac{1}{2\pi} \int \{ \sum_{k=0}^{p} \lambda_k \cos(k\omega) \} f_S(\omega) d\omega \qquad \text{for all } f_S \in L^1$$

The last equality can only be satisfied if

$$f_S^L(\omega) = -\sigma^2 \left| \sum_{k=0}^{q-1} b_k e^{-ik\omega} \right|^2 / \{ \sum_{k=0}^{p} \lambda_k \cos(k\omega) \} \quad .$$

3) Let $p < q$. By, e.g., Theorem II.10 of Hannan (1970), there exists a MA(p)-process with autocovariances $r_0^m, \ldots, r_p^m$. Let $\sigma^2 > 0$, $b_1, \ldots, b_p$ be its parameters, $b_0 = 1$, and

$$f_S^L(\omega) = \sigma^2 \left| \sum_{k=0}^{p} b_k e^{-ik\omega} \right|^2$$

be its spectral density. Choosing $\lambda_0 = 1$, $\lambda_1 = \ldots = \lambda_p = 0$, f_S^L satisfies the condition of 2) and, therefore, is most indeterministic with respect to $\mathcal{D}_{ac}$ for q-step prediction. As H_o is the q-step Wiener predictor with respect to f_S^L it is minimax-robust, by Corollary 3.2.

4) Let $p \geq q$. Let f_S^L be most indeterministic in $\mathcal{D}_{ac}$. By part 2) and Theorem II.10 of Hannan (1970)

$$(3.8) \quad f_S^L(\omega) = \sigma^2 \left| \sum_{k=0}^{q-1} b_k e^{-ik\omega} \right|^2 / \left| \sum_{k=0}^{p} a_k e^{-ik\omega} \right|^2$$

for suitable real-valued $a_0 = 1$, $a_1, \ldots, a_p$ ($a_0 = 1$ follows from the fact that f_S^L is the spectral density of a mixed autoregressive-moving average (ARMA) process and that, by (3.3), σ^2 is already the innovation variance of the process. f_S^L is, however, not an arbitrary ARMA(p,q-1) spectral density. By (3.8), its MA coefficients coincide with its first q-1 impulse response coefficients. Furthermore, by standard results on rational spectral densities, (3.8) implies

$$b(\omega) = \left(\sum_{k=0}^{q-1} b_k e^{-ik\omega} \right) / \left(\sum_{k=0}^{p} a_k e^{-ik\omega} \right) \quad .$$

Inserting this relation into (3.4) implies

$$H_W^L(\omega) = -e^{iq\omega} \left(\sum_{k=1}^{p} a_k e^{-ik\omega} \right) \quad .$$

As $H_W^L \in L_-^2$ the coefficients of trigonometric functions $e^{i(q-k)\omega}$ with positive q-k have to vanish, i.e. we have

$$a_1 = \ldots = a_{q-1} = 0 \; ; \; H_W^L(\omega) = \sum_{k=q}^{p} a_k e^{i(q-k)\omega} \quad .$$

Assertion b) now follows from 2) and 4). QED

In the rest of this section we investigate the problem of existence of a most indeterministic spectral density with respect to autocovariance information sets and the explicit calculation of the minimax-robust q-step predictor. We restrict the discussion to the case where $p \geq q$ and where $\mathcal{R}$ contains only one positive definite autocovariance vector $c = (c_0,\ldots,c_p)'$, i.e. we are considering the spectral information set

$$\mathcal{D}(c) = \{f_S \in L^1;\ \frac{1}{2\pi}\int\cos(k\omega)f_S(\omega)d\omega = c_k,\quad k = 0,\ldots,p\}.$$

If there is an ARMA(p,q-1) process $\{S_n^L\}$ with spectral density $f_S^L \in \mathcal{D}(c)$, given by (3.5), then its parameters are determined by p+1 equations, involving $c_0,\ldots,c_p$ only. By (3.3) and (3.5) we have

$$\text{(3.9a)}\quad S_n^L = \sum_{k=0}^{\infty} b_k\,\varepsilon_{n-k}\ ,\quad \operatorname{Var}(\varepsilon_n) = \sigma^2$$

$$\text{(3.9b)}\quad S_n^L + \sum_{k=q}^{p} a_k S_{n-k}^L = \sum_{k=0}^{q-1} b_k\varepsilon_{n-k}\ ,$$

where here and in the rest of this section $\{\varepsilon_n\}$ is white noise. Let $c_0,\ldots,c_p$ be the first p+1 autocovariance of $\{S_n^L\}$. Then, setting $b_k = 0$ for $k < 0$, we have, using (3.9a)

$$E\{S_n^L S_0^L\} = c_n,\quad n = 0,\ldots,p, \text{ and}$$

$$E\{\varepsilon_n S_0^L\} = \sigma^2\, b_{-n},\quad -\infty < n < \infty\ .$$

Multiplying both sides of (3.9b) by S_0^L and taking expectations for $n = 0,\ldots,p$ we get the following p+1 equations for the p+1 parameters $\sigma^2, b_1,\ldots,b_{q-1}, a_q,\ldots,a_p$:

$$\text{(3.10a)}\quad c_n + \sum_{k=q}^{p} a_k c_{n-k} = \sigma^2 \sum_{k=0}^{q-1} b_k b_{k-n}\qquad n = 0,\ldots,q-1$$

$$\text{(3.10b)}\quad c_n + \sum_{k=q}^{p} a_k c_{n-k} = 0\qquad n = q,\ldots,p\ .$$

Equations (3.10) are a special version, due to the vanishing of the first q-1 AR parameters and to the identity of the MA parameters with the first q-1 impulse response coefficients, of the standard equations for ARMA-parameters, given e.g. by Box and Jenkins (1976, ch. 3.4). It is important to note that the coefficients $a_q,\ldots,a_p$ of the minimax-robust q-step predictor can be calculated easily and efficiently from the linear Yule-Walker type equations (3.10b) alone, which, by positive definiteness of $c_0,\ldots,c_p$, are uniquely solvable. For the purpose of robust prediction, it is not necessary to solve the nonlinear equations (3.10a) for $\sigma^2, b_1,\ldots,b_{q-1}$. One has only to worry about the existence of such a solution, as this is equivalent to the existence of a most indeterministic spectral density for q-step prediction in $\mathcal{D}(c)$, and, therefore, it is needed for applying Theorem 3.1b.

For every time series $\{S_n\}$ with autocovariances $c_0,\ldots,c_p$,

$$S_n^p = -\sum_{k=q}^{p} a_k S_{n-k} \ ,$$

where $a_q,\ldots,a_p$ are the solutions of (3.10b), is the best linear predictor for S_n based on $S_{n-q},\ldots,S_{n-p}$ only. The existence of a most indeterministic spectral density for q-step prediction in $\mathcal{D}(c)$ is equivalent to the existence of a time series $\{S_n^L\}$ with autocovariances $c_0,\ldots,c_p$ for which the best finite q-step predictor $S_n^{L,p}$ is the best q-step predictor for S_n based on the complete past of the process. Based on these considerations, Theorem 3.1 could be proved for the spectral information set $\mathcal{D}(c)$ without recourse to optimization methods by remarking that, for every time series with spectral density in $\mathcal{D}(c)$, the best finite q-step predictor gives rise to the same mean-square error and that for the time series $\{S_n^L\}$ it is the best q-step predictor anyway. This argument was first proposed by Akaike in the context of proving the maximum-entropy property of AR-processes which relates to robust 1-step prediction (see Priestley (1981), p. 604 ff, and Franke (1981a) for the details).

If there exists a purely nondeterministic ARMA(p,q-1) process $\{S_n^L\}$ satisfying (3.9), with autocovariances c_n, $n \geq 0$, then (3.10b) is valid for all $n \geq q$. This follows again from multiplying both sides of (3.9b) by S_0^L and taking expectations. Therefore, if $c_0,\ldots,c_p$ are given and if $a_q,\ldots,a_p$ solve (3.10b) then the autocovariances of $\{S_n^L\}$ for lag $n = p+1,\ldots,p+q-1$ can be calculated from

$$(3.11) \qquad c_n = -\sum_{k=q}^{p} a_k c_{n-k} \ , \quad n = p+1,\ldots,p+q-1 \quad .$$

As autocovariances of a purely nondeterministic time series, $c_0,\ldots,c_{p+q-1}$ are positive definite, the following result implies that the positive definiteness of this sequence is also sufficient for the existence of the desired ARMA (p,q-1) process if an additional weak technical condition is satisfied.

Proposition 3.1: *Let* $c = (c_0,\ldots,c_p)'$ *be positive definite. Let* $c_{p+1},\ldots,c_{p+q-1}$ *be defined by (3.11), where* $a_q,\ldots,a_p$ *are the solutions of (3.10b).*

a) If there exists a purely nondeterministic spectral density in $\mathcal{D}(c)$ *which is most indeterministic with respect to q-step prediction then* $c_0,\ldots,c_{p+q-1}$ *is positive definite.*

b) If $c_0,\ldots,c_{p+q-1}$ *is positive definite then there exist* $\sigma^2 > 0$, $b_1,\ldots,b_{q-1}$ *solving (3.10a), and they can be chosen such that the polynomial*

$$B(z) = \sum_{k=0}^{q-1} b_k z^{q-1-k} \qquad \text{(where } b_0 = 1\text{)}$$

has no zeroes outside of the closed unit circle $\{z; |z| \leq 1\}$.

c) Let $A(z) = \sum_{k=0}^{p} a_k z^{p-k}$ (where $a_0 = 1, a_1 \ldots = a_{q-1} = 0$).

If $c_0,\ldots,c_{p+q-1}$ *is positive definite and if* $A(z)$ *and the polynomial* $B(z)$, *defined in b), have no common zero on the boundary* $\{z; |z| = 1\}$ *of the unit circle then there exists a purely nondeterministic spectral density in* $\mathcal{D}(c)$ *which is most indeterministic with respect to q-step prediction.*

Proof: 1) Part a) follows from the remarks preceding (3.11).
2) Assume $c_0,\ldots,c_{p+q-1}$ to be positive definite. Then they are autocovariances of a purely nondeterministic time series, say $\{Y_n\}$. Let

$$Z_n = Y_n + \sum_{j=q}^{p} a_j Y_{n-j} \quad , \quad -\infty < n < \infty ,$$

be the process of finite q-step prediction errors with respect to $\{Y_n\}$. As a finite linear combination of the purely nondeterministic process $\{Y_n\}$, $\{Z_n\}$ is purely nondeterministic too. Therefore, its autocovariances r_n, $-\infty < n < \infty$, are positive definite. By (3.11), they are given for lag $n < q$ by

$$r_n = E\{Z_n Z_0\}$$

$$= c_n + \sum_{j=q}^{p} a_j c_{n+j} + \sum_{j=q}^{p} a_j c_{n-j} + \sum_{j,k=q}^{p} a_j a_k c_{n+j-k}$$

$$= c_n + \sum_{j=q}^{p} a_j c_{n-j} \quad , \qquad n = 0,\ldots,q-1 \quad .$$

By, e.g., Theorem II.10 of Hannan (1970), there exists a MA(q-1) process $\{X_n\}$ with autocovariances $r_0,\ldots,r_{q-1}$. Its parameters $\sigma^2 > 0, b_1,\ldots,b_{q-1}$ satisfy

$$r_n = E\{X_n X_0\} = \sigma^2 \sum_{j=0}^{q-1} b_j b_{j-n} \ , \quad n = 0,\ldots,q-1,$$

and B(z) has no zeroes outside of the closed unit circle. b) follows.

3) By part b), we know $c_0,\ldots,c_p, b_1,\ldots,b_{q-1}$ and σ^2. If A(z) and B(z) have no common zero on the boundary of the unit circle then, by results of Franke (1984) on the general problem of existence of an ARMA process with prescribed finitely many autocovariances and impulse response coefficients, there exists an ARMA(p,q-1) process $\{X_n\}$ with autocovariances $c_0,\ldots,c_p$, impulse response coefficients $b_1,\ldots,b_{q-1}$ and innovation variance σ^2. Additionally, using (3.10), it has the representation

$$X_n + \sum_{k=q}^{p} a_k X_{n-k} = \sum_{k=0}^{q-1} b_k \varepsilon_{n-k} \ , \qquad \mathrm{Var}(\varepsilon_n) = \sigma^2,$$

i.e., by Theorem 3.1, $\{X_n\}$ is the desired process with spectral density, which lies in $\mathcal{D}(c)$ and which is most indeterministic for q-step prediction. QED

In Appendix 1, we shall discuss an example illustrating the necessity and meaning of the technical condition on A(z), B(z). If $c_0,\ldots,c_p$ are empirical autocovariances then this condition is satisfied with probability 1. In practice, therefore, one has only to check if the extended autocovariance sequence $c_0,\ldots,c_{p+q-1}$ is positive definite. For this purpose, the following well-known result can be used.

Lemma 3.1: *For* $N \geq 1$, *let* $c_0,\ldots,c_N$ *be given. Let* $C_n = (c_{j-k})_{1 \leq j,k \leq n}$ *be the* $n \times n$*-Toeplitz matrix corresponding to* $c_0,\ldots,c_{n-1}$. *If* C_n *is invertible then let*

$\alpha^{(n)} = (\alpha_1^{(n)},\ldots,\alpha_n^{(n)})'$ *denote the solution of the Yule-Walker equation*

(3.12) $$C_n\alpha^{(n)} = -(c_1,\ldots,c_n)' .$$

Let $s_0 = c_0$, *and let* $s_n, n = 1,\ldots,N$, *be defined by*

(3.13) $$s_n = c_0 + \sum_{j=1}^{n} \alpha_j^{(n)} c_j \quad \text{if } C_n \text{ is invertible}$$

$$s_n = 0 \quad \text{if } C_n \text{ is singular.}$$

Then, $c_0,\ldots,c_N$ *is positive definite iff* $s_n > 0$, $n = 0,\ldots,N$.

If, for $n \geq 1$, $c_0,\ldots,c_n$ are positive definite, then s_n is the mean-square error of the best finite linear one-step predictor, depending only on the last n observations, for time series with autocovariances $c_0,\ldots,c_n$, and it is positive. Technically, the lemma follows from observing that the modified Cholesky decomposition, discussed by Wilkinson (1967), of the symmetric Toeplitz matrix C_n is given by

$$C_n = A_n D_n A_n' \quad , \quad n = 1,\ldots,N ,$$

where D_n is a $n \times n$ diagonal matrix with elements $s_0,\ldots,s_{n-1}$ and A_n is a unit lower triangular matrix with k-th row $(\alpha_{k-1}^{(k-1)},\ldots,\alpha_1^{(k-1)},1)$, $k = 1,\ldots,n$. C_n is positive definite iff all elements of D_n are positive.

Let us point out that, for the purpose of checking the assumption of Proposition 3.1b), s_n, $n = p+1,\ldots,p+q-1$, can be calculated recursively and efficiently by solving (3.12) and (3.13) by means of the Levinson-Durbin algorithm (see, e.g., Whittle (1963), p. 37).

It is not possible to weaken the assumption of Proposition 3.1b). For $p \geq q > 1$ one can choose positive definite $c_0,\ldots,c_p$ such that there exists no time series with these autocovariances for which the best finite q-step predictor of length p-q+1 coincides with the best q-step predictor using the whole past. As an example, we consider $p = q = 2$. We fix c_0, c_1 such that $|c_1| < c_0$. Then, using Lemma 3.1, c_0, c_1, c_2 is positive definite iff s_2, given by (3.13), is positive. The latter is equivalent to $|\alpha_2^{(2)}| < 1$. By (3.10b), $a_2 = -c_2/c_0$, and, by (3.11) we calculate $c_3 = -a_2 c_1$. Using again Lemma 3.1, the extended autocovariance sequence $c_0,\ldots,c_3$ is positive definite iff $|\alpha_3^{(3)}| < 1$, where, by (3.12),

$$\alpha_3^{(3)} = -\alpha_2^{(2)} c_1/\{(1-\alpha_2^{(2)})c_0\} .$$

We conclude that $c_0,\ldots,c_3$ is positive definite iff

$$\alpha_2^{(2)} < \{1 + |c_1|/c_0\}^{-1} .$$

Therefore, there exists a region of c_2-values where c_0,c_1,c_2 is still positive definite, but $c_0,\ldots,c_3$ is not.

As we have shown, the best finite q-step predictor for time series with prescribed autocovariances is, under certain conditions, a minimax-robust q-step predictor. In practice, one often is not only interested in q-step prediction for a fixed q but in extrapolation of the whole process up to time q, i.e. in simultaneous n-step prediction for $n = 1,\ldots,q$. It is possible to calculate the q-step prediction coefficients $a_q,\ldots,a_p$, given by (3.10b), and the corresponding q-step prediction errors recursively (in q), similar to the Levinson-Durbin algorithm.

In the following, let $c_0,\ldots,c_p$ be a fixed positive definite sequence, and let q vary in $1,\ldots,p$. Let $\{X_n\}$ denote a purely nondeterministic time series with autocovariances $c_0,\ldots,c_p$. Let $a_k^{(q)}$, $k = q,\ldots,p$, given by (3.10b), and

$$\pi_q = c_0 + \sum_{k=q}^{p} a_k^{(q)} c_k$$

be coefficients and mean-square error of the best predictor for X_0 based on $X_{-q},\ldots,X_{-p}$. Let $\alpha_k^{(n)}$, $k = 1,\ldots,n$, given by (3.12), and s_n, given by the first relation of (3.13), be coefficients and mean-square error of the best predictor for X_0 based on $X_{-1},\ldots,X_{-n}$.

We start the recursion by solving (3.12) for $n = p$. Then, we can calculate $\alpha_k^{(p+1-q)}$, $k = 1,\ldots,p+1-q$, and s_{p+1-q} recursively for $q = 2,\ldots,p$ by using the Levinson-Durbin algorithm (see e.g., Whittle (1963), p. 37) backwards. The recursive procedure is completed by the following relations

$$a_k^{(1)} = \alpha_k^{(p)}, \quad k = 1,\ldots,p; \qquad \pi_1 = s_p.$$

(3.14a) $a_k^{(q+1)} = a_k^{(q)} - a_q^{(q)} \alpha_{k-p}^{(p-q)}$, $k = q+1,\dots,p$, $q = 1,\dots,p-1$,

(3.14b) $\pi_{q+1} = \pi_q + (a_q^{(q)})^2 s_{p-q}$, $q = 1,\dots,p-1$.

To prove (3.14), let P_q denote the projection onto the linear hull of $X_{-q},\dots,X_{-p}$. We have for $q < p$

$$\sum_{k=q+1}^{p} a_k^{(q+1)} X_{-k} = -P_{q+1} X_0$$

$$= -P_{q+1} P_q X_0$$

$$= \sum_{k=q+1}^{p} a_k^{(q)} X_{-k} + a_q^{(q)} P_{q+1} X_{-q}$$

$$= \sum_{k=q+1}^{p} a_k^{(q)} X_{-k} - a_q^{(q)} \sum_{j=1}^{p-q} \alpha_j^{(p-q)} X_{-q-j} .$$

(3.14a) follows. Using this recursion and defining $\alpha_0^{(n)} = 1$, $n = 1,\dots,p$, we get

$$E\{X_{-q}(X_0 + \sum_{k=q+1}^{p} a_k^{(q+1)} X_{-k})\} = c_q + \sum_{k=q}^{p} (a_k^{(q)} - a_q^{(q)} \alpha_{k-q}^{(p-q)}) c_{q-k}$$

$$= -a_q^{(q)} \sum_{j=o}^{p-q} \alpha_j^{(p-q)} c_j$$

$$= -a_q^{(q)} s_{p-q} ,$$

where the second equality follows from (3.10b). Finally, using this relation,

$$\pi_q = E\{X_0(X_0 - P_q X_0)\}$$

$$= E\{(X_0 - P_{q+1} X_0)(X_0 - P_q X_0)\}$$

$$= \pi_{q+1} + E\{(X_0 - P_{q+1} X_0) \sum_{k=q}^{p} a_k^{(q)} X_{-k}\}$$

$$= \pi_{q+1} + a_q^{(q)} E\{(X_0 - P_{q+1} X_0) X_{-q}\}$$

$$= \pi_{q+1} (a_q^{(q)})^2 s_{p-q} .$$

4. A Time-Domain Approach to Minimax Robust Filtering: General Analysis

In Sections 2 and 3 above, a frequency-domain formulation was used to characterize and find solution to the robust Wiener-Kolmogorov problem. It is also interesting to view this problem in the time domain and within the more general framework of (possibly) nonlinear minimum mean-square error (MMSE) estimation.

To consider these issues suppose $(\Omega,\mathcal{F},\mu)$ is a probability space, Y is a real-valued random variable on $(\Omega,\mathcal{F})$ with $\int Y^2 d\mu$ finite, and X is a random variable taking values in a Borel space $(\Sigma,\mathcal{S})$, as defined by Breiman (1968), e.g. in sequence space $(\mathbb{R}^\infty,\mathcal{B}^\infty)$. Let $\mathcal{J}_X$ denote the class of all real-valued $\mathcal{F}(X)$-measurable functions and, for $g \in \mathcal{J}_X$, define

$$(4.1) \qquad \mathcal{E}(g;\mu) = \int |Y-g|^2 d\mu \quad .$$

Note that with μ fixed we have, of course, that

$$(4.2) \qquad E_\mu\{Y|X\} \text{ minimizes } \mathcal{E}(.;\mu) \text{ on } \mathcal{J}_X.$$

If on the other hand μ is known only to lie in some class $\mathcal{M}$ of probability measures on $(\Omega,\mathcal{F})$, we may consider the following alternative to $\min_{g \in \mathcal{J}_X} \mathcal{E}(g;\mu)$.

$$(4.3) \qquad \min_{g \in \mathcal{J}_X} \{ \sup_{\mu \in \mathcal{M}} \mathcal{E}(g;\mu)\} \quad .$$

In the following, we assume:

$$(4.4) \qquad \mathcal{M} \text{ is convex, and } \int Y^2 d\mu \le c \text{ for all } \mu \in \mathcal{M} \quad .$$

In order to study the solutions to (4.3) we first consider the functional

$$(4.5) \qquad C(\mu) \triangleq \min_{g \in \mathcal{J}_X} \mathcal{E}(g;\mu) = \int Y^2 d\mu - \int (E_\mu\{Y|X\})^2 d\mu \quad .$$

Note that $0 \le C(\mu) \le \int Y^2 d\mu \le c$ for all $\mu \in \mathcal{M}$ and that C is concave on $\mathcal{M}$ since it is the pointwise minimum of the linear functions $\int|Y-g|^2 d\mu$. Thus, we see that the condition

$$(4.6) \qquad C(\mu_o) = \max_{\mu \in \mathcal{M}} C(\mu)$$

is equivalent to the condition (see, e.g., Franke (1981a), Lemma 1)

$$(4.7) \qquad C_o'(\mu) \triangleq \lim_{\varepsilon \downarrow 0} \{[C((1-\varepsilon)\mu_0+\varepsilon\mu)-C(\mu_0)]/\varepsilon\} \leq 0, \forall \mu \in \mathcal{M} .$$

Note that $C_o'(\mu)$ is the directional derivative of C at μ_o in direction μ. Its explicit form is provided by the following lemma, where we write $g_o = E_{\mu_o}\{Y|X\}$.

Lemma 3: $C_o'(\mu) = \int(g_o-Y)^2 d\mu = \mathcal{E}(g_o;\mu)$ *for all* $\mu,\mu_o \in \mathcal{M}$.

Proof: Let $\mu_\varepsilon = (1-\varepsilon)\mu_o + \varepsilon\mu$, and let μ^X, μ_ε^X, $0 \leq \varepsilon < 1$, denote the restrictions of μ,μ_ε to $\mathcal{F}(X)$. As μ_o^X,μ^X are absolutely continuous with respect to $\mu_o^X + \mu^X$ with densities φ_o,φ bounded by 1, we have immediately from the definition of conditional expectation

$$E_{\mu_\varepsilon}\{Y|X\} = g_o + (E_\mu\{Y|X\} - g_o)\varepsilon\varphi/\varphi_\varepsilon , \quad \mu_\varepsilon^X - \text{a.s.},$$

where $\varphi_\varepsilon = (1-\varepsilon)\varphi_o + \varepsilon\varphi$. After inserting this relation into the definition of $C(\mu_\varepsilon)$, the lemma follows from a straightforward calculation, using $\varepsilon\varphi/\varphi_\varepsilon \to 0, (\mu_o+\mu)$-a.s. and dominated convergence. QED

The condition (4.6) is thus equivalent to

$$(4.8) \qquad \mathcal{E}(g_0;\mu) \leq \mathcal{E}(g_0;\mu_0), \quad \forall \mu \in \mathcal{M} .$$

From (4.2) and (4.8) it is evident that a measure $\mu_0 \in \mathcal{M}$ and the corresponding conditional-mean estimate of Y, $E_{\mu_0}\{Y|X\}$ form a saddlepoint solution for the minimax problem of (4.3) if and only if μ satisfies (4.6). In view of the definition of $C(\mu)$, such an element μ_0 can be considered to be a member of $\mathcal{M}$ that is least favorable for MMSE estimation. We summarize the above in the following proposition.

Proposition 4.1: *Suppose* $(\Omega,\mathcal{F})$ *is a measurable space and* X *and* Y *are Borel-space-valued variables on* $(\Omega,\mathcal{F})$ *with* Y *real-valued. Let* $\mathcal{M}$ *be a class of probability measures on* $(\Omega,\mathcal{F})$

satisfying (4.4). Then, with $\mathcal{G}_X, \mathcal{E}$*, and* C *defined as above, a measure* $\mu_0 \in \mathcal{M}$ *and the random variable* $E_{\mu_0}\{Y|X\}$ *form a saddlepoint solution to*

$$\min_{g \in \mathcal{G}_X} \{ \sup_{\mu \in \mathcal{M}} \mathcal{E}(g;\mu) \}$$

if and only if

$$C(\mu_o) = \max_{\mu \in \mathcal{M}} C(\mu) \ .$$

Regarding the existence of a maximum of $C(\mu)$ we note that the following result, a proof of which is found in Appendix 3, implies that under weak conditions $C(\mu)$ is weakly upper semi-continuous on $\mathcal{M}$. In this case, a maximum exists if $\mathcal{M}$ is weakly compact.

Proposition 4.2: *The functional* $C(\mu)$ *is weakly upper semi-continuous if* $\mathcal{M}$ *satisfies one of the following two conditions:*

(i) $\int Y^2 d\mu = c$ *for all* $\mu \in \mathcal{M}$.

(ii) There exist $\varepsilon > 0$ *and* c' *such that* $\int |Y|^{2+\varepsilon} d\mu \leq c'$ *for all* $\mu \in \mathcal{M}$.

To consider a specific situation, assume that X assumes values in sequence space $(\mathbb{R}^\infty, \mathcal{B}^\infty)$, and let X_k denote the k-th coordinate of X, $k > 0$. Assume further that the random variables X_k have finite second moment with respect to each $\mu \in \mathcal{M}$, and let $\mathcal{X}(\mu)$ denote the sub-Hilbert space, generated by the X_k, $k > 0$, of the space $L^2(\mu)$ of $\mathcal{F}$-measurable, real-valued, with respect to μ square-integrable random variables. Then, if $\mathcal{M}$ is constrained only in terms of the second-order statistics, the least-favorable distribution μ_o (when it exists) will make (X,Y) jointly Gaussian, and g_o will be linear. This follows since for each $\mu \in \mathcal{M}$,

$$C(\mu) \leq \mathcal{E}(h;\mu) = C(\mu_G) \ , \tag{4.9}$$

where h is the linear projection of Y on $\mathcal{X}(\mu)$ and where μ_G is the measure under which (X,Y) are jointly Gaussian and have the same second-order statistics as under μ. In the context of filtering of time series, these considerations imply that linear filters are minimax-robust under all filters if we have only information about the second moments of the relevant processes. This result gives a precise meaning to the common opinion that one can restrict attention to linear filters if it is not possible to infer more about the processes than second moments only, i.e. if one cannot exclude Gaussian processes from considerations.

Equation (4.8) is necessary and sufficient for (μ_0, g_0) to form a saddlepoint for (4.3). In general this condition would not be very tractable for choosing a solution g_0. However, suppose we consider uncertainty classes of the form

(4.10) $$\mathcal{M} = \{\mu \mid \int g_k(X,Y)d\mu = \gamma_k,\ k = 1,2,\ldots\}$$

where the g_k's are appropriately measurable, linearly independent functions. With a slight abuse of notation, write

$$g_o(x) = E_{\mu_o}\{Y \mid X=x\} \quad , \quad x \in \Sigma ,$$

where $g_o(x)$ now stands for a $\mathcal{J}$-measurable function. Within regularity of the g_k's, a necessary and sufficient condition for (4.8) to hold with $\mathcal{M}$ as in (4.10) is the existence of real Lagrange multipliers $\lambda_1, \lambda_2, \ldots$ such that

(4.11a) $$(g_0(x)-y)^2 + \sum_{k=1}^{\infty} \lambda_k g_k(x,y) \le 0 \quad \text{for all } x \in \Sigma, y \in \mathbb{R}$$

(4.11b) $$\int\{(g_0-Y)^2 + \sum_{k=1}^{\infty} \lambda_k g_k(X,Y)\}d\mu_o = 0 \quad .$$

These equations are more suggestive of the form that the minimax estimators must take on, and we will use them in the following section to find solutions to the robust linear filtering problem. We do not enter into technical details, in particular we do not provide a formal theoretical framework in which relations can be derived exactly. Most of the results in the following section can be derived rigorously quite similarly to the discussion of the q-step prediction problem in Section 3.

5. A Time-Domain Approach to Minimax Robust Filtering: Applications

To illustrate the use of the analysis in the preceding section, consider the q-step noiseless prediction problem, in which the quantities Y and X of Section 4 are given by

(5.1) $$Y = S_{n+q} \text{ and } X = \{S_k\}_{k=-\infty}^{n}$$

where $\{S_k\}$ is a zero-mean unit-variance time series.

An uncertainty class of interest in this situation is that of the form

(5.2) $$\mathcal{M}_o = \{\mu | \int S_\ell S_{\ell+k} d\mu = c_k,\ \ell \in \mathbf{Z}\ ,\ k = 0,\ldots,p\}$$

where the sequence $c_0, c_1, \ldots, c_p$ forms a positive definite Toeplitz matrix. This case corresponds to the situation in which the first p lag correlation coefficients of the observation sequence are known and stationary.

In view of the discussion in the preceding section, the minimax predictor, g_o, for the class $\mathcal{M}_o$ will be linear when it exists since $\mathcal{M}_o$ specifies only the second-order properties of the process $\{S_k\}$. That is, in seeking g_o we need only consider random variables of the form $g_o = -\sum_{i=q}^{\infty} a_i S_{n+q-i}$. The minimax predictor must also satisfy (4.11a) which, for this case, becomes

(5.3) $$\left(\sum_{i=q}^{\infty} a_i s_{n+q-i} + s_{n+q}\right)^2 + \sum_{\ell=-\infty}^{\infty} \sum_{k=0}^{p} \lambda_{\ell,k} s_\ell s_{\ell+k} \leq 0$$

for all real sequences $\{s_i\}$. Note that (5.3) can only be achieved with equality; thus equating terms in (5.3) yields the conditions (for $p \geq q$)

(5.4) $$\begin{aligned} \lambda_{n+q,0} &= 1 \\ \lambda_{n+q-1,i} &= a_i \quad ; \quad i = q,\ldots,p \\ \lambda_{n+q-i,i-j} &= a_i a_j \quad ; \quad i,j = q,\ldots,p \end{aligned}$$

and all the other a_i's and $\lambda_{\ell,k}$'s must be zero.

We see from (5.4) that, if μ_o exists, the minimax robust predictor must be of the form

(5.5) $$\hat{S}_{n+q} = -\sum_{i=q}^{p} a_i S_{n+q-i} \quad ,$$

which is a finite-length filter. Furthermore, the coefficients

$-a_q, \ldots, -a_p$ must be the coefficients for optimum linear estimation of S_{n+q} from $S_{n+q-p}, \ldots, S_n$, and these coefficients are given as solutions of the equations (3.10b). Also, in order for (5.5) to be the optimum predictor corresponding to some measure μ_0, we must have

$$(5.6) \qquad S_{n+q} = \hat{S}_{n+q} + \eta_{n+q}$$

when η_{n+q} is independent $\{S_k\}_{k=-\infty}^{n}$ under μ_o. This implies that μ_o (if it exists) must yield a Gaussian process generated by the equation

$$(5.7) \qquad S_k = -\sum_{i=q}^{p} a_i S_{k-i} + \eta_k \quad , \quad k \in \mathbb{Z}$$

where the residual process $\{\eta_k\}$, is a Gaussian MA(q-1) process. Now one can proceed exactly as in Section 3 to consider the problem of existence of such an ARMA(p,q-1) process with specified autocovariances $c_0, \ldots, c_p$.

The above analysis indicates the solution to the robust Wiener-Kolmogorov pure prediction problem when the uncertainty class is of the form $\mathcal{M}_0$ of (5.2). A more realistic class is one which assumes that the first p c_k's are also uncertain, i.e., we consider the class

$$(5.8) \qquad \mathcal{M}_{\mathcal{C}} = \{\mu \mid \int S_\ell S_{\ell+k} d\mu = c_k \quad \ell \in \mathbb{Z} \ , k = 0, \ldots, p; \ C_p \in \mathcal{C}\ \}$$

where $C_p = \text{Toeplitz}\{c_0, c_1, \ldots, c_p\}$ and $\mathcal{C}$ is an uncertainty class of positive definite $(p+1) \times (p+1)$ Toeplitz matrices.

In view of the preceding analysis, the problem

$$(5.9) \qquad \min_{g \in \mathcal{S}_X} \{ \sup_{\mu \in \mathcal{M}_{\mathcal{C}}} \varepsilon(g;\mu) \}$$

can be approached by first considering the finite-length linear estimation problem

$$(5.10) \qquad \min_{\underline{a} \in \mathbb{R}^{p+1-q}} \{ \sup_{C_p \in \mathcal{C}} \tilde{\mathcal{E}}(\underline{a};C_p) \}$$

where $\tilde{\mathcal{E}}(a;C_p)$ is the mean-square prediction error associated with the predictor $-\sum_{i=q}^{p} a_i S_{n+q-i}$ when C_p is the true covariance matrix of $(S_{n+q-p}, \ldots, S_{n+q})'$. Note that

$$(5.11) \qquad \tilde{\mathcal{E}}(\underline{a};C_p) = c_0 + 2\underline{a}'\begin{pmatrix} c_q \\ \vdots \\ c_p \end{pmatrix} + \underline{a}'\begin{pmatrix} c_0 & \cdots & c_{p-q} \\ \vdots & & \\ c_{p-q} & \cdots & c_0 \end{pmatrix}\underline{a}$$

which is quadratic-linear in $(\underline{a},C_p)$. It is straightforward to show that a pair $(\underline{a}_0,C_p^o)$ with $\underline{a}_o$ optimum for $C_p \in \mathcal{C}$ forms a saddlepoint for (5.10) if and only if C_p^o is least favorable; such a C_p^o must exist if $\mathcal{C}$ is compact. If such an $\underline{a}_o$ does exist, then within the previous restrictions the predictor

$$(5.12) \qquad \hat{S}_{n+q} = -\sum_{i=q}^{p} a_{o,i}S_{n+q-i}$$

is the solution to the overall minimax problem (5.9).

Since $\tilde{\mathcal{E}}(\underline{a};C_p)$ is linear in C_p it can be written as

$$(5.13) \qquad \tilde{\mathcal{E}}(\underline{a};C_p) = \mathrm{tr}\{B_{\underline{a}}C_p\}$$

for a matrix $B_{\underline{a}}$ depending only on q, p, and $\underline{a}$. Also, for fixed C_p the optimum $\underline{a}$ is given by

$$(5.14) \qquad -\underline{a}(C_p) = \begin{pmatrix} c_0 & \cdots & c_{p-q} \\ \vdots & & \\ c_{p-q} & \cdots & c_0 \end{pmatrix}^{-1}\begin{pmatrix} c_q \\ \vdots \\ c_p \end{pmatrix}.$$

These two equations can be used to compute $\underline{a}_0$ and/or C_p efficiently for many uncertainty classes of interest (a similar optimization problem is discussed in Verdú and Poor (1984)). Thus, the computation of minimax robust q-step predictors for uncertainty classes on the covariance structure (i.e., of the form (5.8)) is straightforward.

In a manner similar to that in the above paragraphs, Eqs. (4.14) can be applied to find solutions to other problems of robust filtering within covariance uncertainty. For example, consider the situation in which

$$(5.15) \qquad Y = S_n \text{ and } X = \{X_k\}_{k=-\infty}^{n} \equiv \{S_k+N_k\}_{k=-\infty}^{n}$$

where $\{S_k\}$ and $\{N_k\}$ are as discussed in Section 1 with $\{N_k\}$

being white, and the uncertainty class is given by

$$(5.16)\quad \mathcal{M} = \{\mu \mid \int S_\ell S_{\ell+k} d\mu = c_k, \int S_\ell N_{\ell+m} d\mu = 0, \int N_\ell N_{\ell+m} d\mu = \sigma^2 \delta_m; \\ \ell, m \in \mathbb{Z},\ k = 0,\ldots,p\},$$

where δ_m denotes the Kronecker delta function.

Using (4.14) in conjunction with (5.16) it can be seen that, if g_o exists, it must be of the form

$$(5.17)\qquad g_o = -\sum_{i=0}^{p} a_i X_{n-i} \quad .$$

This in turn, implies that

$$(5.18)\qquad S_n = \sum_{i=0}^{p} a_i X_{n-i} + \varepsilon_n$$

with ε_k independent of $\{X_n, n \le k\}$. Thus, we have that μ_o (if it exists) must be such that $\{X_i\}$ is generated by

$$(5.19)\qquad (1-a_0)X_k = \sum_{i=1}^{p} a_i X_{k-i} + \varepsilon_k + N_k, \quad k \in \mathbb{Z} \quad .$$

with ε_k and N_k Gaussian and independent of the past and of each other. In other words, the least-favorable observation process is an AR(p) process as the prescribed autocovariances of the observation process are positive definite. The regression coefficients must be chosen via the Yule-Walker equations to match the covariance sequence $E\{X_\ell X_{\ell+k}\} = r_k + \sigma^2 \delta_k$, $k = 0,\ldots,p$, and, as in the case of 1-step pure prediction discussed by Franke (1981a), this corresponds to maxentropic observations (but not a maxentropic signal). The least favorable signal is the sum of the AR(p) observation process and of the white noise process $\{-N_i\}$. Therefore, by results of Pagano (1974), it is an ARMA(p,p) process with the same AR coefficients as the least favorable observation process.

For uncertain C_p and σ^2, the computation of saddlepoints is more or less the same for this case as for the case of prediction. (i.e., equations similar to (5.13) and (5.14) apply.)

More generally, for uncertainty classes of the form of (4.13), the necessary and sufficient conditions of (4.14) indicate that certain types of information must be known in order for a minimax-robust estimate to exist. In particular, we must be able to obtain $(g_o(x)-y)^2$ as a linear combination of the g_k's. Examination of (5.3) indicates the fulfillment of this requirement for the pure prediction case. Other covariance and crosscovariance constrained classes will exhibit similar phenomena. However, for uncertainty classes involving only higher order (> 2) moments of Y, the possibility of finding solutions to (4.14) seems doubtful.

As a final note, it is interesting to consider the one-step pure prediction problem (q = 1 in (5.1)) for the uncertainty class

$$(5.20) \quad \mathcal{M} = \{\mu \mid (S_\ell, S_{\ell+1}, \ldots, S_{\ell+p}) \text{ have joint density } f_p; \; \ell \in \mathbf{Z}\}$$

when f_p is consistent with second-order stationarity of $\{S_k\}$. For this problem it is straightforward to verify that a saddlepoint to (4.3) is given by (μ_o, g_o) where

$$(5.21) \quad g_o = E\{S_{n+1} \mid S_{n+1-p}, \ldots, S_n\}$$

and μ_o corresponds to the stationary p^{th}-order Markov process with transition density

$$(5.22) \quad f(s_{n+1} \mid \ldots s_n) = \frac{f_p(s_{n+1-p}, \ldots, s_{n+1})}{\int_{-\infty}^{\infty} f_p(s_{n+1-p}, \ldots, s_{n+1}) ds_{n+1}} \; .$$

It is interesting to note that this process has maximum entropy rate (in the sense of Shannon entropy) among all processes with p^{th}-order block density f_p. This generalizes the property exhibited by the least-favorable minimax 1-step prediction problem with fixed autocorrelations up to lag p.

Note that,as in the covariance-constrained situation, the result of the above paragraph allows for the solution of (4.14) over classes of the form

$$(5.23) \quad \mathcal{M} = \{\mu \mid (S_\ell, \ldots, S_{\ell+p}) \sim f_p, \; \ell \in \mathbf{Z} \; ; \; f_p \in F\},$$

where F is a class of pth-order densities, by first solving the finite-length problem of (minimax) estimating S_{n+1} from $(S_{n+1-p},\ldots,S_n)$. For the class $\mathcal{M}$ of (5.2), this procedure leads to the maxentropic Gaussian process via the argument in the paragraph containing (4.12).

An interesting question is whether a general saddlepoint solution to the robust q-step prediction problem can be given explicitly for the class of (5.20). Of course, the natural conjecture for the estimator is

$$(5.24) \qquad g_o = E\{S_{n+q} | S_{n+q-p},\ldots,S_n\} .$$

The only question is whether there is a legitimate random process having this optimum predictor. This process would have to be generated by

$$(5.25) \qquad S_k = E\{S_k | S_{k-p},\ldots,S_{k-q}\} + \eta_k$$

where $\{\eta_k\}$ is a (q-1)-dependent process, independent of $(S_{k-p},\ldots,S_{k-q})$. Of course $E\{S_k | S_{k-p},\ldots,S_{k-q}\}$ can be obtained from f_p, so that η_k can be written explicitly as a function of S_k and $(S_{k-p},\ldots,S_{k-q})$. However, the question of existence of a process consistent with f_p for which the resulting sequence is (q-1)-dependent is, in general, a difficult one.

Appendix 1

Let $c_0,\ldots,c_p$ be positive definite and satisfy the condition of Proposition (3.1b), i.e. there exist $\sigma^2 > 0$, $a_q,\ldots,a_p$ and $b_1,\ldots,b_{q-1}$ solving (3.10). If one wants to show the existence of an ARMA(p,q-1) process with autocovariances $c_0,\ldots,c_p$ and impulse response coefficients $b_1,\ldots,b_{q-1}$ there are two different ways of constructing this process. First, one can extrapolate the autocovariances, by using (3.11) for all $n > p$, such that they have the correct recursive structure. This procedure results always in a positive definite sequence $c_0, c_1, \ldots$. Then it is easy to show that a time series $\{X_n'\}$ with that autocovariance sequence satisfies (Franke, 1984)

$$(A1.1) \qquad X_n' + \sum_{k=q}^{p} a_k X_{n-k}' = \sum_{k=0}^{q-1} b_k \varepsilon_{n-k} \qquad Var(\varepsilon_n) = \sigma^2 .$$

Secondly, one can extrapolate the impulse response coefficients by means of

$$b_n = - \sum_{k=q}^{p} a_k \, b_{n-k} \quad , \quad n \geq q,$$

which always results in a square-summable sequence. Then, it is easy to show that the time series, given by its MA(∞) representation

$$X_n = \sum_{k=0}^{\infty} b_k \varepsilon_{n-k} \quad , \quad \mathrm{Var}(\varepsilon_n) = \sigma^2,$$

also satisfies (A1.1). However, if the polynomials $A(z)$, $B(z)$ of Proposition 3.1 have common zeros on the boundary of the unit circle, it is possible that $\{X_n'\}$ and $\{X_n\}$ differ by a time series with discrete spectrum concentrated in the common zeros of $A(z)$ and $B(z)$. Then, though there exists the purely nondeterministic time series $\{X_n\}$, satisfying the ARMA relation (A1.1), it has not the required autocovariances $c_0,\dots,c_p$.

As an example consider p=3, q=2. Let $c_0=2$, $c_1=-2/3$, $c_2=5/3$, $c_3=-1$, which are positive definite. By (3.10b), $a_2=-3/4$, $a_3=1/4$. Defining c_4 by (3.11) results in $c_4=17/12$, and the condition of Proposition 3.1b) is satisfied. Solving (3.10a), we get $\sigma^2=1/4$, $b_1=1$. Therefore, $A(z)$ and $B(z)$ have the common zero -1. There exists a purely nondeterministic time series $\{X_n\}$ satisfying (A1.1), namely the AR(2) process, given by

$$X_n - X_{n-1} + X_{n-2}/4 = \varepsilon_n \quad , \quad \mathrm{Var}(\varepsilon_n) = \sigma^2 .$$

However, $c_0,\dots,c_3$ are not autocovariances of $\{X_n\}$, but rather of the time series given by

$$X_n' = X_n + (-1)^n Y,$$

where Y is a random variable independent of $\{X_n\}$. $\{X_n'\}$ satisfies (A1.1) but it has mixed spectrum with a point mass in $\omega = \pi$. This does not conform with the assumption of Section 2. It is possible to extend the theory of robust linear filtering to processes with general spectral measures (Franke, 1980; Vastola and Poor, 1984), but then, other problems arise.

Appendix 2

For spectral densities f_S, f_N and for $H \in L^2$ let $e(f_S, f_N; h)$ be defined as in Section 2. Let $\mathcal{L}$ be given by (2.2), and let

$$R(f_S, f_N) = \min_{H \in \mathcal{L}} e(f_S, f_N; H)$$

be the minimal filtering error for signal and noise spectral densities f_S, f_N which can be achieved with filters in $\mathcal{L}$. Then, an extension of the methods used by Franke (1981a) gives, under suitable weak regularity conditions, that a pair $(f_S^o, f_N^o) \in \mathcal{J}$ is least favorable for the spectral information set $\mathcal{J}$ iff

$$\text{(A2.1)} \qquad R_o'(f_S, f_N) \le R_o'(f_S^o, f_N^o) \text{ for all } (f_S, f_N) \in \mathcal{J} \, .$$

Here, $R_o'(f_S, f_N)$ denotes the directional derivatives of R at (f_S^o, f_N^o) in direction (f_S, f_N) defined analogously to (4.7). The following result relates (A2.1) to the criterion of Corollary 3.1.

Proposition A2.1: *Let* f_S, f_N, f_S^o, f_N^o *be spectral densities, and assume for* $\Delta > 0$

$$\text{(A2.2)} \qquad f_S^o(\omega) + f_N^o(\omega) \ge \Delta > 0 \, .$$

Then,

$$R_o'(f_S, f_N) = e(f_S, f_N; H_W^o)$$

where H_W^o *denotes the Wiener filter with respect to* (f_S^o, f_N^o).

Proof: 1) Let $f_S^\varepsilon = (1-\varepsilon) f_S^o + \varepsilon f_S$, $f_N^\varepsilon = (1-\varepsilon) f_N^o + \varepsilon f_N$, and let H_W^ε denote the Wiener filter with respect to $(f_S^\varepsilon, f_N^\varepsilon)$, which, by (A2.2) is a.e. uniquely defined and is contained in L^2 for all $0 \le \varepsilon < 1$. First, we show that H_W^ε converges weakly in L^2 to H_W^o for $\varepsilon \to 0+$.

Let

$$M = \frac{1}{2\pi} \int (f_S^o(\omega) + f_S(\omega)) d\omega \sup_\omega |D(\omega)|^2 \, .$$

Then, completely analogously to step 1), 2) of the proof of Theorem 2.1, one can conclude that for $\varepsilon_o > 0$ and $\Delta_o = (1-\varepsilon_o)\Delta$ we have

$$\|H_W^\varepsilon\| \le \|D\| + \sqrt{2 M/\Delta_o} \quad , \quad \text{for } 0 \le \varepsilon \le \varepsilon_o \, .$$

As $\{H_W^\varepsilon; 0 \le \varepsilon \le \varepsilon_o\}$ is bounded in Hilbert-space norm it contains a sequence $H_W^{\varepsilon(j)}$ which converges weakly for $\varepsilon(j) \to 0$ (see, e.g., Weidmann (1976), Satz 4.25), say to $H^* \in L^2$. As weak convergence implies convergence of Fourier coefficients we conclude immediately $H^* \in \mathcal{L}$.

Let us for the moment assume that f_S, f_N, f_S^o, f_N^o are bounded. Then, $H_W^{\varepsilon(j)}, D-H_W^{\varepsilon(j)}$ converge weakly to H^*, $D-H^*$ in $L^2(f_N^o)$. Let $\|\cdot\|_S$, $\|\cdot\|_N$ denote the norm in those two Hilbert spaces. We have by definition of e

$$\text{(A2.3)} \quad \begin{aligned} \liminf e(f_S^o, f_N^o; H_W^{\varepsilon(j)}) &\ge \liminf \| D-H_W^{\varepsilon(j)}\|_S^2 \\ &\quad + \liminf \| H_W^{\varepsilon(j)}\|_N^2 \\ &\ge \|D-H^*\|_S^2 + \| H^*\|_N^2 \\ &= e(f_S^o, f_N^o; H^*) \quad , \end{aligned}$$

where the second inequality follows, e.g., from Satz 4.24 of Weidmann (1976).

By definition of H_W^ε we have for $0 \le \varepsilon \le \varepsilon_o$

$$\text{(A2.4)} \quad e(f_S^o, f_N^o; H_W^\varepsilon) + \varepsilon e(f_S, f_N; H_W^\varepsilon) \le e(f_S^o, f_N^o; H_W^o) + \varepsilon\, e(f_S, f_N; H_W^o).$$

As both H_W^ε are bounded in norm uniformly in $0 \le \varepsilon \le \varepsilon_o$, $e(f_S, f_N; H_W^o)$ is also bounded uniformly in $0 \le \varepsilon \le \varepsilon_o$, say by the constant $C > 0$. Therefore, by (A2.3), (A2.4),

$$\begin{aligned} e(f_S^o, f_N^o; H^*) &\le \liminf e(f_S^o, f_N^o; H_W^{\varepsilon(j)}) \\ &\le \liminf \{e(f_S^o, f_N^o; H_W^o) + 2\varepsilon(j) C\} \\ &= e(f_S^o, f_N^o; H_W^o). \end{aligned}$$

As H_W^o is the optimal filter in $\mathcal{L}$ with respect to (f_S^o, f_N^o), we have $H^* = H_W^o$.

2) Now, (A2.3) is satisfied if we replace f_S^o, f_N^o by f_S, f_N, and we already know $H^* = H_W^o$. Therefore,

(A2.5) $\liminf \frac{1}{\varepsilon} \{R(f_S^{\varepsilon},f_N^{\varepsilon}) - R(f_S^{O},f_N^{O})\}$

$$= \liminf \frac{1}{\varepsilon} \{e(f_S^O,f_N^O;H_W^{\varepsilon}) - e(f_S^O,f_N^O,H_W^O) + \varepsilon e(f_S f_N;H_W^{\varepsilon})\}$$

$$\geq \liminf e(f_S f_N;H_W^{\varepsilon})$$

$$\geq e(f_S f_N;H_W^O) \quad .$$

For unbounded spectral densities, the same inequality is satisifed, too. This can be shown directly by first applying (A2.5) to truncated spectral densities, e.g. $f_{S,K} = \min(f_S,K)$, and then letting $K \to \infty$.

Now the proof is completed by noting that for all $\varepsilon > 0$

$$\frac{1}{\varepsilon} \{R(f_S^{\varepsilon},f_N^{\varepsilon}) - R(f_S^O,f_N^O)\} = \frac{1}{\varepsilon} e(f_S^{\varepsilon},f_N^{\varepsilon};H_W^{\varepsilon}) - e(f_S^O,f_N^O;H_W^O)\}$$

$$\leq e(f_S,f_N;H_W^O),$$

which together with (A2.5) implies that $R_O'(f_S,f_N)$ exists and coincides with $e(f_S,f_N;H_W^O)$. QED

Appendix 3: A Proof of Proposition 4.2

We first show that $C(\mu)$ is norm-u.s.c. on $\mathfrak{M}$ and then use the concavity of $C(\mu)$ to deduce weak upper-semicontinuity (see, Barbu and Precupanu (1978), p. 78). Suppose $\{\mu_i\}_{i=1}^{\infty}$ is a sequence in $\mathfrak{M}$ converging in norm to $\mu_0 \in \mathfrak{M}$. As norm on $\mathfrak{M}$ we consider the variational distance

$$\|\mu_1-\mu_2\| = \sup_{A \in \mathfrak{F}} |\mu_1(A)-\mu_2(A)| .$$

Define $\nu = \sum_{i=0}^{\infty} 2^{-i}\mu_i$, which is a finite measure dominating all μ_i, $i \geq 0$. Let φ_i denote the density of μ_i with respect to ν. As a first step, we show that under either of the conditions of the proposition we have

(A3.1) $\int|Y|^n d\mu_i \to \int|Y|^n d\mu_0 \quad$ for $n = 1,2$.

Let us first assume that condition (ii) be satisfied, and let $p = 1 + \varepsilon/2$, and $1/q + 1/p = 1$. Then we have, using Hölder's inequality

$$(A3.2)\quad |\int Y^2 d\mu_i - \int Y^2 d\mu_0| \le \int Y^2 |\varphi_i - \varphi_0| d\nu$$

$$\le \{\int |Y|^{2+\varepsilon} |\varphi_i - \varphi_0| d\nu\}^{1/p} \{\int |\varphi_i - \varphi_0| d\nu\}^{1/q}$$

$$\le (2c')^{1/p} \| \mu_i - \mu_0 \|^{1/q}$$

(A3.1) follows for $n=2$, as, under condition (i) it is trivial. (A3.1) follows for $n=1$ quite similar from the uniform boundedness of the second moments of Y under all $\mu \in \mathcal{M}$.

From (A3.1) we conclude that the functional $\int (Y-g)^2 d\mu$ is norm-continuous on $\mathcal{M}$ for all bounded measurable g. Now

$$C(\mu) = \inf_{M > 0} \inf_{\{g \in \mathcal{B}_X, |g| \le M\}} \int (Y-g)^2 d\mu \ .$$

As the infimum of continuous functionals, $C(\mu)$ is upper semicontinuous (Hewitt-Stromberg, 1965) in the norm topology. As C is concave, it is also weakly upper semicontinuous by the above mentioned result of Barbu and Precupanu. QED

Acknowledgment

This research was supported in part by the Deutsche Forschungsgemeinschaft and in part by the U.S. Office of Naval Research under Contract N00014-81-K-0014.

References

V. Barbu and Th. Precupanu (1978) Convexity and Optimization in Banach Spaces. Sijthoff and Noordhoff, Alphen aan de Rijr.

G. E. P. Box and G. M. Jenkins (1976) Time Series Analysis: Forecasting and Control. Holden-Day, San Francisco.

L. Breiman (1968) Probability. Addison-Wesley, Reading, Mass.

L. Breiman (1973) "A Note on Minimax Filtering". Ann. Probab., Vol. 1, pp. 175-179.

J. Franke (1980) "A General Approach to the Robust Prediction of Discrete Time Series". SFB 123 "Stochastische Mathematische Modelle", Preprint No. 74, Heidelberg.

J. Franke (1981a) "Minimax-Robust Prediction of Discrete Time-Series". Z. Wahrscheinlichkeitstheorie verw. Gebiete.

J. Franke (1981b) "Linear Interpolation and Prediction of Time Series with Partially Known Spectral Density and its Relation to Maximum Entropy Spectral Estimation". Proc. of the First ASSP Workshop on Spectral Estimation, McMaster University, Hamilton/Ontario.

J. Franke (1984) "ARMA-Processes have Maximal Entropy among Time Series with Prescribed Autocovariances and Impulse Response Coefficients". Submitted for Publication.

E. J. Hannan (1970) Multiple Time Series. Wiley, New York.

Y. Hosoya (1978) "Robust Linear Extrapolation of Second-Order Stationary Processes". Ann. Probab., Vol. 6, pp. 574-584.

P. J. Huber (1964) "Robust Estimation of a Location Parameter". Ann. Math. Statist., Vol. 35, pp. 73-104.

S. A. Kassam and T. L. Lim (1977) "Robust Wiener Filters". J. Franklin Inst., Vol. 304, pp. 171-185.

S. A. Kassam (1982) "Robust Hypothesis Testing and Robust Time Series Interpolation and Regression". J. Time Series Analysis, Vol. 3, pp. 185-194.

M. Pagano (1974) "Estimation of Models of Autoregressive Signal Plus White Noise". Ann. Statist., Vol. 2, pp. 99-108.

H. V. Poor (1980) "On Robust Wiener Filtering". IEEE Trans. Autom. Contr., Vol. AC-25, pp. 531-536.

M. B. Priestley (1981) Spectral Analysis and Time Series. Academic Press, London.

M. Taniguchi (1981) "Robust Regression and Interpolation for Time Series". J. Time Series Analysis, Vol. 2, pp. 53-62.

K. S. Vastola and H. V. Poor (1983) "An Analysis of the Effects of Spectral Uncertainty on Wiener Filtering". Automatica, Vol. 19, pp. 289-293.

K. S. Vastola and H. V. Poor (1984) "Robust Wiener-Kolmogorov Theory". IEEE Trans. Inform. Theory, Vol. IT-30 (to appear in March).

S. Verdú and H. V. Poor (1984) "Minimax Linear Observers and Regulators for Stochastic Systems with Uncertain Second-Order Statistics". IEEE Trans. Autom. Control, Vol. AC-31 (to appear in June).

J. Weidmann (1976) Lineare Operatoren im Hilbertraum. Teubner, Stuttgart.

P. Whittle (1963) Prediction and Regulation. The English Universities Press, London.

J. H. Wilkinson (1967) "The Solution of Ill-Conditioned Linear Equations". In: Mathematical Methods for Digital Computers. Eds. A. Ralston and H. S. Wilf, Wiley, New York.

THE PROBLEM OF UNSUSPECTED SERIAL CORRELATIONS

H. Graf, F.R. Hampel, J.-D. Tacier *)
Fachgruppe für Statistik
ETH-Zentrum, CH-8092 Zurich

Abstract

The problem of violation of the assumption of independence for supposedly independent identically distributed data is discussed, and historical observations of the occurring phenomena are cited together with the first attempts to model them with the aid of self-similar processes. Various empirical features of a chemical measurement series given by Student are analysed in detail. A method for estimation and testing of the correlation parameter of self-similar processes is presented and applied to two examples from geophysics and from high-quality physical measurement work.

Acknowledgements

The authors like to acknowledge the help by Professor H.U. Dütsch (LAPETH, Zurich) and the National Bureau of Standards (Washington, D.C.) in providing some long series of measurements for analysis.

1. Introduction and historical background

At present, most research on robust methods in time series analysis is concentrated on the problem of outliers - in a broad sense, including transient stretches of different behaviour and slight

*) Now at Höhere Technische Lehranstalt (HTL) CH-5200 Brugg-Windisch

deviations in the marginal distribution. However, it is normally assumed that the correlation structure of the model holds exactly (apart, perhaps, from some effects caused by the outliers). A conceptually different problem arises if one believes that the assumed correlation structure holds only "approximately" (in a sense or senses to be specified); that the true autocorrelation function differs "slightly" from the assumed one. Probably next to nothing is presently known about this problem in its full generality. But there is a special case which is of considerable practical as well as theoretical interest: the problem of slight serial correlations when independence is believed, or, put differently, the problem of slight deviations from the simplest stochastic process, namely white noise, or the problem of violation of one of the most fundamental assumptions in much of statistics, namely the assumption of independence. This case may also serve as a stepping-stone to the treatment of data whose correlation structure can "almost" be described by, for example, an ARMA process with rather few parameters.

There has already been quite some work on the effects of dependence; see for example the old summary in Scheffé (1959), Ch.10, or the work by Portnoy (1977); cf. also the paper by W. Ploberger at this Heidelberg workshop. Most of the older work considers only ARMA processes with very few parameters as alternatives to independence. The effects of simple Gauss-Markov processes are already quite drastic, not so much on point estimates, but on confidence intervals and related quantities: if independence is assumed, they are wrong by a factor which tends to a constant as the length of the sequence increases, but a constant which can be arbitrarily large.

What are the consequences for statistical practice? Clearly, if lack of independence is ignored, standard errors and tests can be completely misleading. On the other hand, strong short-term dependence - which is needed in ARMA processes with few parameters to create strong effects - can be easily detected in series of moderate length, and in situations where such strong dependence occurs, one will often be inclined already a priori (and otherwise be forced a posteriori) to use an ARMA model rather than the assumption of independence. But the question arises whether the model of ARMA processes with few parameters can capture the essential features of all important deviations from independence occurring in real life.

There is strong empirical evidence to the contrary. Many sequences of supposedly independent high-quality data from various fields of science obtained in order to estimate some absolute constant, look neither independent nor short-term correlated like an ARMA process with few parameters. Rather even the distant past is correlated with the distant future, although individual serial correlations may all be small. The series of deviations from some central value look like something in between independent random errors and constant systematic errors. Several eminent statisticians were clearly aware of their peculiar features long ago although they had no stochastic models which could describe them in a reasonable way. The astronomer Newcomb (1895) found them to be common in astronomy and called them "semi-systematic errors". K. Pearson (1902) did long series of experiments himself with pseudo-astronomical data, which later were reanalysed, together with some other series, by Jeffreys (1939, Ch. 5.6 and 5.7), who spoke of "internal correlation". Student (1927) found the same kind of features in carefully obtained chemical data and called the deviations "semi-constant errors". In order to show what kind of features occur, we shall analyse one of Student's series in depth (Section 2).

In a different field, H. Fairfield Smith (1938) found empirical power laws for the variance of the mean yield over neighbouring plots in agriculture. The mere fact of correlations was not surprising but the form of the law was (cf. also Whittle 1956): the variance of the mean over n equally-sized plots decreases not like n^{-1}, but like $n^{-\alpha}$ where α can be anywhere between 0 and 1. Starting in 1965, B.B. Mandelbrot and co-workers (cf. Mandelbrot and Wallis 1968, 1969a,b,c; Mandelbrot and van Ness 1968; Mandelbrot and Taqqu 1979) made the important step of introducing the model of self-similar processes to hydrological and other geophysical time series which showed similarly

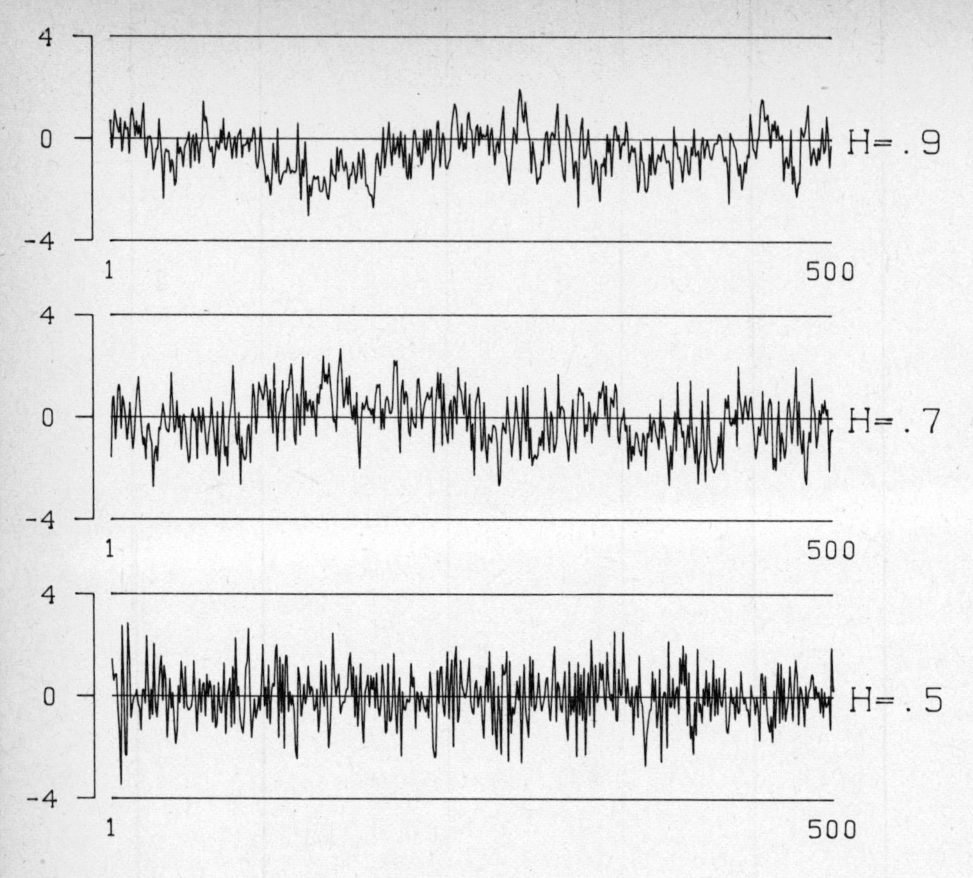

Figure 1:

Simulated realizations of standardized fractional Gaussian noise for three different values of the self-similarity parameter H. Each realization consists of 500 points, which have been connected with straight lines. The case $H = .5$ is Gaussian white noise, while $H > .5$ means long-range dependence.

startling features; this model yields also Smith's result, and its only correlation parameter $H = 1 - \alpha/2$ can also be estimated by a method called R/S analysis; moreover, for many geophysical time series H lies also anywhere between 1/2 (corresponding to independence) and 1 (the border of nonstationarity).

Self-similar processes had been introduced by Kolmogorov (1940) in a rather abstract setting and later found applications in fields like turbulence theory and electrical engineering. Their most important special class are fractional Brownian motions. Loosely speaking, the latter can be derived as infinite moving averages with hyperbolic weight function over the past increments of ordinary Brownian motion (cf. Mandelbrot and van Ness 1968). Their increment processes in discretized time are called (discrete) fractional Gaussian noises; they are stationary, ergodic, weakly mixing and regular, but not strongly mixing, and have the property that the variance of the mean of n successive increments $\mathrm{VAR}\ \bar{X}_n \propto n^{2H-2}$. Furthermore, their serial correlations fall off like $|\mathrm{lag}|^{2H-2}$, and their power spectrum density at frequency $\lambda \varepsilon (0,\pi]$ is approximately proportional to λ^{1-2H} for $H > 1/2$, so that a plot of log power spectrum against log frequency yields approximately a straight line. Hence usual standard errors are wrong by a factor tending to infinity, and although all individual serial correlations may be small and hence hard to detect, their sum still tends to infinity. These increment processes provide the simplest stochastic model which mimics the peculiar features of long series of measurement data mentioned above, and estimation of the self-similarity parameter H (which we shall consider in Section 3) plays a central role as the first step in the quantitative treatment of these correlation effects.

D.R. Cox (cf. Whittle 1956) has independently derived the model of self-similar processes from a physical model for textile data (yarn diameters). Other eminent statisticians show a growing concern about the qualitative dangers of lack of independence (cf. Daniel and Wood 1971, p. 59, Mosteller and Tukey 1977, Ch. 7, Box, Hunter, and Hunter 1978, Ch. 3). Probably also Berkson's (1938) findings can be partly explained by lack of independence. It is one of our main points that in first approximation not only geophysical, but also other scientific time series of supposedly independent data can and should be modeled by increments of self-similar processes (cf. also Hampel 1973, 1978, 1979, Hampel et al. 1982).

2. Student's data on aspartic acid

We shall now analyse under various aspects one of the series given graphically by Student (1927, p. 156), namely 135 determinations of the nitrogen content of pure crystalline aspartic

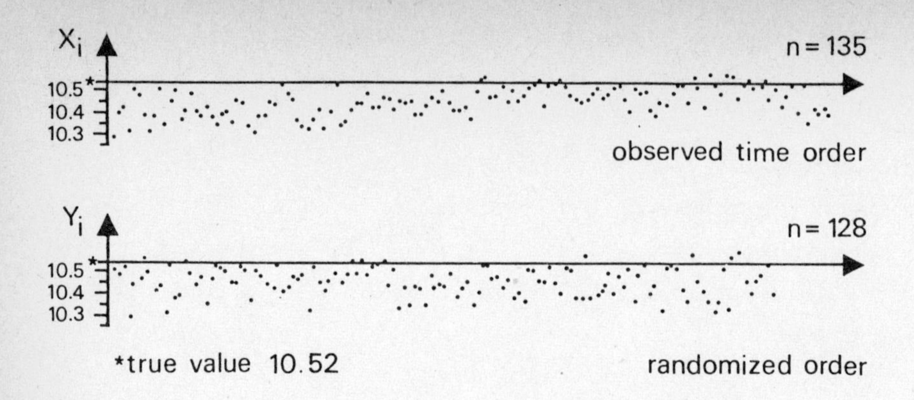

Figure 2:
Nitrogen content of aspartic acid (Student's data).

acid. They are as independent as any series of scientific measurements might be expected to be. Their values are given as X_i at the top of Figure 2. To contrast them with what independent data would look like, a random permutation of the first 128 points is given as Y_i .

When looking at the X_i , one might be tempted to say that a linear trend should be removed before further analysis. This would not make sense in this case (estimation of an absolute constant), but removal of an exponential trend could actually be justified from a remark in Student (1927). Both trend removals were tried; they gave very similar results which still retained the essential qualitative features of the original data except that the estimated H (see below) dropped from 0.83 to 0.74 (in both cases), which is still far above 0.5 . We therefore might as well give the following analysis for the original data.

Figure 3 shows the mean $\bar{X}_n$ of the first n observations together with 99%-confidence intervals based on the assumption of independence. The mean does not converge well, but rather trails off. The horizontal line is the true value which is known in this case; starting with 7 observations, it is not contained in the confidence interval any more. Even worse, 2 of the 99%-confidence intervals based partly on the same data, do not even overlap! By contrast, Figure 4 for the $\bar{Y}_n$ looks like expected

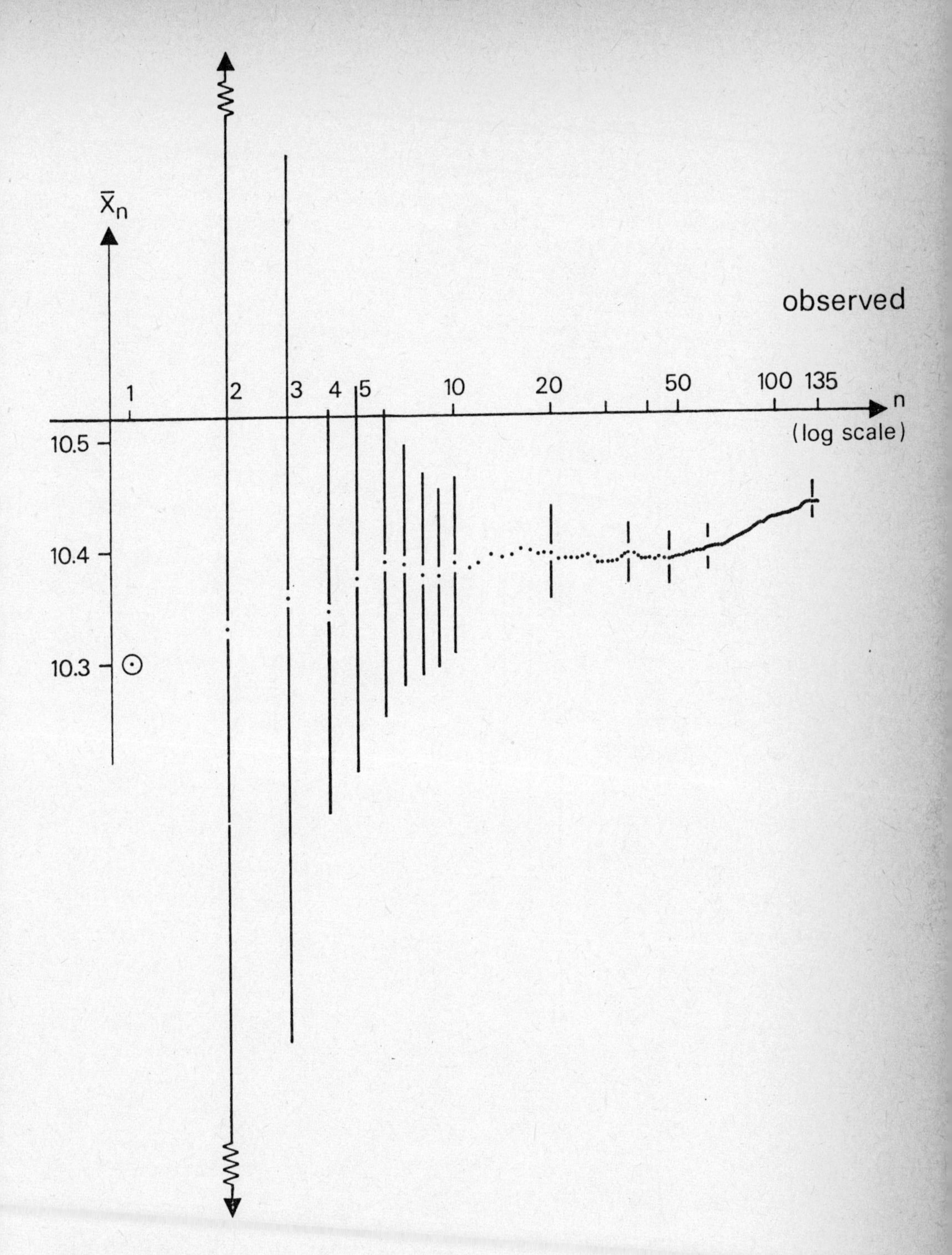

Figure 3:

$\bar{X}_n$ = mean of first n observations against n , with 99%-confidence intervals (Student's data).

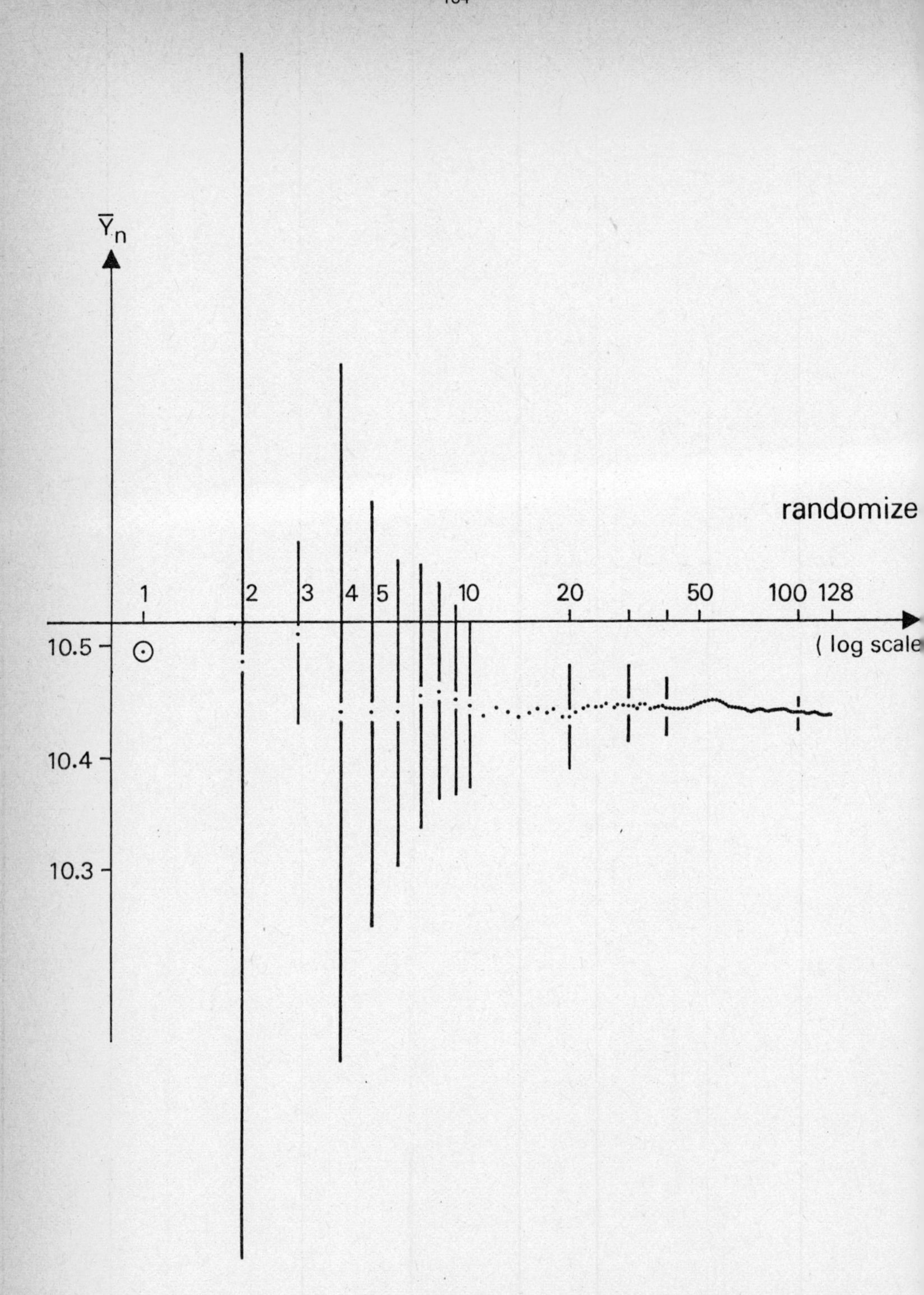

<u>Figure 4:</u>

$\bar{Y}_n$ = mean of first n permuted observations against n, with 99%-confidence intervals (Student's data).

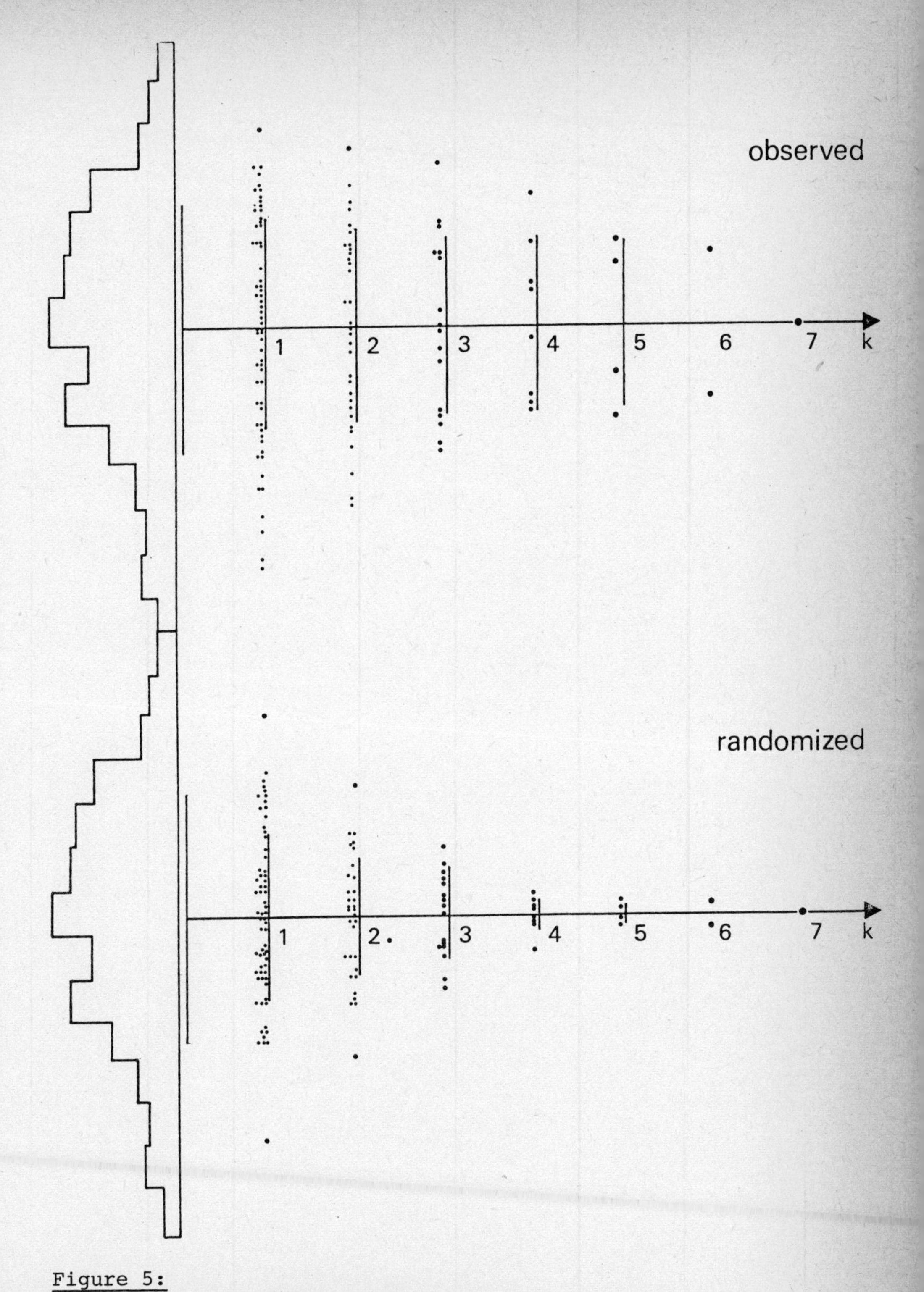

<u>Figure 5:</u>

Sample means of consecutive subgroups of size 2^k against k, with 1-σ-regions and histogram (Student's data).

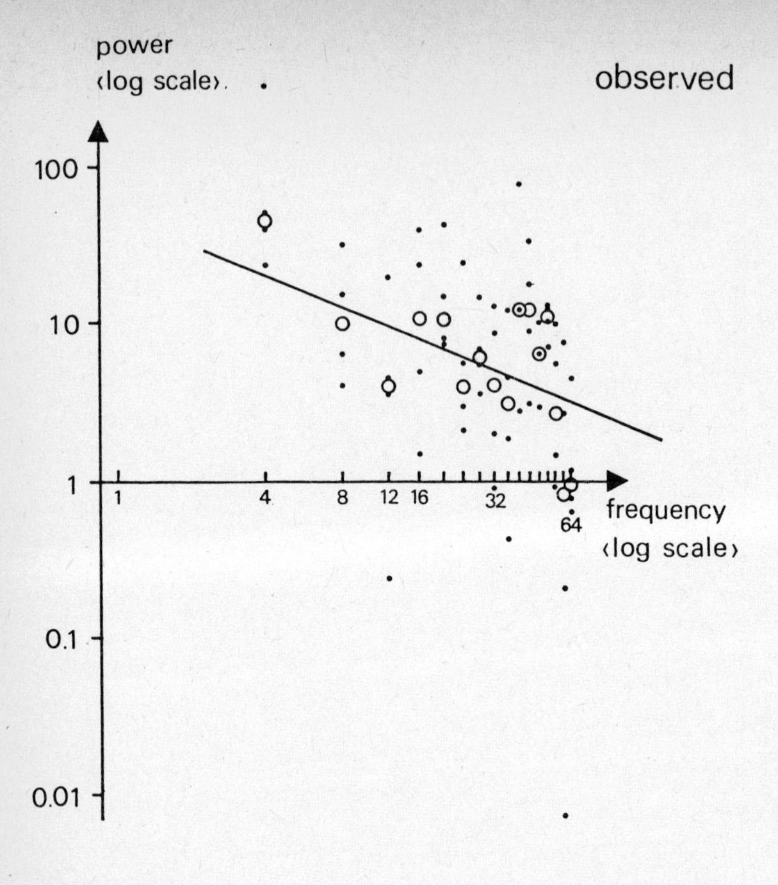

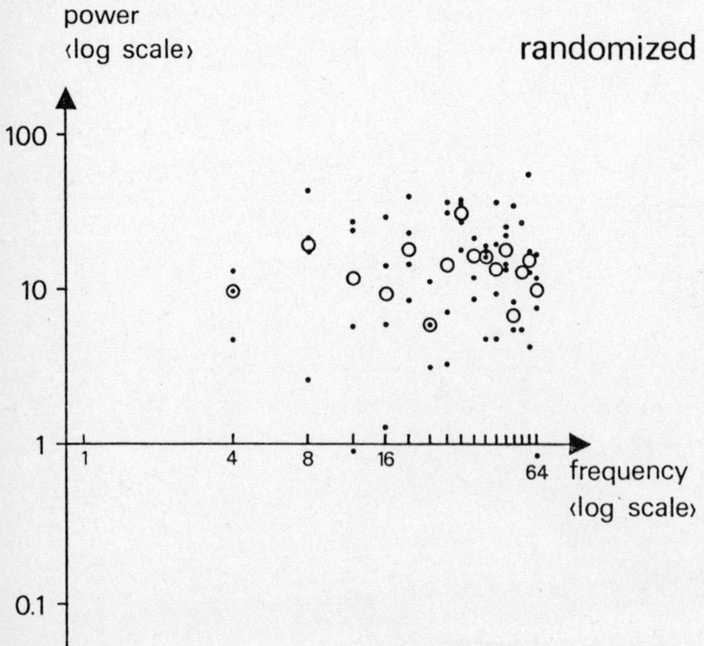

Figure 6: Power spectrum ("pox plot") for Student's data as observed and randomized.

from independence and the law of large numbers, apart from what seems to be a systematic error. Figure 5 contains means of nonoverlapping subsamples of size 2^k together with lines for the region "grand mean $\pm\ \sigma$" . It is striking that the variability of the means decreases much more slowly for the X_i than for the Y_i with increasing sample size. Since the marginal is quite normal, it is possible, with some iteration, to estimate the rate of decrease of the standard error and hence H . This method uses the "variance-time-function" and is related to an analysis-of-variance approach with hierarchical splitting of the time series. Another way of looking at the data is provided by the power spectrum (Figure 6), where circles denote medians over 4 neighbouring frequencies plotted at the same abscissa. The spectrum of white noise should be constant as it is for the Y_i , but for the X_i it clearly shows a linear trend. All pictures based on the X_i agree quite well with the model of increments of a self-similar process with H around .8 . Our best estimate of H is based on the power spectrum (Section 3) and yields $\hat{H} = 0.83 \pm 0.06$. This implies that the naive confidence interval for $n = 135$ has to be multiplied by a factor of about 4.5 ! Put differently, the variance of the mean of such a fairly short series is already 20 times as large as believed under the independence assumption. But there is also a happy observation: the corrected 99%-confidence interval contains the true value again, hence in this case at least, it is not necessary to assume a constant systematic error; all peculiarities can here be explained by the semi-systematic errors.

3. Estimation of H and examples

Having decided to model the correlation structure of a phenomenon by that of fractional Gaussian noise, we immediately face the problem of estimating the self-similarity parameter H , which determines the correlations. Besides the interest in itself, an estimate of H is needed for computing realistic confidence intervals for the mean (cf. Mohr (1981)) and related quantities. Various methods for estimating H , including one based on the periodogram, are investigated in Mohr (1981). A different approach to estimating H via the periodogram is worked out in Graf (1983). The aim was to obtain reasonably efficient and at the same time robust estimators which are not too sensitive to high-frequency deviations from the model.

Before the Fourier transformation, gross errors must be "corrected". This can be done by using methods such as those proposed in Kleiner et al. (1979). The periodogram is then computed for the "cleaned" time series. The distribution of the periodogram ordinates at the Fourier frequencies strictly between 0 and π is approximated by independent exponential distributions. In the FORTRAN subroutines listed in Graf (1983) this has been done by equating expectations under the model. What would be the likelihood equations for the approximating distribution is then derived and robustified by putting a bound on the scores. The effect of, say, a periodicity in the data on the estimated H can be bounded in this way at the cost of a small efficiency loss (say, 5%). In the investigation of long-range properties it is often desirable to rely not too heavily on the high-frequency periodogram ordinates. This can be achieved, at the cost of a moderate efficiency loss (say, 25%), by bounding the effects of high frequencies more severely.

Heuristic considerations regarding the asymptotic properties of the proposed class of estimators led to results which agree well with what one can extrapolate from simulations performed in the Gaussian case. Such simulations were done for series lengths between 64 and 1024. Even for the shortest series, the simulated distributions of the estimators do not differ greatly from normality. Confidence intervals for H which should be reasonably accurate and associated tests are obtained from these simulations.

For fractional Gaussian noise, the approach outlined above appears to be about 10% more efficient than the most efficient estimator investigated in Mohr (1981). In moderately non-Gaussian situations, it appears still to be reasonable.

Example 1: The Arosa $\lambda\lambda$ C total ozone series.

These monthly data are taken from Birrer (1975) for the years 1932 to 1971 and from a personal communication by Prof.H.U.Dütsch,

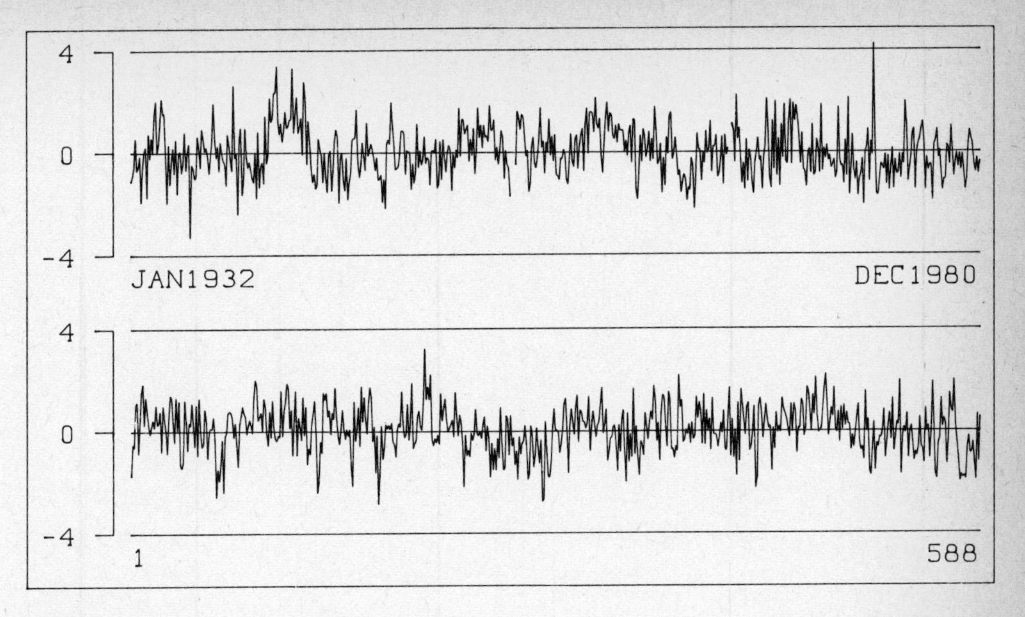

Figure 7:

Upper frieze: series of seasonally adjusted monthly means 1932-1980 of the Arosa λλ C total ozone measurements.
Lower frieze: a realization of standardized fractional Gaussian noise with $H = .7$ for comparison.

Laboratorium für Atmosphärenphysik, ETH Zürich, for the years 1972 to 1980. They exhibit strong seasonal variation in location and scale, which has been removed for our purposes. The seasonally adjusted data, having skewness .341 and kurtosis .405, are shown in the upper frieze of Figure 7.

They contain 5 missing values which were linearly interpolated from their neighbours before the periodogram was computed. The latter is shown in Figure 8 using log-log co-ordinates.

We first apply estimator HUB02%. This estimator puts an upper bound on each periodogram ordinate which equals the 98%-point of the corresponding exponential distribution. Solving the resulting equations and applying a (very small) bias correction yields $\hat{H} = .701$. One-sided lower and upper 97.5%-confidence limits are obtained as .646 and .760, respectively. "Bringing in" before the Fourier transformation the isolated high value for October 1974, which is explained by extreme meteorological conditions, does hardly affect $\hat{H}$. A realization of fractional

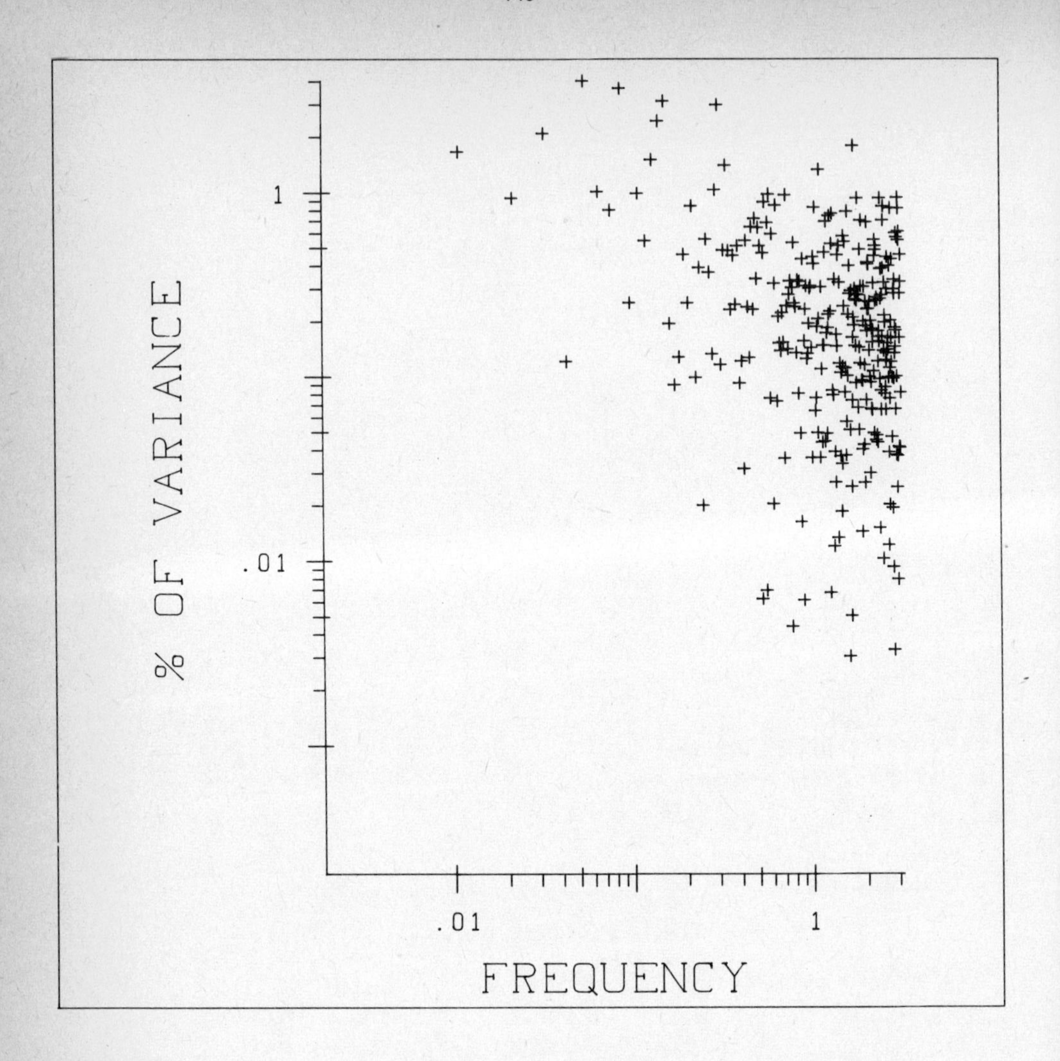

Figure 8:

Periodogram for the monthly means 1932-1980 of the homogeneous Arosa $\lambda\lambda$ C total ozone series after removal of the seasonal variation and interpolation of the missing values (logarithmic scale on both axes).

Gaussian noise with $H = .7$ is shown for comparison in the lower frieze of Figure 7. It mimics the ozone data quite well, although the general impression is that the latter deviate somewhat more from the zero line, most obviously for the two-year period 1940 to 1941.

We now wish to depend less heavily on the high-frequency periodogram ordinates, at the expense of a wider confidence interval. To this end we apply estimator HUBINC, which puts much

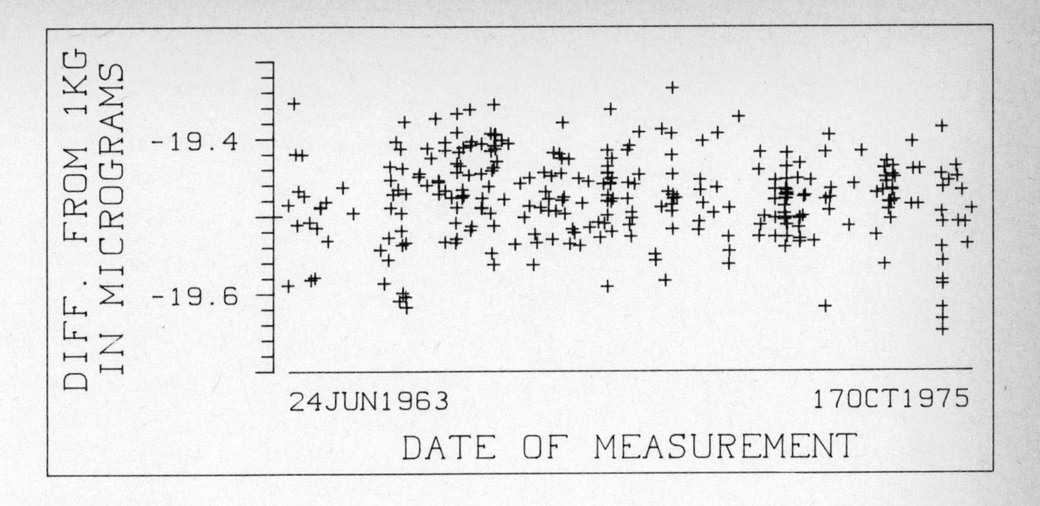

Figure 9:

NBS precision measurements on the 1 kg check standard weight.

smaller bounds on the high-frequency ordinates, yielding the appreciably higher value $\hat{H} = .738$, which is in accordance with our visual impression. The corresponding 95%-confidence interval is (.674, .806) .

Example 2: The NBS measurements on the 1 kg check standard weight.

We know already that correlations are common in routine work such as chemical routine analysis; but they seem to be present also in high precision measurement data. Some argue that this is not surprising since in this kind of data most of the variability can quite likely be attributed to causes such as changes in the atmospheric conditions, and these conditions are known to be correlated at different time points.

A set of high-quality data was supplied to us by the Statistical Engineering Division of the U.S. National Bureau of Standards, Washington, D.C. Figure 9 shows 289 measurements on the 1 kg check standard weight made under conditions as constant as possible. These data have skewness -.306 and kurtosis .363 . The measurements were made between 1963 and 1975 with irregular time intervals separating them. We restrict ourselves to testing the null hypothesis of no correlations with reasonable power

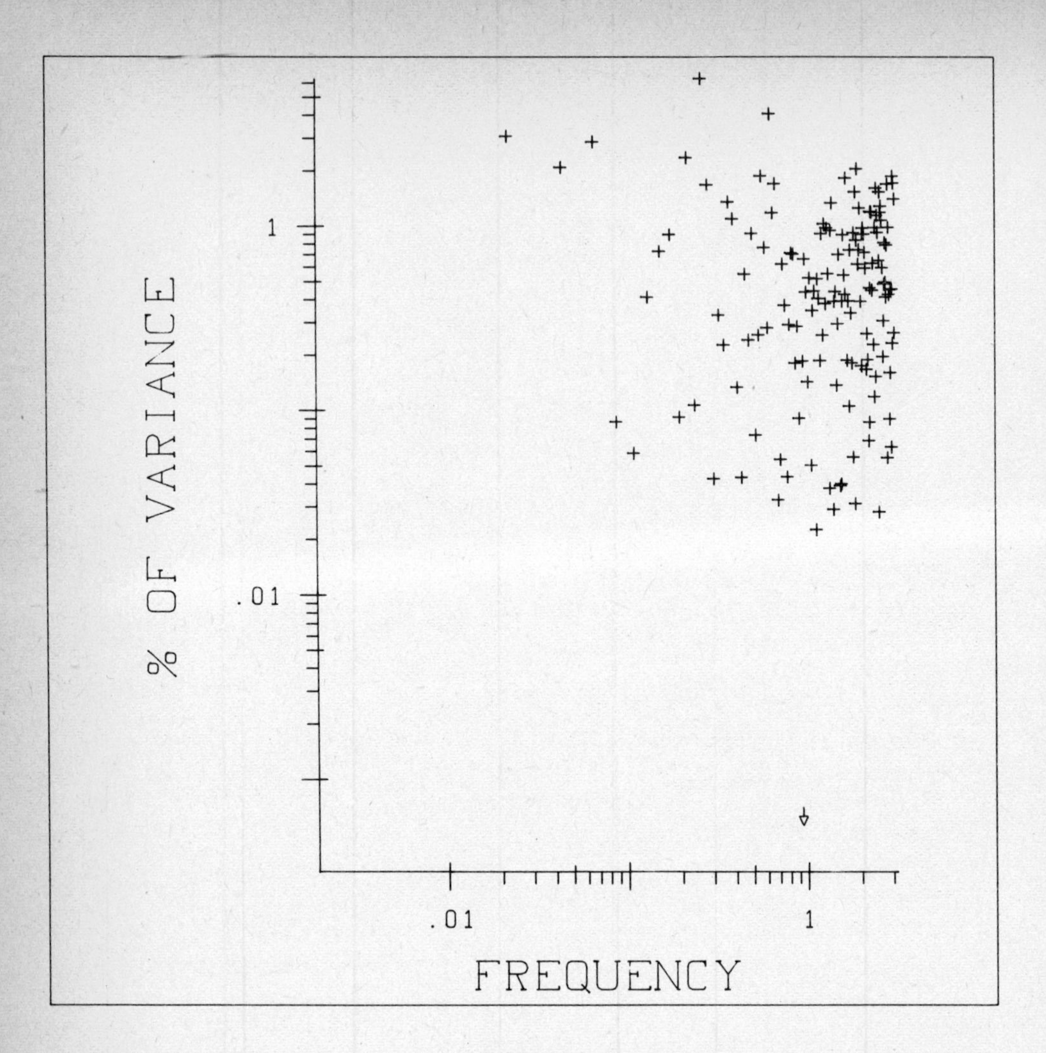

Figure 10:

Periodogram for the NBS precision measurements on the 1 kg check standard weight (logarithmic scale on both axes. The ordinate represented by the arrow is actually slightly smaller than would correspond to the position of the symbol).

against many kinds of long-range correlations. We wish to rely mainly on the low-frequency periodogram ordinates. The periodogram for these data, with only serial order retained, is shown in Figure 10 using log-log co-ordinates. Estimator HUBINC yields $\hat{H} = .605$. With the aid of our simulations a standardized test statistic of 2.26 can be derived, to be compared with the standardized 99%-point of HUBINC for $n = 289$ and $H = .5$, which is derived as 2.23 from the simulations. Thus the 289 measurements considered here just suffice to reject the null hypothesis of no correlations at the 1%-level. Incidentally, the lag-one-correlation, which is often recommended as a test statistic with fair power against a wide range of (short-term correlated) alternatives, happens to be .077 for these data, which is not significant even at the one-sided 5%-level.

References:

Berkson, J.:
Some difficulties of interpretation encountered in the application of the chi-square test.
J. Amer. Statist. Assoc. 33, 526-36 (1938).

Birrer, W.:
Homogenisierung und Diskussion der Totalozon-Messreihe von Arosa 1926-1971.
Laboratorium für Atmosphärenphysik ETH Zürich (1975).

Box, G.E.P.; Hunter, W.G.; Hunter, J.S.:
Statistics for experimenters.
New York: Wiley (1978).

Daniel, C.; Wood, F.S.:
Fitting equations to data (1st/2nd edition).
New York: Wiley (1971/80).

Graf, H.-P.:
Long-range correlations and estimation of the self-similarity parameter.
Ph.D. thesis, Zürich: Eidgenössische Technische Hochschule (1983).

Hampel, F.R.:
Robust estimation: A condensed partial survey.
Z. Wahrscheinlichkeitstheorie verw. Geb. 27, 87-104 (1973).

Hampel, F.R.:
Modern trends in the theory of robustness.
Math. Operationsforschung Statist., Ser. Statist. 9, 425-442 (1978).

Hampel, F.R.:
Discussion of the meeting on Robustness.
42nd Session ISI, Manila, Vol. XLVIII, Book 2, 100-102 (1979).

Hampel, F.R.; Marazzi, A.; Ronchetti, E.; Rousseeuw, P.; Stahel, W.; Welsch, R.E.:
Handouts for the Instructional Meeting on Robust Statistical Methods.
15th European Meeting of Statisticians, Palermo/Italy (1982).

Jeffreys, H.:
Theory of probability.
Oxford: Clarendon Press (1939, 1948, 1961).

Kleiner, B.; Martin, R.D.; Thomson, D.J.:
Robust estimation of power spectra.
J. Roy. Statist. Soc. B 41, 313-351 (1979).

Kolmogorov, A.N.:
Wienersche Spiralen und einige andere interessante Kurven im Hilbertschen Raum.
C.R. (Doklady), Acad. Sci. URSS (N.S.) 26, 115-118 (1940).

Mandelbrot, B.B.; Taqqu, M.S.:
Robust R/S analysis of long run serial correlation.
42nd Session ISI, Manila, Vol XLVIII, Book 2, 69-99 (1979).

Mandelbrot, B.B.; van Ness, J.W.:
Fractional Brownian motions, fractional noises and applications.
SIAM Review 10, 422-437 (1968).

Mandelbrot, B.B.; Wallis, J.R.:
Noah, Joseph and operational hydrology.
Water Resources Research 4, 909-918 (1968).

Mandelbrot, B.B.; Wallis, J.R.:
Computer experiments with fractional Gaussian noises.
Water Resources Research 5, 228-267 (1969a).

Mandelbrot, B.B.; Wallis, J.R.:
Some long-run properties of geophysical records.
Water Resources Research 5, 321-340 (1969b).

Mandelbrot, B.B.; Wallis, J.R.:
Robustness of the rescaled range R/S in the measurement of non-cyclic long run statistical dependence.
Water Resources Research 5, 967-988 (1969c).

Mohr, D.L.:
Modeling data as a fractional Gaussian noise.
Ph.D. thesis, Princeton, N.J.: Princeton University (1981).

Mosteller, F.; Tukey, J.W.:
Data analysis and regression.
Reading, Mass.: Addison-Wesley (1977).

Newcomb, S.:
Astronomical Constants.
p. 103 (1895).

Pearson, K.;
On the mathematical theory of errors of judgment, with special reference to the personal equation.
Phil. Trans. A 198, 235-299 (1902).

Portnoy, S.L.:
Robust estimation in dependent situations.
Ann. Statist. 5, 22-43 (1977).

Scheffé, H.:
The Analysis of Variance.
New York: John Wiley (1959).

Smith, H. Fairfield:
An empirical law describing heterogeneity in the yields of agricultural crops.
J. Agric. Sci. 28, 1-23 (1938).

Student:
Errors of routine analysis.
Biometrika 19, 151-164 (1927).

Whittle, P.:
On the variation of yield variance with plot size.
Biometrika 43, 337-343 (1956).

THE ESTIMATION OF ARMA PROCESSES

E.J. Hannan
Australian National University

Abstract

A description is given of a method for estimating an ARMA process, $y(t)$, from observations for $t=1,2,\ldots T$, and following this a discussion is given of the theory necessary for the validation of the method. The first stage of the method involves the fitting of an autoregression, of order h_T determined by a criterion such as AIC. The asymptotic theory depends on the behaviour of $h_T^{-1} \Sigma_1^T\{\hat{\varepsilon}(t)-\varepsilon(t)\}^2$ where $\hat{\varepsilon}(t)$ is the residual from the autoregression and $\varepsilon(t)$ is the innovation. Theorems relating to this expression are proved.

1.

The Estimation Method

The methods we now describe were introduced in Hannan and Rissanen (1982) and Hannan and Kavalieris (1984a). Let $y(t)$, $t=1,2,\dots,T$ be generated by the ARMA system

$$\sum_0^p \alpha(j)y(t-j) = \sum_0^q \beta(j)\varepsilon(t-j);$$

$$\alpha(0) = \beta(0) = 1; \quad E\{\varepsilon(s)\varepsilon(t)\} = \delta_{st}\sigma^2. \qquad (1.1)$$

Putting $a(z) = \Sigma\alpha(j)z^j$, $b(z) = \Sigma\beta(j)z^j$ and assuming these relatively prime they are required to satisfy

$$a(t),\ b(z) \neq 0 \ , \quad |z| \leq 1 \ . \qquad (1.2)$$

Then the $\varepsilon(t)$ are the linear innovations. For results discussed in detail in section 3 further conditions on the $\varepsilon(t)$ will be needed but we omit discussion of that in this section. When emphasis is needed we indicate true values by a zero subscript e.g. p_0, q_0, $\alpha_0(j)$ etc. The method has 3 stages.

Stage I. An autoregression is fitted by least squares, i.e.

$$\sum_0^h \hat{\phi}(j)y(t-j) = \hat{\varepsilon}(t) \ , \quad \hat{\phi}(0) = 1 \ , \qquad (1.3)$$

where the integer h is chosen to minimise one or other of $AIC(h) = \log\hat{\sigma}_h^2 + 2h/T$, $BIC(h) = \log\hat{\sigma}_h^2 + h\log T/T$, $h \leq (\log T)^a$, $a < \infty$. Here $\hat{\sigma}_h^2$ is the residual variance in (1.3). The upper bound on h is needed for theorems but in practice the actual minimum would be found. Since the minimising value depends on T we call it h_T. It is shown in Hannan and Kavalieris (1984b) that

$h_T = \{\log T/(2\log\rho_0)\}\{1+o(1)\}$. Here ρ_0 is the modulus of a zero of $b_o(z)$ nearest to $|z| = 1$. The term that is $o(1)$ will be shown to be of that order, a.s., under conditions given in section 3 and in probability (when we write $O_p(1)$) under weaker conditions. The same is true of other order results and convergence results mentioned in this section.

Stage II. Carry out the regression

$$y(t) = -\sum_1^p \tilde{\alpha}_p(j)y(t-j) + \sum_1^q \tilde{\beta}_q(j)\hat{\varepsilon}(t-j) + \text{error} \tag{1.4}$$

Choose $\tilde{p}, \tilde{q}$ to minimise

$$\mathrm{BIC}(p,q) = \log \tilde{\sigma}^2_{p,q} + (p+q)\log T/T,$$

$$p \leq P,\ q \leq Q \ . \tag{1.5}$$

Here $\tilde{\sigma}^2_{p,q}$ is the residual variance in (1.4) . It is assumed that $p_o \leq P$, $q_o \leq Q$. In fact P,Q may be allowed to increase slowly to infinity with T but we do not discuss that. We shall write $\tilde{\alpha}(j), \tilde{\beta}(j)$, with no supscripts, for the coefficients in (1.4) when $p = \tilde{p}$, $q = \tilde{q}$. As shown in Hannan and Kavalieris (1984a), under conditions to be relaxed below, for $p \leq p_o$, $q \leq q_o$

$$\tilde{\sigma}^2_{p,q} = \frac{1}{T}\sum_1^T \varepsilon(t)^2 + \sigma_0^2 h_T(\omega_v^2 - 1)/T + o(\log T/T). \tag{1.6}$$

Here ω_v^2 is the (true) mean square of the one step prediction error for $u(t) = \Sigma\beta_o(j)e(t-j)$, $E\{e(s)e(t)\} = \delta_{st}$, from $v = \min(p-p_o, q-q_o)$ previous values. From (1.6) it follows that $\mathrm{BIC}(p_o+1,q_o+1) - \mathrm{BIC}(p_o,q_o) = h_T(\omega_1^2-\omega_0^2)/T + 2\log T/T + o(\log T/T)$, so that if $\omega_o^2 - \omega_1^2 > 2\log T/h_T$ then $\tilde{p}, \tilde{q}$ will tend to be too large. (Asymptotically underestimation occurs with probability zero.) Indeed (1.6) shows that $\tilde{p} \to p_o+v_o$, $\tilde{q} \to q_o+v_o$ where v_o is the smallest integer for which

$$\omega_v^2 - \omega_{v+1}^2 > 4\log\rho_o \ . \tag{1.7}$$

For $p_o = q_o = 1$ then $\rho_o = |\beta_o(1)|^{-1}$ and $v_o = 0$ unless $|\beta_o(1)| > 0.89$.

To correct this overestimation various methods are suggested in Hannan and Kavalieris (1984a) . Here we discuss only one, namely the repetition of the procedure based on (1.4), (1.5) with $\hat{\varepsilon}(t)$ in (1.4) replaced by $\tilde{\varepsilon}(t)$ where

$$\tilde{\varepsilon}(t) = -\sum_{1}^{\tilde{q}} \tilde{\beta}(j)\tilde{\varepsilon}(t-j) + \sum_{0}^{\tilde{p}} \tilde{\alpha}(j)y(t-j), \quad \tilde{\varepsilon}(t) = y(t) = 0, \quad t \leq 0 . \tag{1.8}$$

Now if $\tilde{p}^{(1)}$, $\tilde{q}^{(1)}$ are the values that minimise (1.5) we have $\tilde{p}^{(1)} \to p_o$, $\tilde{q}^{(1)} \to q_o$. The reason for this is as follows. The predictor of $u(t)$ discussed below (1.6) is of the form $\Sigma_o^v \psi_{vj} u(t-j)$ so that

$\omega_v^2 = E[\{\Sigma_o^v \psi_{vj} u(t-j)\}^2]$, $\psi_{vo} = 1$. Let $\psi_v(z) = \Sigma_o^v \psi_{vj} z^j$. Let $\tilde{a}(z)$, $\tilde{b}(z)$ be constructed from the $\tilde{\alpha}(j)$, $\tilde{\beta}(j)$ from (1.4). Then

$$\tilde{a}(z) = \tilde{\psi}_v(z)a_o(z) + O(Q_T), \quad \tilde{b}(z) = \tilde{\psi}_v(z)b_o(z) + O(Q_T) \tag{1.9}$$

where $Q_T = (\log\log T/T)^{\frac{1}{2}}$ and where $\tilde{\psi}_v(z) \to \psi_v(z)$ (i.e. its coefficients so converge). From (1.8) $\tilde{\varepsilon}(t)$ is essentially of the form $\{\tilde{a}(z)/\tilde{b}(z)\}y(t)$, where z is now used as the backwards shift operator. Now $(\tilde{a}/\tilde{b}) - (a_o/b_o) = (\tilde{a}b_o - \tilde{b}a_o)/(\tilde{b}b_o)$ and $\tilde{a}b_o - \tilde{b}a_o = O(Q_T)$ while $\tilde{b}b_o$ eventually has all zeros bounded away from and outside of $|z| = 1$ (by (1.9) and $\tilde{\psi}_v \to \psi_v$). Thus, approximately

$$\frac{1}{T}\sum_{1}^{T}\{\tilde{\varepsilon}(t) - \varepsilon(t)\}^2 = \frac{1}{2\pi}\int I(\omega)\left|\frac{\tilde{a}b_o - \tilde{b}a_o}{\tilde{b}b_o}\right|^2 d\omega = O(Q_T^2) \tag{1.10}$$

wherein

$$I(\omega) = T^{-1}|\Sigma_1^T y(t)e^{it\omega}|^2 .$$

As we shall see in section 3,

$$\frac{1}{T}\sum_{1}^{T}\{\hat{\varepsilon}(t) - \varepsilon(t)\}^2 = O(h_T/T) \tag{1.11}$$

and it is the improvement from (1.11) to (1.10) that ensures convergence for $\tilde{p}^{(1)}$, $\tilde{q}^{(1)}$. However the proof of (1.10) depends on (1.11) as also does the proof of $\tilde{\psi}_v \to \psi_v$. Of course the margin between (1.10) and (1.11) is small. However simulations (see Hannan and Kavalieris

1984a) show that the repeated procedure does improve the estimates.

Stage III. This consists of a maximum likelihood procedure, taking p,q as $\tilde{p}^{(1)}$, $\tilde{q}^{(1)}$ and using the estimates of the $\alpha_o(j)$, $\beta_o(j)$ obtained from the repeated procedure to initiate an iterative maximisation of the likelihood.

2.

Some Convergence Results.

The first result needed for the study of the problems of section 1 is that contained in An, Chen and Hannan (1982). Though a variety of results are presented there only one is mentioned here. We assume the $y(t)$ to satisfy (1.1), (1.2) and also

$$E\{\varepsilon(t)|F_{t-1}\} = 0, \quad E\{\varepsilon(t)^2|F_{t-1}\} = \sigma_o^2 ,$$

$$E\{\varepsilon(t)^4\} < \infty . \tag{2.1}$$

Here F_t is the σ-algebra determined by $\varepsilon(s)$, $s \leq t$. Let

$$c(j) = \frac{1}{T}\sum_{1}^{T-j} y(t)y(t+j), \quad \gamma(t) = E\{y(t)y(t+j)\} .$$

Then under (1.1), (1.2), (2.1)

$$\max_{0\leq j\leq P(T)} |c(j) - \gamma(j)| = O(Q_T), \ P(T) \leq (\log T)^a, \ a < \infty . \tag{2.2}$$

The second part of (2.1) can be replaced by

$$E\{\varepsilon(t)^2|F_{-\infty}\} = \sigma_o^2$$

if $r(j) = c(j)/c(0)$, $\rho(j) = \gamma(j)/\gamma(0)$ replace $c(j)$, $\gamma(j)$, respectively, in (2.2). The result (2.2) is basic because $h_T = O(\log T)$ and because of the narrow margins available for the establishment of results. (See (1.10), (1.11) for example.)

The second result relates to (1.11). Its importance can be seen as follows. We need consider only $p \geq p_o$, $q \geq q_o$ for the reason stated in parenthesis just above (1.7). Then put

$$\chi(z) = (ab_o - ba_o) = \sum_1^r \chi_j z_j \ , \quad r = \max(p+q_o, q+p_o) \ .$$

Observe that $r+v = p+q$. Then

$$a(z) = \nu(z)a_o(z) + \Sigma\, \chi_j e_j(z),$$
$$b(z) = \nu(z)b_o(z) + \Sigma\, \chi_j f_j(z). \quad (2.3)$$

Here the $e_j(z)$, $f_j(z)$ depend only on a_o, b_o and not on a,b. The polynomial $\nu(z)$ is of degree v and satisfies $\nu(0) = 1$. Now the $v+r$ coefficients in $\chi(z)$, $\nu(z)$ may be used to parameterise $a(z)$, $b(z)$ in place of the $\alpha(j)$, $\beta(j)$. The relation between the two sets of parameters is linear and one to one. This parametrisation is useful because the estimates of the χ_j converge satisfactorily to their "true" values (namely zero) and all convergence problems are concentrated in $\nu(z)$. To see this consider (1.4) . Then rearranging $a(z)y(t) - \{b(z)-1\}\hat{\varepsilon}(t)$ we see the estimates are obtained by minimising

$$\frac{1}{T}\sum_1^T[\chi b_o^{-1}y(t)+\varepsilon(t) - (b_o-1)\{\hat{\varepsilon}(t) - \varepsilon(t)\}$$
$$-(b-b_o)\{\hat{\varepsilon}(t) - \varepsilon(t)\}]^2$$

and using (2.3) this is

$$\frac{1}{T}\sum_1^T[\Sigma\chi_j\ \xi_j(t) + \varepsilon(t)-(b_o-1)\{\hat{\varepsilon}(t) - \varepsilon(t)\}$$
$$- (\nu-1)b_o\{\hat{\varepsilon}(t) - \varepsilon(t)\}]^2 \ ,$$

$$\xi_j(t) = b_o^{-1}y(t) + f_j\{\hat{\varepsilon}(t) - \varepsilon(t)\} \ .$$

The regression of $\varepsilon(t) - (b_o-1)\{\hat{\varepsilon}(t) - \varepsilon(t)\}$ on the $\xi_j(t)$ is easily handled because the matrix composed of the $T^{-1}\Sigma\,\xi_j(t)\xi_k(t)$ converges to a non-singular limit and the $\hat{\chi}_j$ converge to zero at the rate determined by (2.2) so that this source of error in $\tilde{\sigma}^2_{p,q}$ is $O(Q_T^2) = o(\log T/T)$ and is negligible. The critical evaluation needed for (1.6) is that involved in the regression of $\varepsilon(t)-(b_o-1)\{\hat{\varepsilon}(t)-\varepsilon(t)\}$ on the $\{\hat{\varepsilon}(t-j) - \varepsilon(t-j)\}$. For this we need evaluations of the type in (1.11) , namely

$$\frac{1}{h_T}\sum_1^T \varepsilon(t)\{\hat{\varepsilon}(t-j) - \varepsilon(t-j)\} = -\delta_{oj}\sigma_o^2 + o(1),\ j=1,\ldots,v+q_o \tag{2.5}$$

$$\frac{1}{h_T}\sum_1^T \{\hat{\varepsilon}(t-j) - \varepsilon(t-j)\}\{\hat{\varepsilon}(t-k) - \varepsilon(t-k)\} = \delta_{jk}\sigma_o^2 + o(1),$$

$$j,k=1,\ldots,v+q_o. \tag{2.6}$$

Using these results (1.6) follows and also (1.9) . These results are also needed for the establishment of the formula $h_T/\{\log T/(2\log\rho_o)\}\to 1$, when h_T is determined by AIC. To see these we point out (see Hannan and Kavalieris, 1984b) that this result follows where h_T is determined by BIC, using (2.2) and that the h_T for AIC is never smaller than that for BIC, i.e. for AIC

$$\liminf_{T\to\infty}\{h_T/\{\log T/(2\log\rho_o)\} \geq 1\ ,\ \text{a.s.}$$

Thus it is sufficient to show that, for all $c > 1$,

$$\limsup_{T\to\infty}\{h_T/\{\log T/(2\log\rho_o)\} \leq c\ . \tag{2.7}$$

However

$$\hat{\sigma}_h^2 = \frac{1}{T}\sum_1^T \hat{\varepsilon}_h(t)^2 = \frac{1}{T}\sum_1^T \varepsilon(t)^2 + \frac{2}{T}\sum_1^T \varepsilon(t)\{\hat{\varepsilon}_h(t) - \varepsilon(t)\}$$
$$+ \frac{1}{T}\sum_1^T\{\hat{\varepsilon}_h(t) - \varepsilon(t)\}^2,$$

where the subscript indicates the order of the autoregression. If (2,5), (2.6) hold, then, for $h = c\log T/(2\log\rho_o)$, $c>1$,

$$\hat{\sigma}_h^2 = \frac{1}{T}\sum_1^T \varepsilon(t)^2 - h\,\sigma_o^2/T + o(\log T/T)$$

so that

$$AIC(h) - AIC([\log T/(2\log\rho_o)) \geq (c-1)\log T/(2\log\rho_o) + o(\log T/T)$$

so that, for $c>1$, $AIC(h)$ is certainly not minimised at h . This essentially establishes (2.7) .

Of course (2.6) for $j=0$ is equivalent to (1.11) and shows just how accurate the $\hat{\varepsilon}(t)$ are.

3.

The Behaviour of the $\hat{\varepsilon}(t)$.

We now assume (1.1), (1.2), (2.1) and for convenience take $\sigma_o^2 = 1$. In this section all order relations hold a.s. unless otherwise indicated. We now require

$$c_1\{\log T/(2\log\rho_o)\} \leq h_T \leq (\log T)^a,\ a < \infty,\ \ c_1 > 1\ ,\ \text{a.s.} \tag{3.1}$$

This does not affect the proof of (2.7) but does mean that for some of the results in sections 1 and 2 to hold we would need to calculate the $\hat{\varepsilon}(t)$ in (1.4) from an autoregression of order c_1 by the order determined by AIC or BIC , (to the nearest unit). This would be of no consequence in practice but it is doubtful, to say the least, if one would bother to make the change.

We first show that the quantities on the left in (2.5), (2.6) may be replaced by

$$\frac{-1}{Th_T}\sum_{u=1}^{h_T}\{\sum_{s=1}^{T}\varepsilon(s)\varepsilon(s-u)\}\{\sum_{s=1}^{T}\varepsilon(s)\varepsilon(s-u-j) + O(\log\log T/h_T),\quad \text{a.s.} \tag{2.5)'$$

$$\frac{1}{Th_T}\sum_{u=1}^{h_T}\{\sum_{s=1}^{T}\varepsilon(s)\varepsilon(s-u-j)\}\{\sum_{s=1}^{T}\varepsilon(s)\varepsilon(s-u-k)\} + O(\log\log T/h_T),\quad \text{a.s.} \tag{2.6)'$$

The first step in establishing (2.5)', (2.6)' is to show that in (2.5), (2.6) we may replace $\varepsilon(t)$ by

$$\varepsilon_T(t) = \sum_{0}^{h_T}\phi_T(j)y(t-j),\ \phi_T(0) = 1$$

where this is the error in the best prediction of $y(t)$ from $y(t-1),\ldots,y(t-h_T)$ i.e., the $\phi_T(j)$ make $E\{\varepsilon_T(t)^2\}$ a minimum. Now

$$\frac{1}{h_T}\sum_{1}^{T}\{\varepsilon(t)-\varepsilon_T(t)\}^2 = \frac{1}{h_T}\sum_{1}^{T}\ [\sum_{1}^{h_T}\{\phi(j)-\phi_T(j)\}y(t-j)$$
$$+ \sum_{h_T+1}^{\infty}\phi(j)y(t-j)]^2$$

where $a_o/b_o = \Sigma\,\phi(j)\,z^j$. From Baxter (1962) it follows that

$$\sum_1^{h_T}\{\phi(j)-\phi_T(j)\}^2 = O(\rho_o^{-2h_T}) \tag{3.2}$$

so that, by (2.2) and (3.2)

$$\frac{1}{h_T}\sum_1^T\Big[\sum_1^{h_T}\{\phi(j)-\phi_T(j)\}y(t-j)\Big]^2 = \frac{1}{h_T}\sum\sum_1^{h_T}\{\phi(j)-\phi_T(j)\}$$

$$\{\phi(k)-\phi_T(k)\}\gamma(j-k)\{1+o(1)\}$$

$$= O(Th_T^{-1}\,T^{-b}),\quad b > 1.$$

Also $\phi(j) = O(\rho_o^{-j})$ so that, for some $a(j)$ decreasing to zero geometrically

$$\frac{1}{h_T}\sum_1^T\{\sum_{h_T+1}^{\infty}\phi(j)y(t-j)\}^2 \le c\frac{T}{h_T}\,T^{-b}\{\frac{1}{T}\sum_1^{\infty}a(j)|y(t-j)|\}^2$$

$$= O(T^{1-b}h_T^{-1}).$$

Thus we can replace $\varepsilon(t)$ by $\varepsilon_T(t)$. Next, from Hannan and Kavalieris (1984a) we have

$$(\hat\phi_T - \phi_T) = -\hat\Gamma_T^{-1}(\frac{1}{T}\sum_1^T \varepsilon(t)y(t-k))_{k=1,\ldots,h_T} + o(T^{-\frac{1}{2}})$$

where $\hat\phi_T$ has $\hat\phi_T(j)$, $j=1,\ldots,h_T$ as its components and similarly for ϕ_T in relation to the $\phi(j)$ while $\hat\Gamma_T$ has $c(j-k)$, $j,k=1,\ldots,h_T$, as its elements. The term that is $o(T^{-\frac{1}{2}})$ is a vector and we use for that the uniform norm, i.e. the maximum of the moduli of its elements. We indicate this by $|x|$, for a vector x. The corresponding operator norm we write as $|A|$ where this is the row sum norm, i.e. the maximum over rows of the sum of the moduli of the elements in the row. Of course $|AB| \le |A|.|B|$. We use $\|\cdot\|$ for the Euclidean norm. It follows that $|\hat\Gamma_T^{-1} - \Gamma_T^{-1}| = O(h_TQ_T)$. Here Γ_T has $\gamma(j-k)$ as typical element. Indeed $|\hat\Gamma_T^{-1} - \Gamma_T^{-1}| \le |\hat\Gamma_T^{-1}||\hat\Gamma_T - \Gamma_T||\Gamma_T^{-1}|$ and Hannan and Kavalieris (1984a,b) show that $|\hat\Gamma_T^{-1}|, |\Gamma_T^{-1}|$ are bounded so that (2.2) completes the proof. Thus, again using results in An, Chen and Hannan (1982),

$$|(\hat\Gamma_T^{-1} - \Gamma_T^{-1})(\frac{1}{T}\sum_1^T \varepsilon(t)y(t-k))_{k=1,\ldots,h_T}| = O(Q_T^2\,h_T)$$

so that, from (3.3)

$$\varepsilon_T(t) - \hat{\varepsilon}(t) = -y_T(t)'\,\Gamma_T^{-1}\,\frac{1}{T}\sum_1^T \varepsilon(s)y_T(s)$$

$$+ y_T(t)'\,O(Q_T^2 h_T)$$

putting $y_T(t)' = (y(t-1),\ldots,y(t-h_T))$. Since

$$\frac{1}{h_T}\sum_1^T \{y_T(t)'O(Q_T^2h_T)\}^2 = O(Q_T^4h_T)\sum_1^T y_T(t)'y_T(t) = O(h_T^2TQ_T^4)$$

then we may replace $\varepsilon_T(t) - \hat{\varepsilon}(t)$ by

$$-\frac{1}{T}\sum_{s=1}^T \varepsilon(s)\,y_T(t)'\,\Gamma_T^{-1}y_T(s)\ .$$

Next we show that in (3.4) we may replace $y_T(t)'\Gamma_T^{-1}y_T(s)$ by

$$\sum_1^{h_T} \varepsilon(t-j)\varepsilon(s-j)\ .$$

Let A_T be the $h_T \times h_T$ lower triangular matrix having $\alpha_o(j-k)$ in row j column k, putting $\alpha_o(k) = 0,\ k < 0$. Define B_T in the same way in terms of the $\beta_o(j-k)$.

Then $A_T\Gamma_TA_T' = B_T\,B_T' + \Delta_T$ where Δ_T is null outside of the first h_T rows and columns. Thus

$$\Gamma_T^{-1} = (B_T^{-1}A_T)'(B_T^{-1}A_T) + \sum\sum_1^p d_{ij}\nu_{T,i}\nu_{T,j}'$$

where $\nu_{T,i} = (B_T^{-1}A_T)'b_{T,i}$ and $b_{T,i}$ is the ith column of B_T^{-1} while the d_{ij} are independent of T . It follows that the elements of $\nu_{T,i}$ decrease to zero at a geometric rate. Thus

$$y_T(t)'\,\Gamma_T^{-1}y_T(s) = y_T(t)'\Phi_T'\Phi_T\,y_T(s)$$

$$+ \sum_1^p\sum d_{ij}(y_T(t)'\nu_{T,i})(y_T(s)'\nu_{T,j})$$

and Φ_T has $\phi(j-k)$ in row j, column k with $\phi(k)$ put as zero for $k < 0$. Thus

$$-\frac{1}{T}\sum_1^T \varepsilon(s)y_T(t)'\Gamma_T^{-1}y_T(s) = -\frac{1}{T}\sum_1^T \varepsilon(s)y_T(t)'\Phi_T'\Phi_Ty_T(s) +$$

$$\sum\sum d_{ij}\Big(\frac{1}{T}\sum_1^T \varepsilon(s)y_T(s)'\nu_{T,j}\Big)y_T(t)'\nu_{T,i}\,.$$

Now $|T^{-1}\Sigma_1^T \varepsilon(s)y_T(s)'\nu_{Tj}| = O(Q_T)$ by results in An, Chen and Hannan, (1982) so that

$$\frac{1}{h_T}\sum_1^T[\{\Sigma\Sigma d_{ij}\ \frac{1}{T}\sum_1^T\varepsilon(s)y_T(s)\nu_{Tj}'\}y_T(t)'\nu_{Tj}]^2 = O(Q_T^2\, h_T^{-1}\, T)$$

which is $O(\log\log T/h_T)$.
Thus we are reduced to considering (2.5), (2.6) with $\hat{\varepsilon}(t) - \varepsilon(t)$ replaced by

$$-\frac{1}{T}\sum_1^T \varepsilon(s)y_T(t)\Phi_T'\Phi_T y_T(s) = -\frac{1}{T}\ \varepsilon(s)y_T(t)\Phi_T\Phi_T' y_T(s)$$

since $\Phi_T'\Phi_T$ is not altered when the orders of both rows and columns are reversed. However

$$\Phi_T y_T(t) = (\varepsilon(t-j))_{j=1,\ldots,h_T} + \theta_T(t)$$

$$\theta_T(t) = (\sum_{k=h_T-j+1}^{\infty} \phi(k)y(t-j-k))$$

Thus, for example,

$$\frac{1}{h_T}\sum_{t=1}^T\{\frac{1}{T}\sum_1^T\ \varepsilon(s)\theta_T\{(t)\Phi_T y_T(s)\}^2 = \frac{1}{h_T}\sum_1^T\{\theta_T(t)'O(Q_T)\}^2$$

$$= \{\frac{1}{h_T}\sum_1^T\ \theta_T(t)'\theta_T(t)\}O(Q_T^2) = o(T\, h_T^{-1}\, Q_T^2) = o(\log\log T/h_T).$$

Thus we are finally able to replace $\hat{\varepsilon}(t) - \varepsilon(t)$ in (2.5), (2.6) by

$$\frac{1}{T}\sum_1^T\ \varepsilon(s)\sum_{j=1}^{h_T}\ \varepsilon(t-j)\varepsilon(s-j)$$

with an error that is $O(\mathrm{loglot}T/h_T)$ so that the left side of (2.5), for example, is replaced by

$$\frac{1}{h_T}\sum\sum_{u,v=1}^T\ [\frac{1}{T^2}\sum\sum_{s,t=1}^T\ \{\varepsilon(s)\varepsilon(t)\varepsilon(s-u)\varepsilon(s-v)\sum_{\tau=1}^T\ \varepsilon(\tau-u-j)\varepsilon(\tau-v-k)\}]$$

and since $\Sigma\,\varepsilon(\tau-u-j)\varepsilon(\tau-v-k) = T\ \delta_{u+j,v+k} + O\{(T\log\log T)^{\frac{1}{2}}\}$ while $T^{-1}\Sigma\varepsilon(s)\varepsilon(s-u)$ are $O(Q_T)$, uniformly then (2.6)' results and, similarly, (2.5)' . We now evaluate these.

For brevity we consider only the central case, i.e. (2.6)' for $j = k = 0$. The others are treated similarly. Thus we wish to show that

$$\frac{1}{Th_T}\sum_1^{h_T}\{\sum_1^T \varepsilon(s)\varepsilon(s-u)\}^2 - 1$$

converges to zero. Put

$$S(T) = \sum_1^{h_T}\{\sum_1^T\varepsilon(s)\varepsilon(s-u)\}^2 - Th_T$$

Then

$$S(T) = S_1(T) + S_2(T) + \sum_1^T \sum_{h_{t-1}+1}^{h_t} [\{\sum_1^{t-1}\varepsilon(s)\varepsilon(s-j)\}^2-(t-1)]$$

$$+ \sum_1^T\sum_1^{h_t}\{\varepsilon(s-j)^2 - 1\}$$

where $S_1(T)$, $S_2(T)$ are martingales with differences, respectively

$$\{\varepsilon(t)^2-1\}\sum_1^{h_t}\varepsilon(t-j)^2 , \quad 2\varepsilon(t)\sum_1^{h_t}\varepsilon(t-j)\sum_1^{t-1}\varepsilon(s)\varepsilon(s-j) .$$

We first show that $(Th_T)^{-1}$ by the first of these and $S_1(T)$ are $O(\log\log T/h_T)$. First

$$\frac{1}{Th_T}\sum_1^T\sum_1^{h_t}\{\varepsilon(t-j)^2-1\} = \frac{1}{Th_T}\sum_{1-h_T}^{T-1}\{\varepsilon(u)^2 - 1\}k_u$$

where k_u is the number of times $t-j=u$, $t=1,\ldots,T$; $j=1,\ldots,h_t$. Of course $k_u \leq h_T$. Since $E\{\varepsilon(t)^4\} < \infty$

$$\left|\frac{1}{Th_T}\sum_{1-h_T}^{0}\{\varepsilon(u)^2-1\}k_u\right| \leq \frac{1}{Th_T}\sum_0^{h_T-1}u^{\frac{1}{2}}k_{-u} = O(T^{-1}h_T^{3/2}) .$$

Thus we may consider

$$\frac{1}{Th_T}\sum_1^T\{\varepsilon^2(u) - 1\}k_u \qquad (3.5)$$

Omitting the factor $(Th_T)^{-1}$ this is a martingale and the increasing process (associated with the square of the martingale is

$$\sum_1^T E[\{\varepsilon^2(u)-1\}^2|F_{u-1}]k_u^2 \leq h_T^2\sum_1^T E[\{\varepsilon^2(u) - 1\}^2|F_{u-1}] = O(Th_T^2)$$

by ergodicity. Thus by Neveu (1975, p151, Proposition VII-2-4) then (3.5) is $o(T^{-a})$ for all $a < \frac{1}{2}$. Since we repeatedly use the result in Neveu (1975) we shall refer to it as N . Next consider $S_1(T)$. This has increasing process

$$\sum_1^T (E[\{\varepsilon(t)^2-1\}^2|F_{t-1}]\{\sum_1^{h_T}\varepsilon(t-j)^2\}^2 \qquad (3.6)$$

Now

$$\sum_1^{h_T}\varepsilon(t-j)^2 = \sum_1^{h_T}\{\varepsilon(t-j)^2-1\}+h_t = \sum_{t-h_t}^{t-1}\{\varepsilon(u)^2-1\} + h_t .$$

Using the method of subsequences it is not difficult to show that the first term is $o(t^a)$, for all $a > \frac{1}{4}$. Indeed putting $q(t) = t^{-a}\Sigma_{t-h_t}^{t-1}\{\varepsilon(u)^2-1\}$ this has variance $O(t^{-2a}h_t)$ so that $q(t^2)$ converges a.s. to zero by Chebyshev's inequality and the Borel-Cantelli lemma. Moreover

$$E\{\max_{s^2<t\leq(s+1)^2}|q(t) - \frac{s^{4a}}{t^{2a}}q(s^2)|^2\}$$

$$\leq \frac{1}{s^{8a}}E[\max|\sum_{s^2}^{t-1}\{\varepsilon^2(u)-1\} + \sum_{t-h_t}^{t-h_{s^2}}\{\varepsilon^2(u)-1\}|^2$$

$$\leq cs^{-8a}((s+1)^2-s^2+[\sum_{t-h_{(s+1)^2}}^{t-h_{s^2}}(E\{|\varepsilon(u)^2-1|^2\})^{\frac{1}{2}}]^2)$$

$$\leq cs^{-8a+1}$$

using Doob's inequality (Hall and Heyde (1980,p15)) for the first term. Here and below we use c for a general positive, constant. This establishes what we wish. Thus (3.6) is dominated by

$$T^{2a}\sum_1^T E[\{\varepsilon(t)^2-1\}^2|F_{t-1}] = O(T^{2a-1}) \quad .$$

Thus, by N , $(Th_T)^{-1}S_{1T} = o(T^{-b})$, $b < \frac{1}{4}$.

Thirdly consider

$$\sum_{s,t=1}^{T} \sum E[\{\sum_{1}^{h_s} \varepsilon(s-j) \sum_{u=1}^{s-j-1} \varepsilon(u)\varepsilon(u-j)\}^2$$

$$\{\sum_{1}^{h_T} \varepsilon(t-j) \sum_{h=1}^{s-j-1} \varepsilon(u)\varepsilon(u-j)\}^2] . \tag{3.11}$$

Now we may show this is $O(T^4 h_T^2)$ by evaluating the cumulant between the four factors under the expectation symbol and showing that to be $o(T^4 h_T^2)$. Now the proof may be completed by the method of subsequences. Thus put $X(T) = Th_T^{-1} S_3(T)$. Then

$$E\{(Th_T)^{-4} S_3(T)^4\} \leq Ch_T^{-2} .$$

Put $T_j = 2^j$. Then $X(T_j)$ converges to zero, a.s. since $h(T_j)^{-2} = cj^{-2}$. Thus we need to show

$$\max_{T_j<T<T_{j+1}} |X(T) - X(T_j)| \to 0 \text{ , a.s.}$$

Since evidently

$$\max_{T_j<T \leq T_{j+1}} |(\frac{T_j h_T}{Th_{T_j}}-1)| \to 0 \text{ , a.s.}$$

we consider

$$E[\max_{T_j<T\leq T_{j+1}} |X(T) - (T_j h_T / Th_{T_j})X(T_j)|^4]$$

$$\leq (T_j h_{T_j})^{-4} E[\max|S_3(T) - S_3(T_j)|^4] .$$

By Doobs inequality (Hall and Heyde, 1980 p.15) applied to the martingale $S_3(T) - S_3(T_j)$, $T_j < T^s \leq T_{j+1}$, we obtain the bound

$$C(T_j h_{T_j})^{-4} E[\{S_3(T_{j+1}) - S_3(T_j)\}^4]$$

and by the same argument as for $S_3(T)$ this is bounded by

$$C(T_j h_{T_j})^{-4} (T_{j+1} - T_j)^4 h_{T_{j+1}}^2 < c\, h_{T_j}^{-2} .$$

This establishes the theorem.

It is clear that Theorem 2 holds under more general conditions. For example it will certainly hold true if the

$$\frac{1}{Th_T}\sum_{1}^{T}\sum_{h_{t-1}+1}^{h_t}\left[\left\{\sum_{1}^{t-1}\varepsilon(s)\varepsilon(s-j)\right\}^2-(t-1)\right] \tag{3.7}$$

where $t_j = \{t|h_{t-1}+1 \le j \le h_t\}$ or $t_j = \{t|h_t+1 \le j \ge h_{t-1}\}$. In the latter case the square bracketed term in (3.7) should be preceded by a minus sign. Now

$$\left\{\sum_{1}^{t_j}\varepsilon(s)\varepsilon(s-j)\right\}^2 \le c\; t_j \log\log t_j \;.$$

Thus if $h_T \ge (\log T)^b$, $b > 1$, then evidently (3.7) converges to zero. If $h_T = [c \log T]$ then it is easily seen that (3.7) is $O(\log\log T/\log T)$ but that does not seem to hold if, merely, $h_T = c \log T\{1+o(1)\}$. However if

$$E\{\varepsilon(t)^b\} < \infty\,, \quad b > 4 \tag{3.8}$$

then we may show that

$$\lim_{T\to\infty}\frac{1}{Th_T}\sum_{1}^{T}\sum_{h_{t-1}+1}^{h_t}\sum_{1}^{t-1}\{\varepsilon^2(s)\varepsilon^2(s-j)-1+2\varepsilon(s)\varepsilon(s-j)^2\varepsilon(s-2j)\} = 0, \text{ a.s.}$$

and then it follows that

$$\frac{1}{Th_T}\sum_{1}^{T}\sum_{h_{t-1}+1}^{h_t}\Sigma'\Sigma'\{\varepsilon(s)\varepsilon(s-j)\varepsilon(u)\varepsilon(u-j)\}$$

converges in mean square to zero, where the double sum $\Sigma'\Sigma'$ is over s,u such that $s \ne u$, $s \ne u-j$, $u \ne s-j$. Finally consider

$$\frac{1}{Th_T}\sum_{t=1}^{T}\varepsilon(t)\left\{\sum_{j=1}^{h_t}\varepsilon(t-j)\sum_{s=1}^{t-1}\varepsilon(s)\varepsilon(s-j)\right\} \tag{3.9}$$

The term

$$\frac{1}{Th_T}\sum_{t=1}^{T}\varepsilon(t)\sum_{j=1}^{h_t}\varepsilon(t-j)^2\varepsilon(t-2j)$$

may be shown to converge a.s. to zero under (3.8). Subtracting this from (3.9) we are left with $(Th_t)^{-1}$ by a martingale

$$\sum_{t=1}^{T}\varepsilon(t)\left\{\sum_{j=1}^{h_t}\varepsilon(t-j)\sum_{\substack{s=1\\ s\ne t-j}}^{t-1}\varepsilon(s)\varepsilon(s-j)\right\} \tag{3.10}$$

This is easily seen to have variance that is $O(T^2h_T)$ so that (3.9) converges in probability to zero.

If $h_T > (\log T)^a$, $a > 1$ then we may show (3.9) converges a.s. to zero. The proof is long and complicated and will not be given here. It is unfortunate that in the important case where h_T is of order $\log T$ an a.s. convergence result cannot be established. The problem lies with the expression (3.7) and the increasing process associated with the square of the martingale (3.10).

We conclude with the statement of these results as a Theorem.

Theorem. *If (1.1), (1.2), (2.1), (3.1), (3.8) hold then (2.5), (2.6) hold with the error of order $O_p(h_T^{-\frac{1}{2}})$. If additionally*

$$\liminf_{T\to\infty} h_T/(\log T)^a > 0 \ , \quad a > 1$$

then the same result holds with the error converging a.s. to zero.

References.

An, Hong-Zhi, Chen, Zhao-Guo and Hannan, E.J. (1982) Autocorrelation, autoregression and autoregressive approximation. Ann.Statist. 10, 926-936.

Baxter, G. (1962) An asymptotic result for the finite predictor. Math.Scand. 10, 137-174.

Hall, P. and Heyde, C.C. (1980) Martingale Theory and its Application. Academic Press, New York.

Hannan, E.J. (1980) The estimation of the order of our ARMA process. Ann.Statis., 8, 1071-1081.

Hannan, E.J. and Kavalieris, L. (1984a) A method for autoregressive-moving average estimation. Biometrika (Forthcoming).

Hannan, E.J. and Kavalieris, L. (1984b) Regression, autoregression models. (Forthcoming).

Hannan, E.J. and Rissanen, J. (1982) Recursive estimation of ARMA order, Biometrika, 69, 81-94.

Neveu, J. (1975) Discrete Parameter Martingales. North Holland, Amsterdam.

How to determine the bandwidth of some nonlinear smoothers in practice *)

Wolfgang Härdle
Fachbereich Mathematik
Johann-Wolfgang Goethe Universität
D - 6000 Frankfurt/M.

Abstract. A nonlinear smoothing procedure which estimates a regression curve is proposed. A kernel operates on data which are first transformed in the way which is familiar in the theory of M-estimators. The bandwidth of the kernel is chosen by a "crossvalidatory" device and asymptotic optimality properties are proven. The proposed method is compared with AIC and FPE and shown to be asymptotically equivalent. An application to Raman-Spectra and a Monte Carlo study show how well our method works in practice.

1. Introduction

Let us assume that we observed a triangular sequence of datapoints

$$Y_t^{(T)} = \mu_t^{(T)} + Z_t^{(T)} \quad , \; t=1,2,\dots,T \tag{1.1}$$

with expectation

$$E\, Y_t^{(T)} = \mu_t^{(T)} = m(t/T) \quad , \; t=1,2,\dots,T$$

*) Research partially financed by the Deutsche Forschungsgemeinschaft, Sonderforschungsbereich 123, "Stochastische Mathematische Modelle".

and independent identically distributed errors $\{Z_t^{(T)}\}_{t=1}^{T}$ with variance σ^2. The unknown function $m \in C^2[0,1]$, the regression curve, is to be estimated from the observations $\{Y_t^{(T)}\}_{t=1}^{T}$. In this paper we propose a nonlinear smoothing procedure. We choose a zero of the function

$$\theta \mapsto \sum_{s=1}^{T} \alpha_s^{(T)}(x)\psi(Y_s^{(T)} - \theta)$$

and call it the <u>M-smoother</u> $S^{(T)}(x)$ derived from ψ and the weights $\alpha_s^{(T)}$. We assume the weights to be given by a kernel function K as follows

$$\alpha_s^{(T)}(x) = T^{-1}h^{-1}(T)\, K((x - s/T)/h(T)). \tag{1.2}$$

ψ is a given monotone and bounded function, $\psi \in C^2$, with $\psi(0)=0$, $E\psi(Z_t^{(T)}) = 0$. The parameter $h = h(T)$ in the weights (1.2) is called <u>bandwidth</u>. Interpreting $\{\alpha_s^{(T)}(x)\}_{s=1}^{T}$ as a window (Brillinger, 1975, chapter 3.3) the bandwidth h regulates the size (or span) of the window. In practice, one must select a particular size of the window. It seems desirable to use a bandwidth which makes the <u>averaged square error</u> (ASE) small

$$ASE(h;T) = T^{-1} \sum_{t=1}^{T} [S^{(T)}(t/T) - \mu_t^{(T)}]^2. \tag{1.3}$$

Denote by $h_A = h_A(T)$ the bandwidth which minimizes ASE. The ASE is a discrete approximation to the mean integrated square error (MISE)

$$MISE\,(h;T) = E\int_0^1 [S^{(T)}(s) - m(s)]^2\, ds$$

of the estimated regression curve $S^{(T)}(.)$. Since the regression curve $m(.)$ is unknown we cannot determine h_A from the data.

We discuss a data-driven procedure for approximating h_A which is based on cross-validation in the sense of Stone (1974). In our case the cross-validatory choice of $h(T)$ is that value $h_c = h_c(T)$ which minimizes

$$CRVD(h;T) = T^{-1} \sum_{s=1}^{T} [S_-^{(T)}(t/T) - Y_t^{(T)}]^2 \tag{1.4}$$

where $S_-^{(T)}(t/T)$ denotes the M-smoother computed from the subsample $\{Y_s^{(T)}\}_{s\neq t}$, i.e. $S_-^{(T)}(t/T)$ is a suitable zero of

$$\theta \mapsto \sum_{s \neq t} \alpha_s^{(T)}(t/T)\psi(Y_s^{(T)} - \theta).$$

This is not exactly Stone's (1974) "leave-one-out" statistic, which would be obtained if we would use the weights $(T/T-1)\alpha_s^{(T)}(.)$ rather than $\alpha_s^{(T)}(.)$. For technical convenience we prefer our modified definition of $S_-^{(T)}$.

We show that

$$(1.5) \qquad CRVD(h;T) = T^{-1} \sum_{t=1}^{T} [Z_t^{(T)}]^2 + ASE(h;T) + R(h;T)$$

where $R(h;T)$ is a remainder term which tends to zero "uniformly in h" (in a sense to be specified later). Remark that the first term on the RHS of (1.5) is independent of h and therefore the task to minimize $CRVD(h;T)$ is similar to the task to minimize $ASE(h;T)$ over h. We shall in fact prove

$$h_{A/h_C} \xrightarrow{p} 1 \qquad \text{as } T \to \infty .$$

From this asymptotic behavior we could expect that $h_C(T)$, the cross-validatory choice of $h(T)$, is a reasonable selection for the bandwidth in practical situations.

A small Monte Carlo study and an application of M-smoothers to Raman-Spectra shows how the method works in practice. We also consider the relationship of CRVD to other devices such as Akaike's (1970, 1974) AIC or FPE and show that they are equivalent to CRVD.

Cross-validation as a method for choosing the degree of smoothing has been proposed by several authors in slightly different situations. Wahba and Wold (1975) discuss spline nonparametric regression; Chow, Geman and Wu (1983) studied kernel density estimators with a bandwidth selected by cross-validation; Wong (1983) showed consistency of the Nadaraya-Watson estimator for regression curves (fixed, equispaced design) with a cross-validatory choice of the bandwidth. Our device is competing with running medians, considered by Tukey (1977), Velleman and Hoaglin (1981). This estimation device admits a similar presentation.

One has to choose

$$\psi(u) = I\ (u > 0) - 1/2,\ K(u) = I\ (|u| \leq 1/2).$$

Our theory does not apply, since these functions do not satisfy our regularity conditions.

The M-smoothers which we investigate here were proposed in a time series setting by Velleman (1980). There are also relations to the work of Mallows (1980) who considered some nonlinear smoothers in the frequency domain but left open the question how the span of such nonlinear smoothers ought to be chosen in practice.
The feature which distinguishes our model from those treated by the authors mentioned above is the possibility of sampling a curve finer and finer. We find this feature in Raman spectroscopy, a field of anorganic chemistry; here indeed the spacings between two successive wavenumbers may be decreased (Bussian and Härdle, 1984). It seems also that our methods can be used in geophysics, in order to identify so-called "nugget-effects" (Cressie, 1983).

For notational convenience we will suppress the index T where it seems unnecessary for the understanding. In particular we shall write $\alpha_s(t)$ instead of $\alpha_s^{(T)}(t/T)$. Similarly S_{t-} instead of $S_-^{(T)}(t/T)$, S_t instead of $S^{(T)}(t/T)$ and Y_t instead of $Y_t^{(T)}$.

2. The bandwidth selection problem

In this section we will show that approximation (1.5) holds under the following assumptions on K, m and $h = h(T)$:

(2.1) The kernel K is twice differentiable, symmetric, integrates to one and vanishes outside $[-1,1]$;

(2.2) the sequence of bandwidth $h = h(T)$ tends to zero such that $Th(T) \to \infty$, as $T \to \infty$;

(2.3) the regression curve $m\colon [0,1] \to \mathbb{R}$ is twice continuously differentiable with

$$\int_0^1 [m''(x)]^2 dx > 0$$

and

$$m^{(p)}(0) = m^{(p)}(1), \quad p=0,1,2.$$

Assumption (2.1) is fulfilled by many kernels K which have been proposed in the literature. A good example is the Bartlett-Priestley window (Priestley, 1981, page 569; Epanechnikov,1969). The assumption (2.3) is introduced for technical convenience. It allows us to treat the problem without modification at the boundary points. In a practical situation where we suspect that (2.3) is not fulfilled we would use a weighted version of ASE(h;T). This is suggested by work of Gasser and Müller (1979) who showed that the rate of convergence of ASE is different at the boundary points.

A linear approximation $\{\tilde{S}_t\}$ to $\{S_t\}$ will be defined in order to simplify technical details. For $\{\tilde{S}_t\}$ an asymptotic relation similar to (1.5) holds. It is then seen that the problem of approximating $h_A(T)$ can be solved via a cross-validation device based on the linear approximation $\{\tilde{S}_t\}$.

S_t is a zero of the equation $\sum_s \alpha_s(t)\psi(Y_s - \theta) = 0$.

Define

$$\tilde{S}_t = \sum_{s=1}^{T} \alpha_s(t)\tilde{Y}_s \,, \tag{2.4}$$

where $\tilde{Y}_s = \mu_s + \psi(Z_s)/q$, $q = E\,\psi'(Z)$. Note that $\tilde{S}_t$ can be interpreted as a classical kernel regression estimate which <u>linearly</u> operates on the non-observable <u>pseudo-data</u> $\{\tilde{Y}_s\}_{s=1}^T$, while the M-smoother $\{S_t\}$ operates <u>nonlinearly</u> on the <u>original data</u> $\{Y_s\}_{s=1}^T$.

In analogy to (1.3) and (1.4) define now the following quantities for $\{\tilde{S}_t\}$.

$$\begin{aligned} \widetilde{ASE}(h;T) &= T^{-1} \sum_{s=1}^{T} [\tilde{S}_t - \mu_t]^2 \\ \widetilde{CRVD}(h;T) &= T^{-1} \sum_{s=1}^{T} [\tilde{S}_{t-} - \tilde{Y}_t]^2 \end{aligned} \tag{2.5}$$

where $\tilde{S}_{t-} = \sum_{s\neq t} \alpha_s(t)\tilde{Y}_s$. Define also

$$\widetilde{MASE}(h;T) = E\{\widetilde{ASE}(h;T)\}.$$

By computations very similar to Parzen (1962) it is seen that

$$(2.6)\qquad \widetilde{MASE}(h;T) = (Th)^{-1}\int K^2(u)du\, E\psi^2(Z)/q^2 + 1/4\, h^4 \int_0^1 [m''(x)]^2 dx \int u^2K(u)du + o(T^{-1}h^{-1}+h^4).$$

Neglecting the third summand on the RHS in the equation above, we see that $h(T) = \eta T^{-1/s}$, $\eta > 0$ balances the trade-off between the variance - and the $\{\text{bias}\}^2$- part. The value η minimizing

$$(2.7)\qquad \tilde{M}(\eta) = \lim_{T\to\infty} T^{4/5}\, \widetilde{MASE}\,(\eta T^{-1/5};\, T)$$

is obviously

$$c_1 = \{\int_{-1}^{1} K^2(u)du\, E\psi^2(Z)/(q^2 \int_0^1 [m''(x)]^2 dx \int_{-1}^{1} u^2K(u)du)\}^{1/5}.$$

Fix now two constants $a < c_1 < b$ and define $\underline{h} = aT^{-1/5}$, $\bar{h} = bT^{-1/5}$. It will be seen in Theorem 2.1, that the remainder terms R_i, $i=1,2$ in the following equation vanish uniformly over $h \in [\underline{h},\bar{h}]$,

$$(2.8)\qquad \begin{aligned} ASE(h;T) &= \widetilde{ASE}(h;T) + R_1(h;T) \\ &= \widetilde{MASE}(h;T) + R_2(h;T). \end{aligned}$$

To be precise, we show that for all $\varepsilon > 0$

$$(2.9)\qquad P\{\sup_{\underline{h}\le h\le \bar{h}} T^{4/5}|R_i(h;T)| > \varepsilon\} \to 0\ ,\ i=1,2,\ \text{as}\ \ T \to \infty .$$

Therefore the problem of finding $h_A \in \arg\min_{h\in[\underline{h},\bar{h}]} ASE(h;T)$ reduces to selecting a bandwidth between $aT^{-1/5}$ and $bT^{-1/5}$ which minimizes

$$\widetilde{MASE}(h;T) + R_2(h;T).$$

The first approximation in (2.8) is a consequence of the following lemma.

<u>Lemma 2.1</u>

<u>Consider a sequence</u> $Y_t^{(T)}$ <u>with the properties specified in</u> (1.1) <u>and assume that</u> (2.1) - (2.3) <u>holds, then for all</u> $\varepsilon > 0$, <u>as</u> $T \to \infty$

$$P\{\sup_{\underline{h}\leq h\leq \bar{h}} T^{4/5} \mid T^{-1}\sum_{t=1}^{T}(S_t - \tilde{S}_t)^2 \mid > \varepsilon\} \to 0.$$

Proof

Consider the following two functions $\Phi : \mathbb{R}^T \to \mathbb{R}^T$, $\Psi : \mathbb{R}^T \to \mathbb{R}^T$

$$\Phi = (\Phi_1,\ldots,\Phi_T) \;;\quad \Psi = (\Psi_1,\ldots,\Psi_T)$$

where

$$\Phi_t(\underline{\xi}) = -q^{-1}\sum_{s=1}^{T}\alpha_s(t)\psi(Y_s - \xi_t),$$

$$\Psi_t(\underline{\xi}) = \xi_t - \sum_{s=1}^{T}\alpha_s(t)\psi(Z_s)/q - \sum_{s=1}^{T}\alpha_s(t)\mu_s,$$

$$\underline{\xi} = (\xi_1,\ldots,\xi_T).$$

By definition of $\underline{S} = (S_1,\ldots,S_T)$, $\tilde{\underline{S}} = (\tilde{S}_1,\ldots,\tilde{S}_T)$ we have

$$\Phi_t(\underline{S}) = 0, \qquad t=1,2,..,T$$
$$\Psi_t(\tilde{\underline{S}}) = 0.$$

Applying Taylor's theorem to Φ_t yields

$$\Phi_t(\underline{\xi}) = -q^{-1}\sum_{s=1}^{T}\alpha_s(t)\psi(Z_s) - q^{-1}\sum_{s=1}^{T}\alpha_s(t)\psi'(Z_s)[\mu_s - \xi_t]$$
$$- 1/2\, q^{-1}\sum_{s=1}^{T}\alpha_s(t)\psi''(Z_s + a_{s,t})[\mu_s - \xi_t]^2,$$

where $a_{s,t}$ is between 0 and $\mu_s - S_t$.

The difference between Φ_t and Ψ_t is then

$$\Phi_t(\underline{\xi}) - \Psi_t(\underline{\xi}) = -q^{-1}\sum_{s=1}^{T}\alpha_s(t)\mu_s[\psi'(Z_s) - q]$$
$$+ q^{-1}\sum_{s=1}^{T}\alpha_s(t)\xi_t[\psi'(Z_s) - q]$$
$$(2.10)\qquad - 1/2\, q^{-1}\sum_{s=1}^{T}\alpha_s(t)\psi''(Z_s + a_{s,t})[\mu_s - \xi_t]^2$$
$$= R_{1,t} + R_{2,t} + R_{3,t}.$$

We investigate now the rates at which these $R_{i,t}$, $i=1,2,3$ tend uniformly (in the sense of (2.9)) to zero as $T \to \infty$.

1) Define $V_s = (\psi'(Z_s) - q)/q$ and note that $\{V_s\}_{s=1}^T$ are i·i·d rv's with mean zero. Summation by parts yields

$$R_{1,t} = \sum_{s=1}^{T} \alpha_s(t)\mu_s V_s = \sum_{s=1}^{T} \Delta w_{s,t}\{T^{-1}\sum_{r=1}^{s}\mu_r V_r\}$$

with $\Delta w_{s,t} = h^{-1}\{K((t-1)/(hT)) - K((t-s-1)/(hT))\}$.
Assumption (2.1) implies

$$\sum_{s=1}^{T} |\Delta w_{s,t}| \leq C_1 T^{-1} h^{-2}$$

with a constant C_1 depending on K, and therefore

$$|R_{1,t}| \leq C_1 T^{-1} \underline{h}^{-2} \, |T^{-1} \sup_{1\leq s\leq T} \sum_{r=1}^{s} \mu_r V_r|.$$

By Kolmogorov's inequality we have with a constant C_2, bounding the variances of $\mu_s V_s$,

$$P\{\sup_{1\leq s\leq T} |T^{-1}\sum_{r=1}^{s}\mu_r V_r| \geq \varepsilon\} \leq \varepsilon^{-2}\mathrm{var}\{T^{-1}\sum_{s=1}^{T}\mu_s V_s\}$$

$$\leq C_2 \varepsilon^{-2}/T,$$

which shows that

$$\sup_{1\leq t\leq T}\ \sup_{\underline{h}\leq h\leq \bar{h}} |R_{1,t}| = o_p(T^{-2/5}).$$

2) The term $R_{2,t}$ is estimated similarly.
3) The third term $R_{3,t}$ splits up into the following three summands.

$$\begin{aligned} R_{3,t} = &\sum_{s=1}^{T}\alpha_s(t)\psi''(Z_s + a_{s,t})[\mu_s - \mu_t]^2 \\ &+ 2\sum_{s=1}^{T}\alpha_s(t)\psi''(Z_s + a_{s,t})[\mu_s - \mu_t][\mu_t - \xi_t] \\ &+ \sum_{s=1}^{T}\alpha_s(t)\psi''(Z_s + a_{s,t})[\mu_t - \xi_t]^2 \\ = &\ U_{1,t} + U_{2,t} + U_{3,t} \end{aligned}$$

If $C_3 \geq 2b$, then as $T \to \infty$

$$P\{\sup_{\underline{h}\leq h\leq \bar{h}} T^{-1}\sum_{t=1}^{T}(\tilde{S}_t - \mu_t)^2 < C_3 T^{-4/5}\} \to 1$$

(Marron and Härdle, 1983).

Define the set

$$\mathcal{F}_T = \{\underline{\xi} \in \mathbb{R}^T : T^{-1} \sum_{t=1}^{T} (\xi_t - \mu_t)^2 \leq C_3 T^{-4/5}\}.$$

Then, if $\underline{\xi} \in \mathcal{F}_T$ it is easily seen that with a constant C_4 bounding K and ψ''

$$U_{3,t} \leq C_4 T^{-3/5}.$$

The Cauchy-schwarz inequality shows that there exist constants C_5, C_6 with:

$$U_{1,t} \leq C_5 T^{-3/5}$$

and

$$U_{2,t} \leq C_6 T^{-3/5}.$$

Putting these statements together we finally have that for $\underline{\xi} \in \mathcal{F}_T$

$$\text{(2.11)} \quad \sup_{\underline{h} \leq h \leq \bar{h}} T^{-1} \sum_{t=1}^{T} (\Phi_t(\underline{\xi}) - \Psi_t(\underline{\xi}))^2 = o_p(T^{-4/5}).$$

Now the triangle inequality yields

$$\sup_{\underline{h} \leq h \leq \bar{h}} T^{-1} \sum_{t=1}^{T} (\Phi_t(\underline{\xi}) - (\xi_t - \mu_t))^2 = o_p(T^{-4/5}).$$

Therefore the function $\underline{\eta} \mapsto \underline{\eta} - \Phi(\underline{\eta} + \mu)$ maps the compact, convex set $\mathcal{F}_T - \underline{\mu}$, $\underline{\mu} = (\mu_1, \ldots, \mu_T)$ into itself and by a suitable fixed-point theorem there exists a fixed point $\hat{\underline{\eta}}$ in $\mathcal{F}_T - \underline{\mu}$. Setting $\underline{S} = \hat{\underline{\eta}} + \underline{\mu}$ we see that $\Phi(\underline{S}) = 0$.
We furthermore have by (2.11)

$$T^{-1} \sum_{t=1}^{T} [\Phi_t(S_t) - \Psi_t(S_t)]^2$$
$$= T^{-1} \sum_{t=1}^{T} (S_t - \tilde{S}_t)^2 = o_p(T^{-4/5})$$

with the "o_p" denoting a rv which uniformly in $h \in [\underline{h}, \bar{h}]$ tends to zero. This proves the lemma.

With this lemma we obtain that the difference between ASE and $\widetilde{\text{ASE}}$ is of smaller order than $T^{-4/5}$ uniformly over $h \in [\underline{h}, \bar{h}]$.

This result, together with the second equality in (2.8), yields that $h_A T^{1/5} \overset{p}{\to} C_1$ as $T \to \infty$.

Theorem 2.1

Consider the sequence $Y_t^{(T)}$, as defined in (1.1), and assume that (2.1) - (2.3) hold. Then for all $\varepsilon > 0$, as $T \to \infty$,

$$P\{\sup_{h \in [\underline{h},\bar{h}]} T^{4/5}|ASE(h;T) - \widetilde{ASE}(h;T)| \geq \varepsilon\} \to 0 \tag{2.12}$$

and $h_A \in \arg\min_{h \in [\underline{h},\bar{h}]} ASE(h;T)$ satisfies $T^{1/5} h_A(T) \overset{p}{\to} c_1$, with

$$c_1 = \{\int_{-1}^{1} K^2(u)du E\psi^2(Z)/(q^2 \int_0^1 [m''(x)]^2 dx \int_{-1}^{1} u^2 K(u)du)\}^{1/5}. \tag{2.13}$$

Proof

Statement (2.12) follows by lemma 2.1 and an application of the Cauchy-Schwarz-inequality. Note that the remainder term in (2.6) is tending to zero uniformly in $h \in [\underline{h},\bar{h}]$. Therefore $\tilde{M}(\ )$, as defined in (2.7) reads as

$$\tilde{M}(\eta) = \eta^{-1} \int K^2(u)du\, E_F \psi^2(Z)/q^2 + 1/4\, \eta^4 \int_0^1 [m''(x)]^2 dx$$

which is a continuous and convex function for $\eta \in [a,b]$ and has its unique minimum at $c_1 = \arg\min_{h \in [\underline{h},\bar{h}]} \tilde{M}(hT^{1/5})$. Now

(2.12) and Theorem 1 of Marron and Härdle (1983) yield that, as $T \to \infty$

$$\begin{aligned}
&\sup_{\eta \in [a,b]} |T^{4/5}\, ASE(\eta T^{-1/5};T) - \tilde{M}(\eta)| \\
\leq\ &\sup_{\eta \in [a,b]} |T^{4/5}\{ASE(\eta T^{-1/5};T) - \widetilde{ASE}(\eta T^{-1/5};T)\}| \\
+\ &\sup_{\eta \in [a,b]} |T^{4/5}\{\widetilde{ASE}(\eta T^{-1/5};T) - \widetilde{MASE}(\eta T^{-1/5};T)\}| \\
+\ &\sup_{\eta \in [a,b]} |T^{4/5}\, \widetilde{MASE}(\eta T^{-1/5};T) - \tilde{M}(\eta)| \overset{p}{\to} 0.
\end{aligned}$$

The following arguments are as in Rice (1983). For any $\delta > 0$ define

$$D(\delta) = \inf_{|\eta - c_1| > \delta} (\tilde{M}(\eta) - \tilde{M}(c_1)).$$

Then

$$P\{|h_A T^{1/5} - c_1| > \delta\}$$

$$\leq P\{\ \tilde{M}(h_A T^{1/5}) - \tilde{M}(c_1) > D(\delta)\}$$

$$\leq P\{\ \tilde{M}(h_A T^{1/5}) - T^{4/5}ASE(h_A;T) + T^{4/5}ASE(c_1 T^{-1/5};T) - \tilde{M}(c_1) > D(\delta)\}$$

$$\leq P\{\ \tilde{M}(h_A T^{1/5}) - T^{4/5}ASE(h_A;T) \geq D(\delta)/2\}$$

$$+ P\{T^{4/5}ASE(c_1 T^{-1/5};T) - \tilde{M}(c_1) \geq D(\delta)/2\}$$

$$\to 0\ ,\ \text{which proves (2.13).}$$

Recall now the definition of S_{t-} and of CRVD(h;T). The next theorem shows that (1.5) holds. Therefore, for large T, instead of minimizing the (unknown) function ASE(.;T), we may minimize CRVD(.;T).

Theorem 2.2

Consider $\{Y_t^{(T)}\}$, as defined in (1.1) and assume that (2.1)-(2.3) holds.
Then, for all $\varepsilon > 0$,

$$\text{(2.14)} \quad P\{\sup_{h\in[\underline{h},\bar{h}]} T^{4/5}|CRVD(h;T) - T^{-1}\sum_{t=1}^{T} Z_t^{(T)^2} - ASE(h;T)| \geq \varepsilon\} \to 0,\ \text{as}\ T \to \infty.$$

and $h_C \in \arg\min_{h\in[\underline{h},\bar{h}]} CRVD(h;T)$ satisfies

$$\text{(2.15)} \quad h_C T^{1/5} \overset{p}{\to} c_1$$

where c_1 is the same constant as in Theorem 2.1.

Proof

Consider the following decomposition

$$\begin{aligned}(\tilde{Y}_t - \tilde{S}_{t-})^2 &= (Y_t - S_{t-})^2 + (S_{t-} - \tilde{S}_{t-})^2 + (\psi(Z_t)/q - Z_t)^2 \\ &+ 2(Y_t - S_{t-})(\psi(Z_t)/q - Z_t + S_{t-} - \tilde{S}_{t-}) \qquad \text{(2.16)} \\ &+ 2(S_{t-} - \tilde{S}_{t-})(\psi(Z_t)/q - Z_t)\end{aligned}$$

where $\widetilde{S}_{t-}$ is the "leave-one-out" statistic based on the pseudo-data $\{Y_s\}_{s\neq t}$. From Härdle and Marron (1983), Theorem 1 we have that, uniformly over $h\in[\underline{h},\bar{h}]$

$$\widetilde{\mathrm{CRVD}}(h;T) = T^{-1}\sum_{t=1}^{T}(\widetilde{Y}_t - \widetilde{S}_{t-})^2$$

(2.17)

$$= T^{-1}\sum_{t=1}^{T}(\psi(Z_t)/q)^2 + \widetilde{\mathrm{ASE}}(h;T) + o_p(T^{-4/5}).$$

Now by Theorem 2.1 and the Cauchy-Schwarz inequality we have

$$\sup_{h\in[\underline{h},\bar{h}]} T^{-1}\sum_{t=1}^{T}(S_t - \widetilde{S}_{t-})^2 = o_p(T^{-4/5}).$$

In view of (2.16) it remains therefore to show that the sum $T^{-1}\sum_{t=1}^{T}$ {of the following terms}

$$(\psi(Z_t)/q - Z_t)^2 + 2(\mu_t - S_{t-})(\psi(Z_t)/q - Z_t + S_{t-} - \widetilde{S}_{t-})$$
$$+ 2\, Z_t(\psi(Z_t)/q - Z_t + S_{t-} - \widetilde{S}_{t-})$$
$$+ 2(S_{t-} - \widetilde{S}_{t-})(\psi(Z_t)/q - Z_t) - \psi(Z_t)/q$$

equals

$$T^{-1}\sum_{t=1}^{T} Z_t^2 + o_p(T^{-4/5})$$

uniformly over $h\in[\underline{h},\bar{h}]$. Observing that some terms cancel each other, we have to show that the "$T^{-1}\sum_{t=1}^{T}$" sum of

$$(\mu_t - S_{t-})(\psi(Z_t)/q - Z_t) + (\mu_t - S_{t-})(S_{t-} - \widetilde{S}_{t-})$$
$$+(\psi(Z_t)/q)(S_{t-} - \widetilde{S}_{t-})$$
$$= W_{1,t} + W_{2,t} + W_{3,t} = o_p(T^{-4/5}).$$

By Theorem 2.1 and (2.6) we have that

$$T^{-1}\sum_{t=1}^{T} W_{2,t} \le (\mathrm{ASE}(h;T))^{1/2}\left(T^{-1}\sum_{t=1}^{T}(S_{t-} - \widetilde{S}_{t-})^2\right)^{1/2}$$
$$= o_p(T^{-4/5}), \text{ uniformly over } h\in[\underline{h},\bar{h}].$$

The third term is estimated as in the proof of Lemma 2.1 by setting $\xi_t = S_{t-}$ in (2.10) and observing that S_{t-} and $\widetilde{S}_{t-}$ are independent of $\psi(Z_t)/q$. It remains to show that

$\sup_{h\in[\underline{h},\bar{h}]} T^{-1/5} \sum_{t=1}^{T} (\mu_t - S_{t-})(\psi(Z_t)/q) = o_p(1)$, since the analysis of the term where $\psi(Z_t)/q$ is replaced by Z_t is the same. Adding and subtracting $\widetilde{S}_{t-}$ and repeating the argument for $W_{3,t}$ it remains to show that

$$\sup_{h\in[\underline{h},\bar{h}]} T^{-1/5} \sum_{t=1}^{T} (\mu_t - \sum_{s\neq t} \alpha_s(t)\mu_s - \sum_{s\neq t} \alpha_s(t)\psi(Z_s)/q) \cdot(\psi(Z_t)/q)$$

$$= o_p(1)$$

Consider the bias term

$$\sup_{h\in[\underline{h},\bar{h}]} T^{-1/5} \sum_{t=1}^{T} (b_T(t)\psi(Z_t)/q)$$

where $b_T(t) = \mu_t - \sum_{s\neq t} \alpha_s(t)\mu_s = O(T^{-4/5})$ in the range $h\in[\underline{h},\bar{h}]$.

This shows that the bias term is $o_p(1)$. Using now the independence of $\psi(Z_t)/q$ from $\sum_{s\neq t} \alpha_s(t)\psi(Z_s)/q$ it follows by similar calculations as in the proof of Lemma 2.1 that $T^{-1} \sum_{t=1}^{T} W_{1,t} = o_p(T^{-4/5})$, uniformly over $h\in[\underline{h},\bar{h}]$. This proves (2.14).

We show now (2.15). Recall the definition of $\widetilde{M}(\eta)$ and $D(\delta)$ then with (2.14) and $\hat{\sigma}_T^2 = T^{-1} \sum_{t=1}^{T} Z_t^{(T)2}$ we have,

$$P\{|T^{1/5}h_c - c_1| > \delta\} \le P\{\widetilde{M}(T^{1/5}h_c) - \widetilde{M}(c_1) > D(\delta)\}$$

$$\le P\{\widetilde{M}(T^{1/5}h_c) - CRVD(h_c;T) - \hat{\sigma}_T^2 + CRVD(T^{-1/5}h_c;T) + \hat{\sigma}_T^2 - \widetilde{M}(c_1) > D(\delta)\}$$

$$\le P\{\widetilde{M}(T^{1/5}h_c) - ASE(h_c;T) \ge D(\delta)/4\}$$

$$+P\{ASE(T^{-1/5}c_1;T) \quad \widetilde{M}(c_1) \ge D(\delta)/4\}$$

$$\to 0 \quad \text{by Theorem 2.1.}$$

3. Relations to other devices for selecting a bandwidth

In section 2 we studied the selection of the bandwidth $h_c \in \arg\min_{\underline{h} \le h \le \bar{h}} \text{CRVD}(h;T)$ on the basis of a modified form of Stone's (1974) crossvalidation function. This was mainly done for historical reasons, since Wahba and Wold (1975) introduced the crossvalidation method as a device to pick up "asymptotically correct" sequences of bandwidth in the setting of regression function estimation. Stone (1977) showed an asymptotic equivalence of the crossvalidation method and Akaike's information criterion (AIC) in the context of model selection. It is therefore of interest to study the equivalence of other devices, such as AIC, FPE, to cross-validation in our context.

Note that in the proof of Theorem 2.2 we have essentially shown that

$$\text{CRVD} = \widetilde{\text{CRVD}} - T^{-1} \sum_{t=1}^{T} \psi^2(Z_t)/q^2 + T^{-1} \sum_{t=1}^{T} Z_t^2 + o_p(T^{-4/5}).$$

Since the two middle terms on the RHS do not depend on h and the last term vanishes uniformly in $h \in [\underline{h}, \bar{h}]$, we conclude with the techniques developed in section 2, that the minima of CRVD approximate asymptotically the minima of $\widetilde{\text{CRVD}}$. We therefore consider only $\widetilde{\text{CRVD}}$ in the following.

Let us rewrite $\widetilde{\text{CRVD}}$:

$$\begin{aligned} \widetilde{\text{CRVD}}(h;T) &= T^{-1} \sum_{t=1}^{T} (\tilde{Y}_t(1 + T^{-1}h^{-1}K(o)) - \tilde{S}_t)^2 \\ &= T^{-1} \sum_{t=1}^{T} (\tilde{Y}_t - \tilde{S}_t)^2 + T^{-1} \sum_{t=1}^{T} (T^{-1}h^{-1}K(o)\tilde{Y}_t)^2 \qquad (3.1) \\ &\quad + 2T^{-1} \sum_{t=1}^{T} (\tilde{Y}_t - \tilde{S}_t)(T^{-1}h^{-1}K(o)\tilde{Y}_t). \end{aligned}$$

It is easy to see that $\sup_{h \in [\underline{h}, \bar{h}]} \left| T^{-1} \sum_{t=1}^{T} (T^{-1}h^{-1}K(o)\tilde{Y}_t)^2 \right|$

$$= o_p(T^{-4/5}).$$

The third term is equal to

$$2T^{-1}h^{-1}K(o)E_F\psi^2(Z)/q^2 + o_p(T^{-4/5})$$

uniformly over $h \in [\underline{h}, \bar{h}]$.

Define the residual sum of squares

$$\widetilde{RSS}(h;T) = T^{-1} \sum_{t=1}^{T} (\tilde{Y}_t - \tilde{S}_t)^2.$$

Then as we have shown above

$$\widetilde{CRVD} = \widetilde{RSS} + 2T^{-1}h^{-1}K(o)V(\psi,F) + o_p(T^{-4/5})$$

where $V(\psi,F) = E_F\psi^2(Z)/q$.

Define the leading term

$$C^*(h;T) = \widetilde{RSS}(h;T) + 2T^{-1}h^{-1}K(o)V(\psi,F) - V(\psi,F). \tag{3.2}$$

We will see in Theorem 3.1 that the minima of $C^*(.;T)$ approximate asymptotically the minima of the following functions.

$$\exp(AIC(h;T)) = \widetilde{RSS}(h;T) \exp(2T^{-1}h^{-1}K(o)) \tag{3.3}$$

$$AIC(h;T) = \log(\widetilde{RSS}(h;T)) + 2T^{-1}h^{-1}K(o)$$

(Akaike, 1974),

$$FPE(h;T) = (1 - T^{-1}h^{-1})/(1 - T^{-1}h^{-1})\ \widetilde{RSS}(h;T) \tag{3.4}$$

(Akaike,1970),

$$SHI(h;T) = \widetilde{RSS}(h;T)(1 + 2T^{-1}h^{-1}K(o) \tag{3.5}$$

(Shibata, 1981).

This list may be extended to GXV (generalized cross-validation, Craven and Wahba, 1979) or $FPE(\alpha)$, a modified FPE criterion from Bhansali and Downham (1977).

Note that all the devices listed from (3.2) to (3.5) carry the same structure. They contain a term involving $\widetilde{RSS}$ which is decreasing as $h \downarrow 0$ and a penalty term getting bigger if h is two small. The next theorem states that a small random or nonrandom disturbance $\delta(h,T)$ of $C^*(h;T)$ does not affect the asymptotic optimality of h.

Theorem 3.1

Suppose that for all $\varepsilon > 0$

$$P\{ \sup_{h\in[\underline{h},\bar{h}]} T^{4/5} | \delta(h;T) | > \varepsilon\} \to 0 \quad , \text{ as } T \to \infty.$$

Then a sequence of bandwidth $h_{c,\delta}(T)$ chosen so as to minimize

$$C^*_\delta(.;T) = C^*(.;T) + \delta(.;T)$$

approximates asymptotically $h_c \in \arg \min_{h\in[\underline{h},\bar{h}]} \text{CRVD}(h;T)$, i.e.

$$h_{c,\delta} - h_c \xrightarrow{p} 0 \quad , \text{ as } T \to \infty.$$

The proof of this theorem follows closely the arguments that were used in the proof of Theorem 2.2.

Shibatas criterion function (3.5) may be written as

$$\begin{aligned} \text{SHI}(h;T) &= (C^*(h;T) + V(\psi,F) - 2T^{-1}h^{-1}K(o)V(\psi,F)) \\ &\qquad (1 + 2T^{-1}h^{-1}K(o)) \\ &= C^*(h;T) + \delta(h;T) + V(\psi,F) \end{aligned}$$

where $\delta(h;T) = o_p(T^{-4/5})$ uniformly over $h\in[\underline{h},\bar{h}]$.
The other functions may be expanded in Taylor-series to see that they are asymptotically equivalent to SHI(h;T).

4. An example and a Monte Carlo study

We report here the results of an application and of a small Monte Carlo simulation. M-smoothers of the function $m(s) = \sin(2\pi s)$ were computed from a sample of $T=100$ equispaced data points t/T, $1 \le t \le T$. The residuals $\{Z_t\}$ were generated according to the pdf

$$(4.1) \qquad g(z) = 9\phi(10z) + 1/90\, \phi(z/9)$$

where ϕ denotes the pdf of a standard normal distribution. By direct computation one sees $\sigma^2 = 8.19$. The kernel we implemented was the so-called Bartlett-Priestley window (Epanechnikov, 1969)

$$(4.1) \qquad \begin{aligned} K(u) &= .75(1-u^2) \qquad & |u| \le 1 \\ &= 0 & |u| > 1 \quad . \end{aligned}$$

The IMSL routine GGNPM was used to generate the Gaussian pseudo random numbers. For each of the 160 Monte Carlo runs, the functions CRVD(h) and ASE(h) for $h = i/200$, $i = 3,5,\dots,15$ were computed. We used Huber's ψ-function

$$\psi(u) = \max(-\varkappa, \min(u,\varkappa)) , \varkappa > 0 \tag{4.3}$$

for $\varkappa = 1.2, 1.5, 3$. The mean and the standard deviation of CRVD and ASE for different bandwidth h and tuning parameter $\varkappa$, together with the correlation between ASE and CRVD, are shown in Table 1. The numbers shown there are consistant with the theory: the averaged CRVD and ASE curves have both their minimum at .065 for $\varkappa = 1.2, 1.5, 3$.

An application of M-smoothing to Raman spectroscopic data was also carried out. For various reasons spiky outliers may corrupt the recorded Raman spectrum. Intermittent high frequency signals, bubbles in the sample, furthermore shock waves within the optical instrumentation may introduce absurd spikes (Bussian and Härdle, 1984). In Figure 1 a typical data sequence, $T = 330$, together with the smoothed series $\{S_t\}_{t=1}^{T}$ is shown. Huber's ψ-curve (4.3) was used and S_t was computed by the Newton-Raphson algorithm. In Figure 2 a batch of CRVD curves for different levels of $\varkappa$ is shown. To simplify the interpretation, on the horizontal axis the scale 2hT is used rather than h itself. The solid line in Figure 2 corresponds to $\varkappa = .2$ and the finest dotted line belongs to $\varkappa = .4$; the three other graphs were computed for $\varkappa = .25, .3, .35$ respectively. The five curves have their minimum all in the range between 6 and 8. Selecting $2hT = 7$ and $\varkappa = .25$ gives the smooth curve of Figure 1. There the M-smoother $\{S_t\}$, overlaid with the original data $\{Y_t\}$, is shown. Obviously $\{S_t\}$ is not affected by the spurious observations at $t \approx 200$ and $t \approx 310$. We tested our assumption on the noise sequence $\{Z_t\}$ by means of Bartlett's test (Priestley, 1981). The test did not reject the white noise hypothesis at a 5% significance level. The programs, written in FORTRAN, can be obtained from the author.

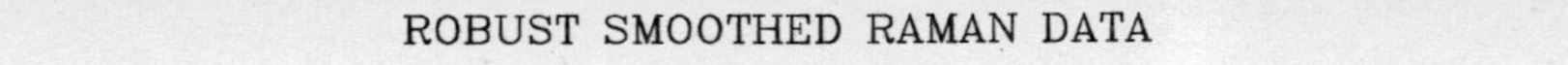

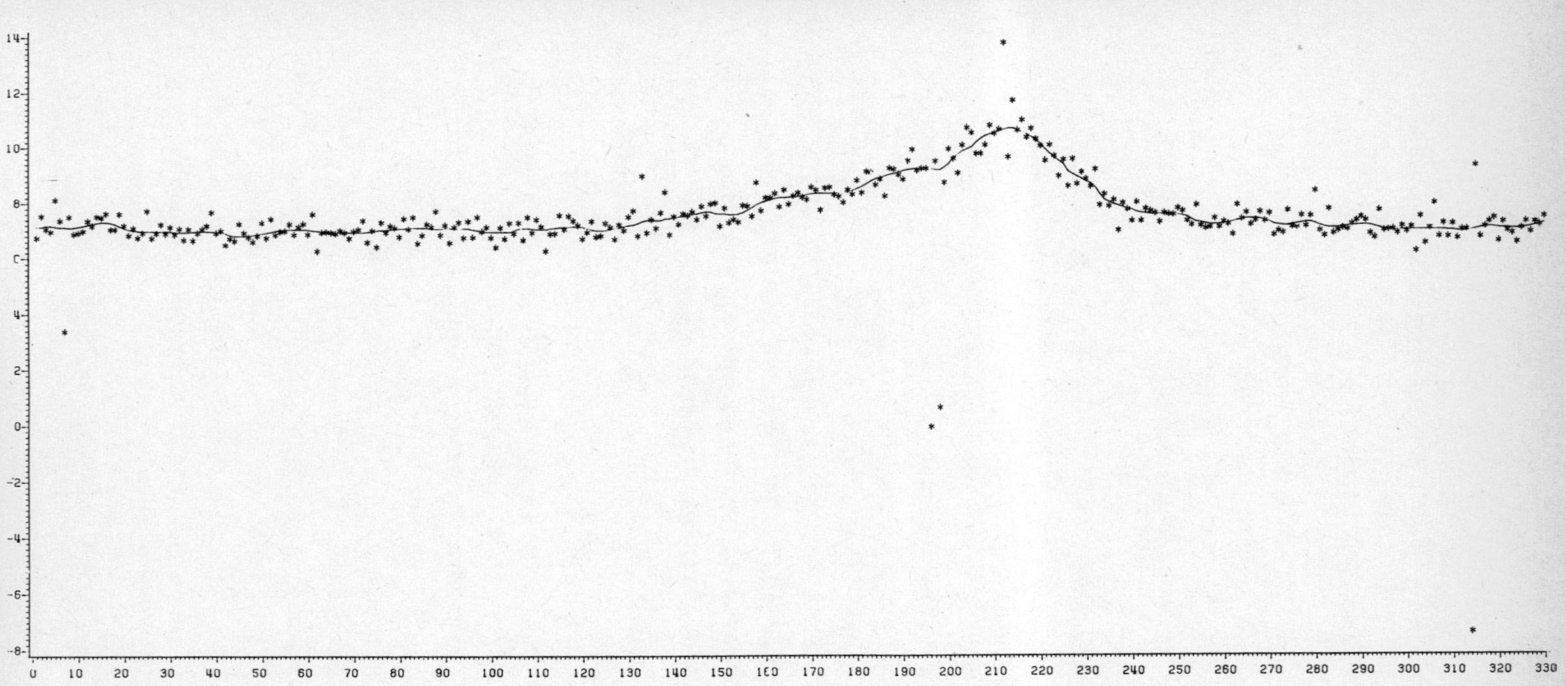

T = 330,K = .25, span = 13

Figure 1

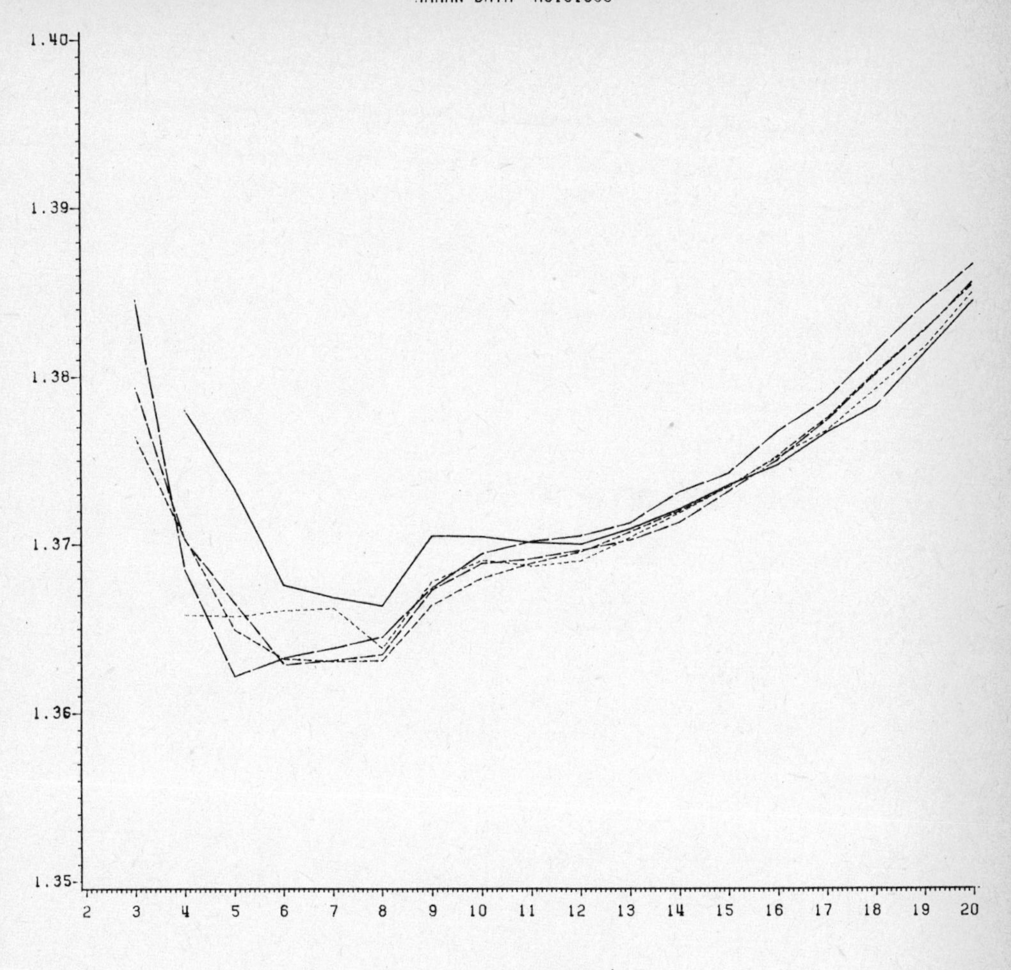

(span+1)/2

κ = .2 to .4 by .05

Figure 2

$(200h+1)/2$		2			3			4			5			6			7			8		
$\varkappa$		1.2	1.5	3.	1.2	1.5	3.	1.2	1.5	3.	1.2	1.5	3.	1.2	1.5	3.	1.2	1.5	3.	1.2	1.5	3.
mean	CRVD	205	207	186	166	122	23	60	37	11	40	27	9.7	47	21	8.6	27	18	7.9	32	20	8
	ASE	198	200	179	159	115	16	53	29	3.7	33	20	2.1	40	13	1.0	20	10	.3	24	12	.4
std	CRVD	172	197	219	169	141	70	117	85	28	78	64	23	88	57	10	68	50	4	77	50	4.5
	ASE	170	196	218	168	141	70	118	85	27	77	62	21	87	57	9	67	49	.1	75	48	.2
corr ASE,CRVD		.997	.997	.996	.994	.991	.735	.852	.8	.637	.771	.698	.585	.74	.661	.534	.603	.645	.492	.604	.673	.449

Table 1

The curves CRVD and ASE averaged over 160 Monte Carlo experiments.

$\mu_t = \sin(2\pi t/T), \quad 1 \le t \le T=100.$

$z_t \quad 9\phi(10x) + 1/90\ \phi(x/9).$

Acknowledgement

I would like to thank my colleagues H. Dinges, J.B. Ferebee, J. Franke, G. Kersting and all other patient souls for many helpful discussions.

References

Akaike H (1970). Statistical predictor identification. Ann. Inst. Stat. Math. 22, 203 - 217

Akaike H (1974). A new look at statistical model identification. IEEE Trans. Auto. Cont. 19, 716 - 723

Bhansali R J and Downham D Y (1977). Some properties of the order of an autoregressive model selected by a generalization of Akaike's FPE criterion, Biometrika, 64, 547-551

Brillinger D R (1975). Time Series, Data Analysis and Theory; Holt, Rinehard and Winston, New York

Bussian B and Härdle W (1984). Robust smoothing applied to white noise and single outlier contaminated Raman spectra. Applied Spectroscopy, 38, 309 - 313

Chow Y S , Geman S and Wu L D (1983). Consistent cross-validated density estimation. Ann Stat. 11, 25 - 38

Craven P and Wahba G (1979). Smoothing noisy data with spline functions. Numerische Mathematik, 31, 377 - 403

Cressie N (1983). Personal communication.

Epanechnikov V A (1969). Nonparametric estimation of a multivariate probability density. Theory Probab. Appl. 14, 153 - 158

Gasser T and Müller H G (1979). Kernel estimation of regression functions. in "Smoothing Techniques for Curve Estimation". Springer Lecture Notes 757

Härdle W and Marron S (1983). Optimal bandwidth selection in nonparametric regression function estimation. Inst. of Stat. Mimeo Series #1530, Chapel Hill, N.C.

Härdle W and Gasser T (1984). Robust nonparametric function fitting. J. Royal Stat. Soc. B 46, 42 - 51

Mallows C (1980). Some theory of nonlinear smoothers. Ann. Stat. 8, 695 - 715

Marron S and Härdle W (1983). Random approximations to an error criterion of nonparametric statistics. Inst. of Stat. Mimeo Series #1538, Chapel Hill, N.C.

Parzen E (1962). On the estimation of a probability density and mode. Ann. Math. Stat. 33, 1065 - 1076

Priestley M B (1981). Spectral analysis and time series. Academic Press, London

Rice J (1983). Bandwidth choice for nonparametric kernel regression. Unpublished manuscript

Shibata R (1981). An optimal selection of regression variables. Biometrika, 68, 45 - 54

Stone M (1974). Crossvalidatory choice and assessment of statistical predictions (with discussion). J. Royal Stat. Soc. B, 36, 111 - 147

Stone M (1977). An asymptotic equivalence of choice of model by crossvalidation and Akaike's criterion. J. Royal Stat. Soc. B, 39, 44 - 47

Tukey (1977). Exploratory data analysis. Addison-Wesley, Reading Massachusetts

Velleman P F (1980). Definition and Comparison of Robust Non-linear Data Smoothing Algorithms. J. Amer. Stat. Ass., 75, 609 - 615

Velleman P F, Hoaglin D C (1981). Applications, Basics, and computing of Exploratory Data Analysis. Duxburry Press, Boston Massachusetts

Wahba G and Wold S (1975). A completely automatic French curve: fitting splines functions by cross-validation. Communications in Statistics 4, 1 - 17

Wong W H (1983). On the consistency of cross-validation in kernel nonparametric regression. Ann. Stat, 11, 1136 - 1141

REMARKS ON NONGAUSSIAN LINEAR PROCESSES WITH ADDITIVE GAUSSIAN NOISE*

K. S. Lii
University of California, Riverside
Riverside, California 92502
and
M. Rosenblatt
University of California San Diego
La Jolla, California 92093

The basic model. In a number of papers [1,2,4,5,6] nonGaussian linear processes are considered as the basic model. Questions relating to the estimation of coefficients and deconvolution were dealt with. We give the assumptions here. Let $\{v_t\}$ be a sequence of independent, identically distributed random variables with $E\, v_t \equiv 0$, $E\, v_t^2 \equiv 1$ and some higher order cumulant $\gamma_s \neq o$ $(s > 2)$. The real coefficients $\{\alpha_j\}$ are in ℓ^2

(1) $$\sum \alpha_j^2 < \infty .$$

It's assumed that one observes the linear process

(2) $$X_t = \sum_j \alpha_j v_{t-j}$$

without knowledge of the α_j's or the v_t's. The object was to *determine procedures for the estimation of the coefficients* α_j *whether or not the system is minimum phase* and *effect deconvolution to estimate the* v_t's.

We shall still look at aspects of the problem just described. But our main concern will be with the modified problem in which we observe *only*

(3) $$Y_t = X_t + \eta_t$$

* Research supported in part by ONR Contract N00014-81-K-0003 and NSF Grant DMS83-12106.

where X_t has the structure given in (2) and $\{\eta_t\}$ is a Gaussian noise process independent of the process $\{X_t\}$.

In many cases the spectral range of the X and η processes will be disjoint. In such a case direct linear filtering will let us get the process $\{X_t\}$. This happens, for example, when the spectrum of the η's is in the high frequency range and that of the X process in the low frequency range. Suppose the interval $[-a,a]$ contains the X spectral mass and that of the η process is outside. Further, let us assume that $\pm a$ are continuity points of the spectral distribution function F of Y. The indicator function of $[-a,a]$ has the Fourier representation

$$\sum_j \frac{1}{\pi} \frac{\sin ja}{j} e^{-ij\lambda}$$

and this implies that

$$X_t = \sum_j \frac{1}{\pi} \frac{\sin ja}{j} Y_{t-j} .$$

Dwyer [3] has been concerned with techniques aimed at gauging the Gaussian or nonGaussian character of additive components of the process Y corresponding to different spectral ranges. He tries to assess the third and fourth order moment properties of such components.

We shall be interested in seeing what happens when the spectral ranges of the X and η processes overlap. If the η process is small compared to the X process, one can still try to deconvolve approximately even by proceeding naively as if η weren't there. An example of such a naive convolution is given below. All this is independent of whether the system is minimum phase.

In Figures 1 and 2 the process Y_t is generated by a Monte Carlo simulation where

$$X_t = v_t - 3.5\, v_{t-1} + 1.5\, v_{t-2}$$

and v_t is a sequence of exponential independent random variables of variance one. Here the roots of the polynomial $1 - 3.5z + 1.5z^2$ are 2 and $1/3$. The additive noise η_t consists of independent Gaussian variables of mean zero and variance $\sigma^2 = 1$ in the case of Figure 1 and $\sigma^2 = 2$ for Figure 2. The first line of the figure graphs the sequence

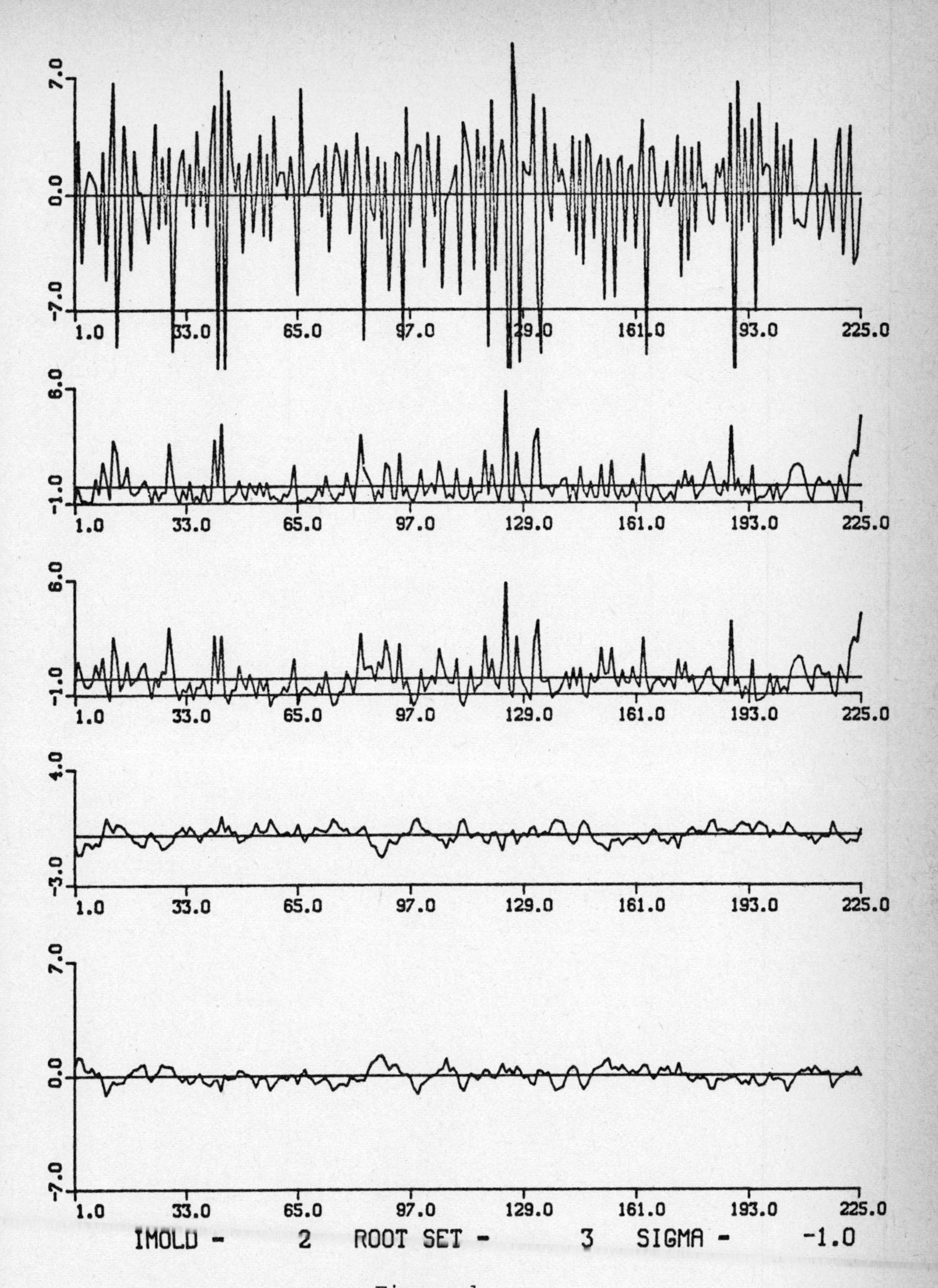
7.0
0.0
-7.0
1.0 33.0 65.0 97.0 129.0 161.0 193.0 225.0
6.0
-1.0
1.0 33.0 65.0 97.0 129.0 161.0 193.0 225.0
6.0
-1.0
1.0 33.0 65.0 97.0 129.0 161.0 193.0 225.0
4.0
-3.0
1.0 33.0 65.0 97.0 129.0 161.0 193.0 225.0
7.0
0.0
-7.0
1.0 33.0 65.0 97.0 129.0 161.0 193.0 225.0
IMOLD - 2 ROOT SET - 3 SIGMA - -1.0

Figure 1.

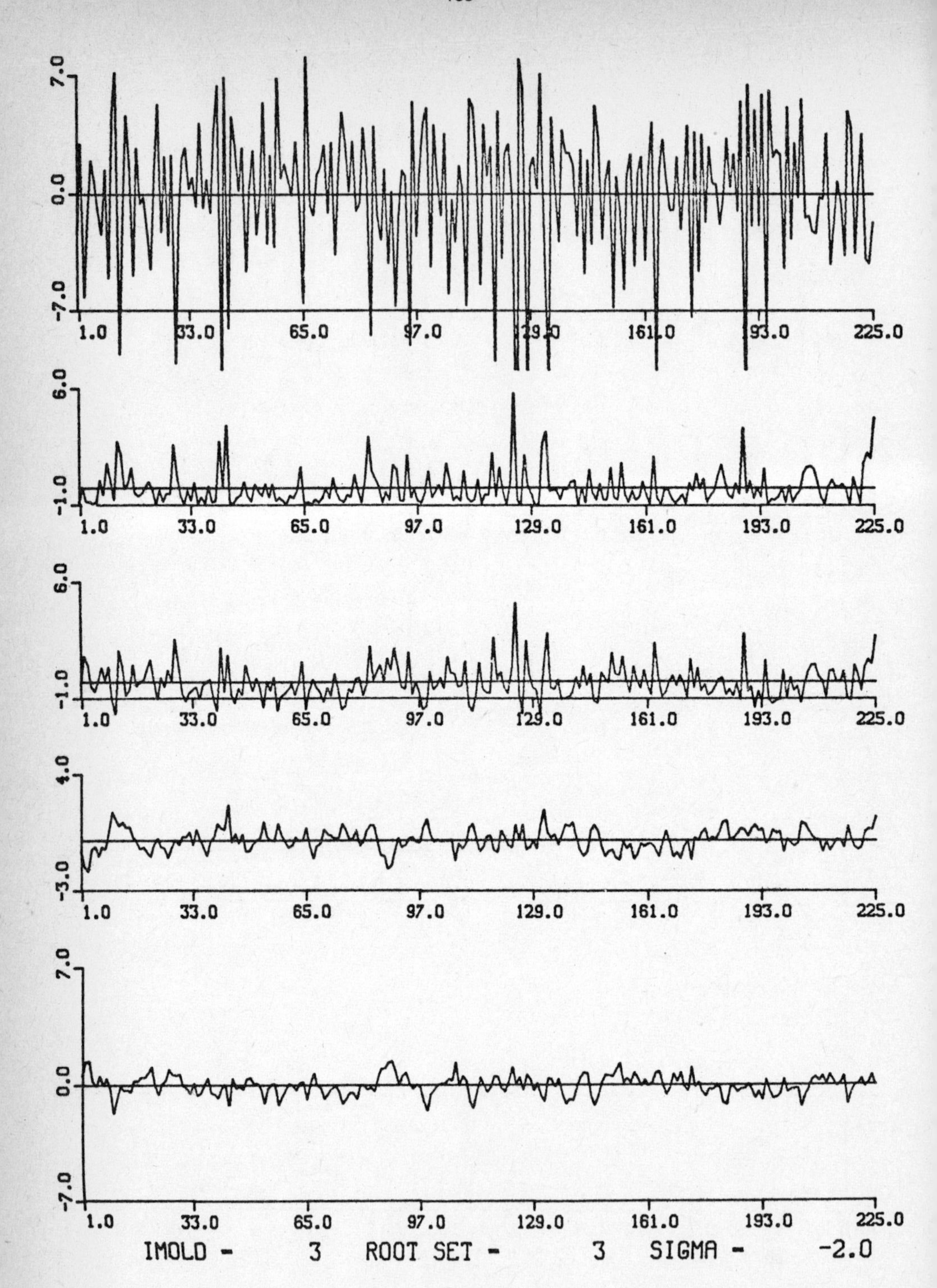
7.0
0.0
-7.0
1.0
33.0
65.0
97.0
129.0
161.0
193.0
225.0
6.0
-1.0
4.0
-3.0
IMOLD - 3 ROOT SET - 3 SIGMA - -2.0

Figure 2.

Y_t. The second line graphs the generating v_t sequence. The third line gives the result of our naive deconvolution neglecting η. The fourth line gives the error in our deconvolution. The last line has the result of a minimum phase deconvolution which is naturally off because one of the roots is inside the unit disc in the complex plane. Clearly our naive deconvolution gets worse as σ^2 increases. But even for $\sigma^2 = 2$, a moderate amount of noise, one can still recognize certain broad features of the v sequence. If η is large one won't be able to deconvolve but one can still hope to estimate a good deal about the α_j's under appropriate conditions.

A simple model. To give some idea of the difficulties that can arise, let us consider the case in which

$$Y_t = a\, v_t + b\, u_t \tag{4}$$

with the $\{v_t\}$ sequence nonGaussian independent, identically distributed with

$$E\, v_t \equiv 0, \quad E\, v_t^2 \equiv 1, \quad E\, v_t^3 \equiv \gamma \neq 0$$

and $\{u_t\}$ an independent, identically distributed sequence of $N(0,1)$ random variables. The $\{v_t\}$ and $\{u_t\}$ sequences are assumed to be independent. It is assumed that $|a|$, $|b|$ and γ are unknown. The counterpart of the problem mentioned above is that of estimating $|a|$ and $|b|$. $|a|$ and $|b|$ are not identifiable in terms of the problem as specified here since v_t might be decomposable

$$v_t = v_t^{(1)} + v_t^{(2)} \tag{5}$$

in terms of two nontrivial independent summands, one of which, say $v_t^{(2)}$, is Gaussian. The problem can be normalized by insisting that v_t be indecomposable in the sense that a representation of the form (5) be impossible. We shall call a random variable v_t with a nontrivial decomposition of the form (5), in that a Gaussian summand with positive variance exists, reducible. This could be expressed in terms of distribution functions. A distribution will be called reducible if it has a nontrivial Gaussian component.

PROPOSITION. *A reducible distribution has a maximal decomposition relative to its Gaussian component.*

Let φ be the characteristic function of the reducible distribution. Then there are constants $c_n > 0$ and corresponding characteristic functions ψ_n such that

$$\varphi(t) = \psi_n(t) \exp\left(- c_n \frac{t^2}{2}\right) .$$

Then

$$\mathrm{Re}\ \varphi(t) \le \mathrm{Re}\ \psi_n(t) \le 1 .$$

Let the F_n be the distributions corresponding to the characteristic functions ψ_n. We have

$$\int_{|x| \ge 1/u} d\, F_n(x) \le \frac{7}{u} \int_0^u \{1 - \mathrm{Re}\ \psi_n(v)\} dv$$

$$\le \frac{7}{u} \int_0^u \{1 - \mathrm{Re}\ \varphi(v)\} dv .$$

We have then uniform bounds on the tails of the distributions F_n. One can therefore choose a subsequence F_{n_k} with $c_{n_k} \uparrow c = \sup_n c_n$ that converges weakly. Let the limiting distribution be F with corresponding characteristic function ψ. It then follows that

$$\varphi(t) = \psi(t) \exp\left(- c \frac{t^2}{2}\right) \tag{6}$$

which corresponds to the unique maximal decomposition.

The problem of estimating $|a|$, $|b|$ now in (4) where v_t is irreducible is one in which $|a|$, $|b|$ are identifiable. However, the problem of deconvolution which is that of estimating v_t is clearly not meaningful.

The representation (6) implies that any distribution with a characteristic function $\varphi(t)$ that doesn't decrease to zero as fast as the Gaussian as $|t| \to \infty$ must be irreducible. This means that all discrete distributions and all gamma distributions must be irreducible.

Let's consider the question of predicting a v_t given Y_t. We shall put this in the form

$$Y = \xi + \eta$$

where ξ and η are independent with means zero and variances σ_1^2

and σ_2^2 respectively. η is normal and ξ is nonnormal with density $g(\cdot)$. We are typically interested in the case in which σ_2^2 is small compared to σ_1^2. The best linear predictor of ξ in mean square is αY where (assuming σ_1^2 and σ_2^2 known)

$$\alpha = \frac{\sigma_1^2}{\sigma_1^2 + \sigma_2^2}$$

and the variance of the prediction error is

$$\frac{\sigma_1^2 \, \sigma_2^2}{\sigma_1^2 + \sigma_2^2} .$$

The best predictor of ξ in mean square is $E(\xi|Y)$. Now

$$E(\xi^k|Y) = \frac{\int \xi^k g(\xi) \frac{1}{\sqrt{2\pi\sigma_2^2}} \exp\left[-\frac{1}{2\sigma_2^2}(Y-\xi)^2\right] d\xi}{\int g(\xi) \frac{1}{\sqrt{2\pi\sigma_2^2}} \exp\left[-\frac{1}{2\sigma_2^2}(Y-\xi)^2\right] d\xi}$$

$$= \frac{\frac{1}{\sqrt{2\pi}} \int (Y-u\sigma_2)^k g(Y-u\sigma_2) \exp\left(-\frac{1}{2}u^2\right) du}{\frac{1}{\sqrt{2\pi}} \int g(Y-u\sigma_2) \exp\left(-\frac{1}{2}u^2\right) du} .$$

Let

$$m_k(Y) = \frac{\int u^k g(Y-u\sigma_2) \exp\left[-\frac{1}{2}u^2\right] du}{\int g(Y-u\sigma_2) \exp\left[-\frac{1}{2}u^2\right] du} .$$

Then

$$E(\xi^k|Y) = \sum_{j=o}^{k} \binom{k}{j} Y^{k-j} \sigma_2^j (-1)^j m_j(Y) .$$

Assuming sufficient smoothness and using a Taylor expansion we have

$$g(Y-u\sigma_2) = g(Y) - u\sigma_2 g'(Y) + \frac{(u\sigma_2)^2}{2!} g''(Y) + \ldots$$

and consequently the best predictor is

$$E(\xi|Y) = Y + \sigma_2^2 \frac{g'(Y) + \frac{\sigma_2^2}{2} g^{(3)}(Y) + \ldots}{g(Y) + \frac{\sigma_2^2}{2} g''(Y) + \ldots}$$

while

$$E(\xi^2|Y) = Y^2 + 2Y\sigma_2^2 \frac{g'(Y) + \frac{\sigma_2^2}{2} g^{(3)}(Y) + \dots}{g(Y) + \frac{\sigma_2^2}{2} g''(Y) + \dots}$$

$$+ \sigma_2^2 \frac{g(Y) + \frac{3}{2}\sigma_2^2 g''(Y) + \dots}{g(Y) + \frac{\sigma_2^2}{2} g''(Y) + \dots} .$$

We should like to compare

$$E((\xi-\alpha Y)^2|Y) = E(\xi^2|Y) - 2\alpha Y E(\xi|Y) + \alpha^2 Y^2$$

with

$$E((\xi-E(\xi|Y))^2|Y) = E(\xi^2|Y) - E(\xi|Y)^2 .$$

It is clear that

$$E((\xi-E(\xi|Y))^2|Y) \le E((\xi-\alpha Y)^2|Y)$$

and the difference is given by

$$(\alpha Y-E(\xi|Y))^2 = ((1-\alpha)Y + \sigma_2^2 \frac{g'(Y)}{g(Y)} + o(\sigma_2))^2 .$$

Such a comparison can give us some idea of the effectiveness of the best predictor versus the linear predictor in the tail region of the g distribution. This tail region may be of greatest interest in certain deconvolution problems (see Wiggins [6]).

Up to this point we have considered prediction in mean square. Suppose we consider instead trying to minimize

$$E|\xi-f(Y)|^4$$

for some appropriate f when $Y = \xi+\eta$ with η $N(o,\sigma_2^2)$ and ξ irreducible nonnormal with mean zero and variance σ_1^2. An analysis can be carried out by considering

$$E(\xi-f(Y))^4 = E(E[(\xi-f(Y))^4|Y]) .$$

We can obviously minimize this if we can minimize the conditional fourth moment on the right side for each Y. This suggests that we consider minimizing

$$E(Z-c)^4$$

in c where Z is a random variable with $E Z^4 < \infty$ and c is a constant. Let m be the mean of Z with $\alpha = m - c$. Then

$$(7) \qquad E(Z-c)^4 = E(Z-m+\alpha)^4 = \alpha^4 + b\sigma^2\alpha^2 + 4\mu_3\alpha + \mu_4$$

with σ^2 the variance of Z and μ_j the jth central moment. Differentiating this with respect to α we get

$$(8) \qquad 4\alpha^3 + 12\sigma^2\alpha + 4\mu_3 .$$

If $\mu_3 = 0$ the unique real zero is $\alpha = 0$ and the minimum is attained by $c = m$. If $\mu_3 \neq 0$ we have a unique real zero since the derivative

$$12\alpha^2 + 12\sigma^2$$

of (8) is positive. The zero will be negative if $\mu_3 > 0$ and positive if $\mu_3 < 0$. Set (8) equal to zero and solve for α. By Cardano's formula one obtains

$$(9) \qquad \alpha = 2^{-1/3} [- \mu_3 + \{\mu_3^2 + 4\sigma^6\}^{1/2}]^{1/3} - 2^{-1/3} [\mu_3 + \{\mu_3^2 + 4\sigma^6\}^{1/2}]^{1/3} .$$

This means that the function $f(Y)$ minimizing (7) is

$$f(Y) = m(Y) - \alpha(Y)$$

where

$$m(Y) = E(\xi|Y)$$

and $\alpha(Y)$ is given by expression (9) with

$$\mu_3 = E((\xi-E(\xi|Y))^3|Y)$$

and

$$\sigma^2 = E((\xi-E(\xi|Y))^2|Y) .$$

Estimation of coefficients. Let $\{Y_t\}$ be a process of type indicated in formula (3), that is, the sum of a linear non-Gaussian process and an independent Gaussian noise process. We shall indicate in a simple way that the coefficients α_j of the linear nonGaussian process can be estimated up to an undetermined multiplier and an undetermined time shift. This is an asymptotic argument.

PROPOSITION. *Let* $\{Y_t\}$ *be a linear nonGaussian process perturbed by independent Gaussian noise. Suppose that* $\sum |k||\alpha_k| < \infty$, $\alpha(i^{-i\lambda}) \neq 0$ *for all* λ *and* $\gamma_3 \neq 0$. *Then the coefficients* α_k *can be consistently estimated up to an undetermined multiplier* c *and an unspecified time shift of the index set.*

The bispectral density of the process $\{Y_t\}$ is

$$b_3(\lambda_1,\lambda_2) = \frac{\gamma_3}{(2\pi)^2}\alpha(e^{-i\lambda_1})\alpha(e^{-i\lambda_2})\alpha(e^{i(\lambda_1+\lambda_2)}) .$$

As before one can estimate the phase of $\alpha(e^{-i\lambda})$ consistently up to an undetermined additive term $ik\pi$ with k integral. This can be accomplished by using bispectral estimates as in Lii and Rosenblatt [4]. Notice that

$$b_3(\lambda,0) = \frac{\gamma_3}{(2\pi)^2}\alpha(1) \; |\alpha(e^{-i\lambda})|^2 . \tag{10}$$

Bispectral estimates (using (10)) then allow us to estimate the spectral density of the process $\{X_t\}$ up to an undetermined constant. The proposition follows immediately from these remarks.

<u>Corollary.</u> *Let the assumptions of the proposition be satisfied. Suppose that the Gaussian noise is white and that* $|\alpha(e^{-i\lambda})| \not\equiv$ *constant for all* λ. *One can then estimate the multiplier* c. *There is then only an unspecified time shift of the index set. The variance of* η *can also be estimated.*

Since the assumptions of the proposition are satisfied the conclusions hold. We have only to estimate the multiplier c. Both the spectral density

$$f_1(\lambda) = \frac{1}{2\pi}|\alpha(e^{-i\lambda})|^2 + \frac{\sigma_\eta^2}{2\pi}$$

of the Y process and

$$b(\lambda,0) = \frac{\gamma_3}{(2\pi)^2}\alpha(1) \; |\alpha(e^{-i\lambda})|^2$$

can be estimated consistently. Clearly

$$f_1(\lambda) = \frac{2\pi}{\gamma_3\alpha(1)} b(\lambda,0) + \frac{\sigma_\eta^2}{2\pi}$$

$$= A\, b(\lambda,0) + B .$$

The coefficients A, B can be consistently estimated given consistent estimates of $f_1(\lambda)$, $b(\lambda,0)$. The conclusion then follows.

<u>The argument of a polynomial transfer function</u>. In many cases it seems reasonable to assume that $\alpha(Z)$ is a polynomial

$$\alpha(Z) = \sum_{j=0}^{p} \alpha_j Z^j .$$

Because the coefficients are assumed real, the roots Z_j are real or occur in complex conjugate pairs. The argument of $\alpha(e^{-i\lambda})$ is the sum of the contributions from the factors corresponding to the roots. We shall consider the contribution from a root Z_j with $|Z_j| \neq 1$. Suppose $Z_j = re^{-i\theta}$ with $0 < r < 1$ and $-\pi < \theta \leq \pi$. Then

$$\log (e^{-i\lambda}-Z_j) = -i\lambda + \log (1-e^{i\lambda}Z_j)$$

$$= -i\lambda - \sum_{k=1}^{\infty} (e^{i\lambda}Z_j)^k/k$$

and this implies that

$$\text{(11)} \qquad \operatorname{Im} \log (e^{-i\lambda}-Z_j) = -\lambda - \sum_{k=1}^{\infty} \frac{r^k \sin k(\lambda-\theta)}{k}$$

for $|Z_j| < 1$.

Similarly for $Z_j = re^{-i\theta}$ with $r > 1$ one has

$$\log (e^{-i\lambda}-Z_j) = \log (-Z_j) + \log (1-Z_j^{-1}e^{-i\lambda})$$

$$= i\pi + \log Z_j - \sum_{k=1}^{\infty} (Z_j^{-1}e^{-i\lambda})^k/k$$

and this implies that we have a representation

$$\text{(12)} \qquad \operatorname{Im} \log (e^{-i\lambda}-Z_j) = \pi - \theta - \sum_{k=1}^{\infty} \frac{r^{-k} \sin k(\theta-\lambda)}{k}$$

for $|Z_j| > 1$. Let us formally consider the difference between (11) and (12) for $r = 1$, that is

$$\text{(13)} \qquad \theta - \lambda - \pi + 2 \sum_{k=1}^{\infty} \frac{\sin k(\theta-\lambda)}{k} .$$

Notice that the Fourier series of $x - \pi \operatorname{sgn} x$ is

$$-2 \sum_{k=1}^{\infty} \frac{2 \sin kx}{k} .$$

This implies that the expression (13) equals zero for $0 < \theta - \lambda < \pi$ and -2π for $-\pi < \theta - \lambda < 0$. This is consistent with the indeterminacy of the argument up to a multiple of 2π. For a real root $|Z_j| < 1$ expression (11) becomes

$$- \lambda \pm \sum_{k=1}^{\infty} \frac{r^k \sin k\lambda}{k}$$

according as to whether $\theta = 0$ or π. In the case of complex conjugate roots of absolute value less than one the sum of the contributions is

$$- 2\lambda - 2 \sum_{k=1}^{\infty} \frac{r^k \sin k\lambda \cos k\theta}{k} .$$

The corresponding remarks for roots greater than one read as follows. If $|Z_j| > 1$ is real expression (12) is

$$(1 - \delta_{\theta,\pi})\pi \pm \sum_{k=1}^{\infty} \frac{r^{-k} \sin k\lambda}{k}$$

with $\pm$ according as to whether $\theta = o$ or π. If there are complex conjugate roots one obtains

$$2\pi + 2 \sum_{k=1}^{\infty} \frac{r^{-k} \sin k\lambda \cos k\theta}{k} .$$

In our case we try to estimate the sum of the contributions to the argument from all zeros and then as a convention readjust the value at zero of the sum so that it is zero there.

This discussion of the relationship between the argument of the transfer function and zeros of $\alpha(Z)$ is natural when $\alpha(Z)$ is a polynomial. A corresponding discussion could be carried out if $\alpha(Z)$ is a rational function in terms of the zeros and poles of $\alpha(Z)$. However, one can easily give examples of analytic functions $\alpha(Z)$ with no zeros such as, for example,

$$\alpha(Z) = \exp \{\sin Z\} .$$

Here, there would be no meaning to such an analysis.

References

1. Beneviste, A., Goursat, M. and Ryget, G. (1980). "Robust identification of a nonminimum phase system," IEEE Trans. on Automatic Control, 25, no. 3, 385-399.

2. Donoho, D. (1981). "On minimum entropy deconvolution," in *Applied Time Series Analysis II* (ed. D. F. Findley) Academic Press, New York, 565-608.

3. Dwyer, R. F. (1982). "Arctic ambient noise statistical measurement results and their implications to sonar performance improvements," NUSC reprint report 6739.

4. Lii, K. S. and Rosenblatt, M. (1982). "Deconvolution and estimation of transfer function phase and coeffients for nonGaussian linear processes," Ann. Statist., 10, 1195-1208.

5. Rosenblatt, M. (1980). "Linear processes and bispectra," J. Appl. Probab. 17, 265-270.

6. Wiggins, R. A. (1978). "Minimum entropy deconvolution," Geoexploration 17.

Gross-Error Sensitivies of GM and RA-Estimates

R. Douglas Martin

University of Washington
Seattle, WA 98195 U.S.A.

V. J. Yohai

Univ. Nac. de Buenos Aires
1426 Buenos Aires, Argentina

1. INTRODUCTION

In a highly important paper for the subject of robust parameter estimation, F. Hampel (1974) introduced the *influence curve* and the *gross-error sensitivity.* Hampel's gross-error sensitivity (GESH) is the supremum of the norm of Hampel's influence curve (ICH) with respect to an argument which represents an observation, or contamination, whose influence is being assessed. The GESH is a measure of the asymptotic bias caused by a vanishingly small fraction of contamination. This interpretation of the GESH has stimulated research on *optimal bounded influence* estimates (see Hampel, 1978; Krasker and Welsch, 1982; Huber, 1983; and the references cited therein).

The considerable interest in Hampel's influence curve is reflected in the large number of papers in the literature which apply and extend ICH to various contexts such as censored data (Samuels, 1978), testing (Lambert, 1981; Ronchetti, 1982), errors-in-variables (Kelly, 1984), and so on.

In spite of the large number of such papers on influence curves, they all share the common base of presuming independent observations. The few papers in existence which deal with influence curves in the time series context (see, for

This work was supported in part by the U.S. Office of Naval Research under contract N00014-84-C-0169.

example, Martin and Jong, 1977; Martin, 1980; Chernick, Downing and Pike, 1982; Kuensch, 1983; Basawa, 1984) use Hampel's ICH, which was defined in the context of independent observation.

Recently, Martin and Yohai (1984) have introduced a new definition of influence curve which is explicitly geared for the time series setting. This influence curve we abbreviate IC in order to have a convenient labeling distinction relative to Hampel's ICH.

The main purpose of this paper is to provide a general definition of gross-error-sensitivity (GES) based on our time-series influence curve (IC), and to provide some explicit calculations of GES's for some particular robust parameter estimates, namely GM-estimates and RA-estimates for first-order autoregression, AR(1), parameters.

In Section 2 we briefly state our definition of IC for time series, and mention some results which have been obtained in Martin and Yohai (1984). However, the bulk of Section 2 is spent discussing a general replacement model for time series (and its various specializations) which is needed for the definition of the IC. Section 3 defines GM and RA-estimates for an AR(1) parameter, and additive-outliers IC's for these estimates are given in Section 4. In Section 5 we give a general definition of GES for time series. We also compute a special case of GES for GM and RA-estimates of the AR(1) model, and draw some general conclusions about the relative behaviors of these estimates for independent versus patchy outliers.

2. DEFINITION OF INFLUENCE CURVE

The General Replacement Model

The following component processes and associated stationary and ergodic marginal measures on $(R^{\infty}, \beta^{\infty})$ are used to construct the contaminated process:

$x_i \sim \mu_x$,	the nominal or core process, often Gaussian
$w_i \sim \mu_w$,	a contaminating process
$z_i^{\gamma} \sim \mu_z^{\gamma}$,	a 0−1 process

where $0 \leqq \gamma \leqq 1$, and

$$P\,(z_i^{\gamma} = 1) = g\,(\gamma) = \gamma + o\,(\gamma) \tag{2.1}$$

for some function g. The contaminated process y_i^{γ} is now obtained by the general *replacement* model

$$y_i^{\gamma} = (1 - z_i^{\gamma})x_i + z_i^{\gamma} w_i \tag{2.2}$$

where $y_i^{\gamma} \sim \mu_y^{\gamma}$ and $\mu_y^0 = \mu_x$, i.e., zero contamination results in perfect observations of x_i. In general we may wish to allow dependence between the z_i^{γ}, x_i and w_i processes in order to model certain kinds of outliers. Correspondingly, the measures μ_y^{γ}, $0 \leqq \gamma \leqq 1$ are determined by the specification of the joint measures μ_{xwz}^{γ}, $0 \leqq \gamma \leqq 1$.

The Pure Replacement Model

Here z_i^{γ}, x_i and w_i are mutally independent processes, i.e.,

$$\mu_{xwz}^{\gamma} = \mu_z^{\gamma} \mu_x \mu_w \,.$$

The AO Model

Allowing dependence between x_i and w_i means that the additive outliers (AO) model

$$y_i = x_i + v_i^* \tag{2.3}$$

used in previous studies (Denby and Martin, 1979; Bustos and Yohai, 1983) may in some situations be obtained as a special case of (2.2). This is the case for example when the v_i^* have for marginal distribution the contamination distribution $F_v = (1-\gamma)\delta_0 + \gamma H$ with degenerate central component δ_0. Just set $w_i = x_i + v_i$ with v_i having marginal distribution H, let $g(\gamma) = \gamma$ in (2.2), and let z_i^γ be independent of x_i and v_i. Here $\mu_{xwz}^\gamma = \mu_z^\gamma \mu_{xw} = \mu_z^\gamma \mu_{x,x+v}$. Throughout the remainder of the paper we use the version of the AO model obtained from (2.2) with $w_i = x_i + v_i$.

The two main *time-configurations* for outliers are (a) isolated outliers, and (b) outliers occuring in *patches* or *bursts.* The need for modeling the latter behavior is well recognized by those who have dealt with real time series problems. It may also be desirable to combine these two situations in order to adequately model some time series data.

Independent Outliers

Since isolated outliers are typically produced by independence in the w_i or v_i, we shall use the terms "independent" and "isolated" as interchangeable adjectives.

Situations in which the outliers are mainly isolated are easily manufactured from either the pure replacement or AO form of (2.2) by letting z_i^γ be i.i.d. with $g(\gamma) = \gamma$ and w_i an appropriately specified process. For example, w_i could be an i.i.d. Gaussian process with mean zero and suitably large variance, or the w_i could be identically equal to a constant value ζ.

Patchy Outliers

Patches of approximately fixed length can be arranged in the following way. Let w_i and v_i be highly correlated processes. In case these processes are identically equal to a constant ζ, they will be regarded as highly correlated. Now let $\tilde{z}_i^p$ be an i.i.d. binomial $B(1,p)$ sequence, and set

$$z_i^\gamma = \begin{cases} 1 \text{ if } \tilde{z}_{i-l}^p = 1 \text{ for some } l = 0,1,\cdots,k-1 \\ 0 \qquad \text{else} \end{cases} \tag{2.4}$$

Here we have set $\gamma = kp$, with k fixed and p variable. Then since

$$P(z_i^\gamma = 1) = 1 - (1-p)^k = kp + o(p)$$

we have

$$g(\gamma) = \gamma + o(\gamma) \tag{2.5}$$

and the average patch length is k for γ small. We denote the probability measure of the process $\{z_i^\gamma\}$ by $\mu_z^{k,\gamma}$.

Definition (Time-Series Influence Curve): Suppose the estimator sequence $\{\mathbf{T}_n\}$ is specified asymptotically by a functional $\mathbf{T} = \mathbf{T}(\mu)$ defined on a subset $\mathbf{P}_o$ of the set $\mathbf{P}_{se}$ of all stationary and ergodic measures, and suppose that μ_y^γ is the measure for y_i^γ defined by (2.1)-(2.2). Then the *influence curve* **IC** of **T** is defined as

$$\mathbf{IC}\,(\mu_w, \mathbf{T}, \{\mu_y^\gamma\}) = \lim_{\gamma \to 0} \frac{\mathbf{T}(\mu_y^\gamma) - \mathbf{T}(\mu_y^o)}{\gamma} \tag{2.6}$$

provided the limit exists.

3. RA AND GM-ESTIMATES OF AR(1) PARAMETER

Let $x_1, x_2, \ldots, x_n$ be observations corresponding to a stationary AR(1) model, i.e.,

$$x_i = \varphi x_{i-1} + u_i \tag{3.1}$$

where the u_i's are i.i.d. random variables and $|\varphi| < 1$. In this paper we will study the gross error sensitivity for two classes of estimates for the AR(1) model: generalized M-estimates (M-estimates) and residual autocovariance estimates (RA-estimates).

The GM-estimates for the AR(1) model are defined by

$$\sum_{i=2}^{n} \eta(r_i(\varphi)\,,\ \ x_{i-1}(1-\varphi^2)^{1/2}) = 0 \tag{3.2}$$

where $r_i(\varphi) = x_i - \varphi x_{i-1}$ and $\eta : R^2 \to R$ is a conveniently chosen bounded function.

The GM-estimates were first proposed for regression by Hampel (1975) and Mallows (1976). GM-estimates were studied for autoregressive models by Denby and Martin (1979), Martin (1980) and Bustos (1982).

The RA-estimates were introduced by Bustos and Yohai (1983). For the AR(1) model the RA-estimate $\hat{\varphi}$ is defined as follows. Let

$$\gamma_l(\varphi) = \frac{1}{n} \sum_{i=l+2}^{n} \eta(r_i(\varphi), r_{i-l}(\varphi)) \qquad 0 \le l \le n-2$$

denote a robust lag-l autocovariance estimate for the residuals with robustifying function $\eta : R^2 \to R$. Then $\hat{\varphi}$ is a solution of the estimate equation

$$\sum_{l=1}^{n-2} \varphi^{l-2} \gamma_l(\varphi) = 0 . \tag{3.3}$$

The two main variants of the function η for GM- and RA-estimates are as follows.

Mallows Variant

$$\eta(u_1,u_2) = \psi(u_1)\psi(u_2)$$

for some bounded robustifying psi-function ψ. This type of estimate was suggested by Mallows (1976) for regression.

Hampel-Krasker-Welsch (HKW) Variant

$$\eta(u_1,u_2) = \psi(u_1,u_2)$$

for some bounded robustifying psi-function ψ.

In the case of a Gaussian AR(1) model and under general conditions on the function η it may be shown that the RA- and GM-estimates have asymptotical normal distribution with the same variance

$$V = \frac{A}{B^2} \tag{3.4}$$

where

$$A = E\eta^2(u,v) \tag{3.5}$$

and

$$B = Eu\ (\partial/\partial u)\ \eta(u,v) \tag{3.6}$$

where u,v are independent $N(0,1)$ random variables. See Bustos (1982) for GM-estimates and Bustos, Fraiman and Yohai (1984) for RA-estimates.

4. SOME AR MODEL RESULTS

AR(1) Models

In this section we state results concerning the IC-function of GM and RA-estimates for the AR(1) model. These results may be found in Martin and Yohai (1984). The following assumptions will be needed.

(A1) $\eta(\cdot\,,\cdot)$ is odd in each variable

(A2) $\eta(u_1,u_2)$ is continuous and $|\eta(u_1,u_2)| \leqq K|u_1|^{k_1}|u_2|^{k_2}$, where k_1 and k_2 are either 0 or 1

(A3)

$$\eta_i(u_1,u_2) = \frac{\partial\eta(u_1,u_2)}{\partial u_i}\ , i = 1,2$$

are continuous and

$$|\eta_1(u_1,u_2)| \leqq K|u_2|^{h_1}, \qquad |\eta_2(u_1,u_2)| \leqq K|u_1|^{h_2}$$

where h_1 and h_2 are either 0 or 1

(A4)

$$B = E\,\{v \cdot (\partial/\partial x)\eta(x,v)|_{x=u}\} \neq 0 \tag{4.1}$$

where u and v are independent $N(0,1)$ random variables.

(A5) The GM- and RA-estimates are asymptotically normal with asymptotic variance given by (3.4), (3.5) and (3.6).

Theorem: Suppose that the AO model holds, i.e., $w_i = x_i + v_i$ in (2.3), with v_i independent of x_i. Assume A1-A4 and that $E\,|v_1|^{k_1+k_2} < \infty$, where k_1 and k_2 are as in A2.

(i) For independent outliers we have

$$IC_{AO,1}(\mu_v, T^{GM}, \varphi)$$

$$= \frac{(1-\varphi^2)^{1/2}}{B} E\ \eta(u_1 - \varphi v_0, (x_0 + v_0)(1-\varphi^2)^{1/2}) \tag{4.2}$$

and

$$IC_{AO,1}(\mu_v, T^{RA}, \varphi) = \frac{1-\varphi^2}{B} E\ \eta(u_1 - \varphi v_0, u_0 + v_0)\,. \tag{4.3}$$

(ii) For patches of outliers of length $k \geqq 2$ we have

$$IC_{AO,k}(\mu_v, T^{GM}, \varphi)$$

$$= \frac{(1-\varphi^2)^{1/2}}{kB} \Big[(k-1)\, E\ \eta(u_1 + v_1 - \varphi v_0, (x_0 + v_0)(1-\varphi^2)^{1/2})$$

$$+ E\ \eta(u_1 - \varphi v_0, (x_0 + v_0)(1-\varphi^2)^{1/2})\Big] \tag{4.4}$$

and

$$IC_{AO,k}(\mu_v, T^{RA}, \varphi)$$

$$= \frac{1-\varphi^2}{kB}\left[\sum_{h=1}^{k-2}(k-h-1)\varphi^{h-1}E\,\eta(u_1+v_1-\varphi v_0,\, u_{1-h}+v_{1-h}-\varphi v_{-h})\right.$$

$$+\sum_{h=1}^{k-1}\varphi^{h-1}E\,\eta(u_1+v_1-\varphi v_0,\, u_{1-h}+v_{1-h})$$

$$+\sum_{h=1}^{k-1}\varphi^{h-1}E\,\eta(u_1-\varphi v_0,\, u_{1-h}+v_{1-h}-\varphi v_{-h})$$

$$\left.+\,\varphi^{k-1}E\,\eta(u_1-\varphi v_0,\, u_{1-k}+v_{1-k})\right]. \tag{4.5}$$

5. GROSS-ERROR SENSITIVITY

In this section we give a general definition of gross-error sensitivity (**GES**) based on the **IC**. Specific results are then given for the **GES**'s of GM and RA estimates of the first-order autoregression parameter.

The Gross Error Sensitivity

Suppose that the contamination process y_i^γ is given by (2.2) and its related assumptions. Then for a fixed set of measures $\{\mu_{xwz}^\gamma\} = \{\mu_{xwz}^\gamma : 0 \leq \gamma < 1\}$, we have a particular arc $\{\mu_y^\gamma\} = \{\mu_y^\gamma : 0 \leq \gamma < 1\}$ of contaminated process measures in $\mathbf{P}_{se}$. Suppose that the **IC** (2.6) exists for a given *family* $\mathbf{P}$ of arcs $\{\mu_y^\gamma\}$ with each arc in $\mathbf{P}_{se}$. This family $\mathbf{P}$ will be generated by letting the contamination process measure μ_w vary over a family $\mathbf{P}_w$, in a manner consistent with the dependency structure $\{\mu_{xwz}^\gamma\}$, while $\{\mu_z^\gamma\} = \{\mu_z^\gamma : 0 \leq \gamma < 1\}$ and μ_x are fixed. Often $\{\mu_z^\gamma\}$ will be independent of μ_{xw}, and then $\mathbf{P}$ will be generated by letting μ_w vary over $\mathbf{P}_w$ in a manner consistent with the dependency structure of μ_{xw}. For the pure replacement model given in Section 2, $\mu_{xw} = \mu_x \mu_w$ so P is then determined by simply letting μ_w range over $\mathbf{P}_w$. In the AO model where $w_i = x_i + v_i$ with x_i and v_i independent, the measures $\mu_{xw} = \mu_{x,x+v}$ are specified by μ_x and μ_v, and P is generated by letting μ_v range over a prescribed family $\mathbf{P}_v$.

Definition: The gross-error sensitivity (GES) of an estimate T at the family P of arcs $\{\mu_y^\gamma\}$ is

$$\mathbf{GES}(\mathbf{P}, \mathbf{T}) = sup_{\{\mu_y^\gamma\} \in \mathbf{P}} \mid \mathbf{IC}\,(\mu_w, \mathbf{T}, \{\mu_y^\gamma\}) \mid . \tag{5.1}$$

In the interesting cases of pure replacement models and AO models, it would be convenient to modify the notation as follows: In the former case the supremum would be over all $\mu_w \in \mathbf{P}_w$, with $\mathbf{P}_w$ and $\{\mu_z^\gamma \mu_x\}$ replacing P as arguments of **GES**. In the latter case the supremum would be over all $\mu_v \in \mathbf{P}_v$, with μ_v replacing μ_w as argument of **IC**, and $\mathbf{P}_v, \{\mu_z^\gamma \mu_{x,x+v}\}$ replacing P as arguments of **GES**. However, an even simpler notation will be used for the special cases treated below.

Comment: In either a pure replacement model or an AO model where the family of arcs P is generated by $\mathbf{P}_w$ or $\mathbf{P}_v$, respectively, the leading case of **GES** is obtained when $w_i \equiv v_i \equiv \zeta$, and correspondingly, $\mathbf{P}_w = \mathbf{P}_v = \{\delta_\zeta\}$ where δ_ζ is the point mass on R^∞. In this case we replace the **IC** argument μ_w by ζ, take the supremum over all ζ, and replace the **GES** argument P by $\{\mu_z^\gamma \mu_x\}$. It is shown below that for AO models the supremum over larger families $\mathbf{P}_v$ are attained at $\{\delta_\zeta\}$.

GES's for GM and RA-Estimates at AO Models

Let x_i be a first-order Gaussian autoregression, and suppose that the AO model holds. Correspondingly, the notation $GES_{AO,k}(\mathrm{P}_v, \varphi, T)$ will be used where $T = T^{GM}$ or $T = T^{RA}$ and k indicates the patch length, with $k = 1$ for independent outliers. In this case the subscripts AO and k, along with the argument φ, serve to replace the argument $\{\mu_z^\gamma \mu_{x,x+v}\}$ suggested above. The results to follow help characterize the differences between GM and RA-estimates at isolated outliers and at patches of outliers.

Theorem 5.1: Suppose that A1-A4 hold, and that u_1, u_2 are independent $N(0,1)$ random variables. Then for independent outliers we have

$$GES_{AO,1}(\mathrm{P}_v, \varphi, T^{GM}) = \frac{(1-\varphi^2)^{1/2}}{B} \, sup_c \, E \, | \, \eta(u_2 - \varphi c \, , \, u_1 + c \, (1-\varphi^2)^{1/2}) | \tag{5.2}$$

$$GES_{AO,1}(\mathrm{P}_v, \varphi, T^{RA}) = \frac{1-\varphi^2}{B} \, sup_c \, E \, | \, \eta(u_1 - c\,\varphi, \, u_2 + c \,) \, | \tag{5.3}$$

Proof: Since v_0 is independent of x_0 and u_0, we have

$$E\eta(u_1 - \varphi v_0, (x_0 + v_0)(1-\varphi^2)^{1/2}) = \int E\eta\,(u_0 - \varphi c\,, (x_0 + c\,)(1-\varphi^2)^{1/2}) d\mu_v(c\,)$$

and the GES is obtained by taking the supremum over all $\mu_v \in \mathrm{P}_v$ of the absolute value of the right-hand side above. But the right-hand side is bounded above in magnitude by $sup_c E \, | \, \eta(u_1 - \varphi c\,, (x_1 + c\,)(1-\varphi^2)^{1/2}) \, |$, and this supremum is achieved by restricting μ_v to point masses. This proves (5.2); (5.3) is proved similarly.

Corollary 5.1: Suppose that in addition to A1-A4, we have

(A6) $\eta(u,v)$ is monotone in each variable and $uv > 0$ implies $\eta(u,v) \geqq 0$.

Then

$$GES_{AO,1}(\mathrm{P}_v, \varphi, T^{GM}) = \frac{(1-\varphi^2)^{1/2}}{B} \, sup_{u,v} \, |\, \eta(u,v) \,| \tag{5.4}$$

$$GES_{AO,1}(\mathrm{P}_v, \varphi, T^{RA}) = \frac{1-\varphi^2}{B} \, sup_{u,v} \, |\, \eta(u,v) \,| \; . \tag{5.5}$$

Remark: In the above corollary, note that by (5.2) the GES for T^{GM} is the supremum of $|ICH(\mathbf{y}_1, T^{GM}, \varphi)|$, but that by (5.3) the GES for T^{RA} is *not* the supremum of $|ICH(\mathbf{y}_1, T^{RA}, \varphi)|$.

Theorem 5.2: Suppose that A1-A4 and A6 hold. Then

(i) For AO patches of length 2 we have

$$GES_{AO,2}(\mathrm{P}_v, \varphi, T^{GM}) = \frac{(1-\varphi^2)^{1/2}}{B} \, sup_{u,v} \, |\, \eta(u,v) \,| \tag{5.6}$$

$$GES_{AO,2}(\mathrm{P}_v, \varphi, T^{RA}) \leq \frac{1-\varphi^2}{B} \cdot \frac{2+|\varphi|}{2} \, sup_{u,v} \, |\, \eta(u,v) \,| \tag{5.7}$$

(ii) For AO patches of length k, with $k \to \infty$, denote $\lim_{k\to\infty} GES_{AO,k}(\mathrm{P}_v, \varphi, T) = GES_{AO,\infty}(\mathrm{P}_v, \varphi, T)$, with $T = T^{GM}$ or T^{RA}.

$$GES_{AO,\infty}(\mathrm{P}_v, \varphi, T^{GM}) = \frac{(1-\varphi^2)^{1/2}}{B} \, sup_{u,v} \, |\, \eta(u,v) \,| \tag{5.8}$$

$$GES_{AO,\infty}(\mathrm{P}_v, \varphi, T^{RA}) = \frac{1+|\varphi|}{B} \, sup_{u,v} \, |\, \eta(u,v) \,| \tag{5.9}$$

Before proving Theorem 5.2, we need the following Lemma which we state without proof.

Lemma: Assume A1 and A5. Let (u_n, v_n) be a sequence of bivariate normal random vectors such that (a) $E(u_n) = E(v_n) = 0$, (b) $\lim_{n\to\infty} var(u_n) = \lim_{n\to\infty} var(v_n) = \infty$ and (c) $\lim_{m\to\infty} \rho(u_n, v_n) = \pm 1$. Then $\lim_{n\to\infty} E\eta(u_n, v_n) = \pm sup_{u,v} \mid \eta(u,v) \mid$.

Proof: We will assume throughout the proof that $\varphi > 0$. The case $\varphi < 0$ is proved similarly.

(i) From (4.2) we have

$$GES_{AO,2}(\mathrm{P}_v, \varphi, T^{GM}) \leq \frac{(1-\varphi^2)^{1/2}}{B} sup_{u,v} \mid \eta(u,v) \mid \tag{5.10}$$

Let v_i^n be a sequence of stationary AR(1) Gaussian processes, $v_i^n = \theta_n v_i^n + h_i^n$, where for n fixed, the h_i^n's are i.i.d. with distribution $N(0, \sigma_n^2)$. Take $\sigma_n^2 \to \infty$ and $\theta_n \to -1$. Then by (4.4) we have

$$\mathbf{IC}(\mu_{v^n}, T^{GM}, \varphi) \mid$$

$$= \frac{(1-\varphi^2)^{\frac{1}{2}}}{2B} \Big[E\eta(u_1 + (\theta_n - \varphi)v_0^n + h_0^n, (x_0 + v_0^n)(1-\varphi^2)^{1/2})$$

$$+ E\eta(u_1 - \varphi v_0^n, (x_0 + v_0^n)(1-\varphi^2)^{1/2}) \Big].$$

It is straightforward to show that we can apply the Lemma to both expectations, and this gives

$$\lim_{n\to\infty} IC(\mu_{v^n}, T^{GM}, \varphi) = \frac{-(1-\varphi^2)^{1/2}}{B} sup_{u,v} \mid \eta(u,v) \mid .$$

Then (5.10) implies (5.6), and (5.7) follows immediately from (4.5).

(ii) Let $\varepsilon > 0$, then by (4.4) we can find k_0 such that for any process (v_i) and for any $k > k_0$

$$| IC_{AO,1}(\mu_v, T^{GM}, \varphi) | \leq \frac{(1-\varphi^2)^{1/2}}{B} \, sup_{u,v} \, | \eta(u,v) | + \varepsilon \; .$$

Therefore

$$\lim_{k \to \infty} GES_{AO,k}(\mathrm{P}, \varphi, T^{GM}) \leq \frac{(1-\varphi^2)^{1/2}}{B} \, sup_{u,v} \, | \eta(u,v) | \; . \tag{5.11}$$

Let μ_c be the degenerate measure corresponding to the constant process $v_i \equiv c$. Then since by A6

$$\lim_{c \to \infty} E\eta(u_1 + c(1-\varphi, (x_0+c)(1-\varphi^2)^{1/2})) = sup_{u,v} \, \eta(u,v)$$

(5.8) implies that for any $\varepsilon > 0$ there exist k_0 such that $k \geq k_0$ implies

$$IC(\mu_c, T^{GM}, \varphi) \geq \frac{(1-\varphi^2)^{1/2}}{B} \, sup_{u,v} \eta(u,v) - \varepsilon$$

for $k \geq k_0$. Thus

$$GES_{AO,\infty}(\mathrm{P}, \varphi, T^{GM}) \geq \frac{(1-\varphi^2)^{1/2}}{B} \, sup_{u,v} \eta(u,v) \tag{5.12}$$

and so (5.11) and (5.12) imply (5.8). The proof of (5.9) is similar.

Dropping the argument P_v from GES's for simplicity, we have the following immediate corollaries to parts (i) and (ii) of Theorem 5.2.

Corollary 5.2: For patches of length 2

$$sup_{-1<\varphi<1} \frac{GES_{AO,2}(\varphi, T^{RA})}{GES_{AO,2}(\varphi, T^{GM})} \leq 1.1 \tag{5.13}$$

and

$$inf_{-1<\varphi<1} \frac{GES_{AO,2}(\varphi, T^{RA})}{GES_{AO,2}(\varphi, T^{GM})} = 0 \ . \tag{5.14}$$

Corollary 5.3: For long patches of length $k \to \infty$

$$sup_{-1<\varphi<1} \frac{GES_{AO,\infty}(\varphi, T^{RA})}{GES_{AO,\infty}(\varphi, T^{GM})} = \infty \tag{5.15}$$

$$inf_{-1<\varphi<1} \frac{GES_{AO,\infty}(\varphi, T^{RA})}{GES_{AO,\infty}(\varphi, T^{GM})} = 1 \ . \tag{5.16}$$

Comment: Corollary 5.1 shows that for independently located outliers, and η monotone in each variable, the RA-estimates are better than the GM-estimates since by (5.5) they have the same asymptotic variance, and at the same time T^{RA} has a smaller GES. Part (i) of Theorem 5.2 and Corollary 5.2 show that for AO patches of length 2 and η monotone in each variable, the GM-estimates are at most slightly better than the RA-estimates, while the RA-estimates may be much better than the GM-estimates with regard to GES as $\varphi \to 1$. On the other hand part (iii) of Theorem 5.2 and Corollary 5.3 shows that for long AO patches and η monotone in each variable, the GM-estimates are always at least as good as RA-estimates, and much better than the RA-estimates as $\varphi \to 1$. In both cases, the relative differences in GES's for GM and RA-estimates is tempered by the fact that the GES's $\to 0$ as $\varphi \to 1$.

Comment: Suppose that η is the HKW type, where $\eta(u,v) = \psi(uv)$, or of the Mallows type, where $\eta(u,v) = \psi(u)\psi(v)$. Then if ψ is a redescending type we will have

$$sup_c \; E \mid \eta(u_2 - \varphi c \, , \, u_1 + c \; (1-\varphi^2)^{1/2}) \mid \; < sup_{u,v} \mid \eta(u,v) \mid \tag{5.17}$$

and

$$sup_c \; E \mid \eta(u_1 - cu_2 - \varphi c \, , \, u_2 + c \,) \mid \; < sup_{u,v} \mid \eta(u,v) \mid . \tag{5.18}$$

Then in view of Theorem 5.1 and Corollary 5.1 we may expect that redescending ψ functions will lead to smaller GES's, and hence smaller biases, than monotone ψ functions having the same supremum in absolute value. In fact, our computations below show that a bisquare psi-function $\psi_{BS,a}$, with a *larger* supremum than a Huber psi-function $\psi_{H,a}$, can yield the smaller *GES* among the two choices.

GES Computations

In Table I we give GES's of GM and RA-estimates based on Mallows type η-functions at the AR(1) model with AO for both independent and patchy outliers. Two psi-functions are used, namely the Tukey redescending bisquare function

$$\psi_{BS,a}(u) = \begin{cases} u\left[1 - \left[u/a\right]^2\right]^2, & |u| \leq a \\ 0 & |u| > a \end{cases}$$

and the Huber function

$$\psi_{H,a}(u) = min(a, \max(u, -a)) .$$

The estimates are matched by the choice of tuning constants a to have the same asymptotic efficiencies at the Gaussian model. This means that $sup_r \mid \psi_{BS,a_1}(r) \mid \; > sup_r \mid \psi_{H,a_2}(r) \mid$, and correspondingly, $sup_{u,v} \mid \eta_{BS,a_1}(u,v) \mid \; > sup_{u,v} \mid \eta_{H,a_2}(u,v) \mid$. Yet, as Table I shows, the GES's for the bisquare psi-function are often smaller than those for the Huber psi-function.

The *GES*'s for the HKW estimates are smaller than those of the Mallows estimates, but the differences are only slight.

Table I. AR(1) GES's.

	$\varphi=.5$		$\varphi=.9$	
Estimator	$k=1$	$k=20$	$k=1$	$k=20$
LS	∞	∞	∞	∞
GM-H	-2.8	2.6	-1.4	1.3
RA-H	-2.5	4.3	-.6	3.3
GM-BS	-1.9	1.6	-1.8	.4
RA-BS	-1.5	3.1	-1.5	2.5

REFERENCES

Basawa, I.V (1984). "*Neyman-LeCam tests based on estimating functions,*" Proc. Neyman Kiefer Conference, June 1983, University of California, Berkeley (to appear).

Bustos, O., (1982). "General M-estimates for contaminated *pth*-order autoregressive processes: consistency and asymptotic normality", *Z. Wahrsch.*, **59,** 491-504.

Bustos, O., Fraiman, R. and Yohai, V. (1984). "Asymptotic behaviour of the estimates based on residual autocovariances for ARMA models." In *Robust and*

Nonlinear Time Series, edited by Franke, J., Hardle, W. and Martin, R.D. Springer-Verlag, New York.

Bustos, O. and Yohai, V.J. (1983). "Robust estimates for ARMA models," submitted to JASA.

Chernick, M.R., Downing, D.J. and Pike, D.H. (1982). "Detecting outliers in time series data", *Journal Amer. Stat. Assoc.*, **77**, 743-747.

Denby, L. and Martin, R.D. (1979). "Robust estimation of the first order autoregressive parameter", *Jour. Amer. Stat. Assoc.*, **74**, No. **365**, 140-146.

Hampel, F.R. (1974). "The influence curve and its role in robust estimation", *Jour. Amer. Stat. Assoc.*, **69**, 383-393.

Hampel, F.R. (1975). "Beyond location parameters: robust concepts and methods", *Proceedings of I.S.I. Meeting, 40th Session*, Warsaw.

Hampel, F.R. (1978). "Optimality bounding the gross-error-sensitivity and the influence of position in factor space", *1978 Proceedings of the A.S.A. Statistical Computing Section*, Amer. Stat. Assoc., Washington, D.C.

Huber, P.J. (1983). "Minimax aspects of bounded influence regression," *Jour. Amer. Statist. Assoc.*, **78**, 66-72.

James, B.R. and James, K.L. (1982). "On the influence curve for quantal bioassay," MS #Z7217 I.M.P.A., Rio de Janeiro, Brazil.

Kelly, G.E. (1984). "The influence function in the errors in variables problem," *Annals of Statistics* **12**, 87-100.

Krasker, W.S. and Welsch, R.E. (1982). "Efficient bounded-influence regression estimation", *Jour. Amer. Stat. Assoc.*, **77**, No. **379**, 595-604.

Kuensch, H. (1983). "Infinitesimal robustness for autoregressive processes." Report No. 83, ETH, Zurich.

Lambert, D. (1981). "Influence functions for testing," *Journal Amer. Statist. Assoc.*, **76**, 649-657.

Mallows, C.L. (1976). "On some topics in robustness", Bell Labs. Tech. Memo, Murray Hill, New Jersey. (Talks given at NBER Workshop on Robust Regression, Cambridge, MA, May 1973, and at ASA-IMS Regional Meeting, Rochester, N.Y., May 21-23, 1975.)

Martin, R.D. (1980). "Robust estimation of autoregressive models" (with discussion), in *Directions in Time Series*, edited by Brillinger, et al., Instit. of Math. Statistics Publication.

Martin, R.D. and Jong, J. (1977). "Asymptotic properties of robust generalized M-estimates for the first-order autoregressive parameter," Bell Labs, Tech. Memo, Murray Hill, New Jersey.

Martin, R.D. and Yohai, V.J. (1984). "Influence curves for time series," Tech. Report No. 51, Department of Statistics, University of Washington, Seattle.

Ronchetti, E. (1982). "Robust testing in linear models: the infinitesimal approach," Ph.D. dissertation, ETH, Zurich.

Samuels, S.J. (1978). "Robustness of survival estimators," Ph.D. dissertation, Dept. of Biostatistics, University of Washington.

SOME ASPECTS OF QUALITATIVE ROBUSTNESS IN TIME SERIES

P. Papantoni-Kazakos
The University of Connecticut
Storrs, Connecticut 06268, U.S.A.

Abstract

Qualitative robustness in time series is considered. The appropriate stochastic distance measures for its definition are presented and discussed. The breakdown point and the sensitivity of robust operations are defined accordingly. Finally, saddle-point game theoretic formalizations are presented, that are consistent with the theory of qualitative robustness, and a class of robust operations is briefly discussed.

1. Introduction

The last twenty years considerable research effort has been devoted to developing theory and methodology of robust statistical procedures for memoryless stationary processes; that is stochastic processes that generate sequences of independent and identically distributed data. Hampel (1968, 1971, 1975) developed the concepts of qualitative robustness, for the above models, while Huber (1964, 1981) used saddle-point game (or min-max) approaches, to develop and evaluate "robust" procedures for memoryless processes. Hampel's qualitative robustness basically corresponds to local performance stability for infinitesimal deviations from a nominal parametric stochastic model. Huber's min-max robustness can not be satisfied, unless qualitative robustness exists. Thus, the satisfaction of qualitative robustness can be considered as a necessary criterion, for "robustness" within some contamination class of stochastic processes. Appropriate performance criteria for the quantitative analysis of "robust" operations are the breakdown points, the influence curves and the sensitivities and efficiencies at selected stochastic processes (or distributions). Those criteria provide the means for comparative evaluation among

---

This work was supported by the U.S. Air Force Office of Scientific Research under grant AFOSR-78-3695.

different robust operations, and their formulation in the memoryless process set up can be found in the works by Hampel (1968, 1971, 1975), by Maronna and Yohai (1981), and by Beran (1977, 1978).

The developed robustness theory and methodologies, for memoryless processes set ups, do not apply to models of stochastic processes with memory; that is, stochastic processes that generate sequences of dependent data. Then, different models of contamination are needed, and different performance criteria are relevant. The first effort for the formulation and analysis of qualitative robustness, for stochastic stationary processes with memory, can be found in the paper by Papantoni-Kazakos and Gray (1979). This effort was followed by the works of Cox (1978), Boente, Fraiman, and Yohai (1982), and Papantoni-Kazakos (1981b, 1983). In spite of those efforts, a complete theory for robust operations in time series is still lacking. The appropriate robust models need to be analyzed carefully, and appropriate performance criteria such as breakdown points, sensitivities, efficiencies, and influence curves, need to be developed and evaluated.

In this paper, we will first present the principles and characteristics of qualitative robustness in time series, within a generalized framework. Such a generalized formalization provides the theory of qualitative robustness with especially attractive characteristics. As compared to previous studies, we will use different, improved, contamination and stability stochastic distances in the formalization of the theory. We will also provide definitions for the breakdown point and the sensitivity, in the time series set up. We will finally sketch an approach for the design of robust operations, such as filters, predictors, and interpolators.

2. Qualitative Robustness Principles

Using the same terminology as in Papantoni-Kazakos (1979, 1981a, 1981b, 1983), let $[\mu, X, R]$ denote some discrete-time, scalar, stationary process, whose data take values on the real line R; where μ is the measure of the process, and X is its name. Let x denote an infinite data sequence, let x^{ℓ} denote a given sequence of ℓ consecutive data, and let X_i^j, x_i^j; $j \geq i$, X^{ℓ} denote respectively the sequence $X_i,\ldots,X_j$ of random variables, the sequence $x_i,\ldots,x_j$ of observed data, and a sequence of ℓ consecutive random variables generated by the underlying process. Let $\gamma(\cdot,\cdot)$ be some metric on the real line, R. Let g_{ℓ} denote a scalar stationary, or time invariant, operation on blocks of ℓ consecutive data.

Operating on a given data sequence x^{ℓ}, the operation g_{ℓ} may either provide a single number, or it may provide a whole range of numbers that is controlled by some distribution. In the former case, g_{ℓ} is a deterministic operation. In the latter case, g_{ℓ} is a stochastic operation, and for each given sequence x^{ℓ}, it induces a measure. This measure is then a conditional random variable, and it is denoted by $g_{\ell,x^{\ell}}$. Stochastic g_{ℓ} operations appear in source encoding, and they may be useful (as we will see) in robust filtering, smoothing, interpolation, and prediction. It is well-known that, if an either deterministic or stochastic stationary operation g_{ℓ} operates sequencially on data blocks $x_j^{\ell+j-1}$; j=...,-1,0,1,.. ., that are generated by a stationary process [μ, X, R], then, the induced process is also stationary. We will denote this induced process by [$\mu o g_{\ell}$, Y, R]. The operation g_{ℓ} is called <u>zero-memory</u>, if its values depend only on the data sequence x^{ℓ}, on which g_{ℓ} operates, and not on previous or future g_{ℓ} mappings; that is, if:

$$g^n_{\ell,\, x_o^{\ell+n-2}}\,(Y_o^{n-1} \in R^n) = \prod_{j=0}^{n-1} g_{\ell,\, x_j^{\ell+j-1}}\,(Y_j \in R) \tag{1}$$

; where $g^n_{\ell,x}$ denotes the measure induced by the infinite data sequence x on the outputs from 0 to n-1.

The objectives of stationary operations g_{ℓ}, as above, may vary from parameter estimation, to filtering and smoothing, to prediction or interpolation. The length, ℓ, of the data blocks on which g_{ℓ} operates may vary from finite to infinitely long values, determining this way small-sample or asymptotic operations. Given some stationary process [μ_o, X, R], given ℓ, given some performance criterion, the operation g_{ℓ} can be selected, in general, optimally at the process [μ_o, X, R]. Then, the parametric parameter estimation, filtering, prediction, interpolation, etc. problems are present. When deviations from the nominal process [μ_o, X, R] are incorporated into the model, the robustness formulation is necessary. Qualitative robustness is a local stability property, and it is necessary for more global robustness as well. Given two stochastic distances d_1 and d_2, a stationary operation g_{ℓ} is called <u>qualitatively robust at some given process [μ_o, X, R]</u>, if given $\varepsilon > 0$, there exists $\delta > 0$, such that,

$$d_1(\mu_o,\mu) < \delta \rightarrow d_2(\mu_o o g_{\ell},\, \mu o g_{\ell}) < \varepsilon \;;\; \mu \in M_s \tag{2}$$

; where M_s the class of stationary processes.

Qualitatively speaking, an operation is qualitatively robust at some given stochastic process, if for small deviations from this process it induces stability, in some appropriate performance sense. The selection of the distances d_1 and d_2 is the crucial element in the meaningful definition of qualitative robustness, and in the subsequent constructive analysis, for the development of qualitatively robust operations. The distance d_1 must reflect data contaminations appearing in the real world, it must be meaningful for the time series set up (stationary processes with memory), and it must be as least restrictive as possible. The distance d_2 must reflect an appropriate performance criterion for the problem of interest, it must be also meaningful for the time series set up, and it must be a strong or relatively restrictive measure, to make the definition of qualitative robustness strong. In the set up of memoryless processes, Hampel (1971) selected as d_1 and d_2 the Prohorov distance between marginal statistics. Hampel's d_1 selection is meaningful in this case. It reflects contamination by occassional outliers on i.i.d. data, and it fully represents deviations from a nominal memoryless process. We believe, however, that his d_2 selection is too weak. Indeed, it does not impose the necessity for bounded estimation operations, whose good qualities have been observed in practice. On the other hand, if d_2 is the Vasershtein distance, in addition to Hampel's (1968, 1971) continuity conditions, boundness is also imposed on the robust estimators. In the set up of stationary processes, Papantoni-Kazakos and Gray (1979) proposed the selection of the rho-bar distance for d_1, and the selection of the Prohorov distance for d_2. The rho-bar distance is an appropriate distance for stationary processes with memory, and it bounds from above the Prohorov distances of all orders. However, it is too strong as a contamination criterion in qualitative robustness, and (as observed by Cox (1978)) it may result in conclusions against intuition. Also, the Prohorov distance as a d_2 selection is characterized by the same weakness, as in Hampel's model. It is important to point out, however, that the sufficient conditions for qualitative robustness provided by Papantoni-Kazakos and Gray (1978) are strong and flexible, they correspond to pointwise strong convergence, and they overcome the weakness presented by their d_1 and d_2 selections. This is also prointed out by Boente, Fraiman, and Yohai (1982), who also propose an alternative for the d_1 selection. Their proposal is based on one of the intermediate stochastic distances used by Papantoni-Kazakos and Gray for the derivation of sufficient conditions for robutness. However, as pointed out by Papantoni-Kazakos (1983), the supremum of Prohorov distances on empirical measures used by Boente, Fraiman, and Yohai may not exist. In addition, their d_1 selection is, in general, not appropriate for nonasymptotic data sequences (small-sample cases), and their d_2 selection is still the marginal Prohorov distance. We point out here that selecting the marginal Prohorov distance as d_2, in addition to not inducing boundness for robust operations, it is also meaningless in problems such as filtering, smoothing, prediction, and interpolation. Indeed, stochastic processes are then induced by the stationary operations, rather than random variables, as in parameter estimation. Then, the marginal Prohorov distance does not reflect the overall induced stochastic process. It only reflects the marginal, first-order distribution induced by this process. To overcome all the problems mentioned above, in the generalized qualitative robustness framework for time series, we propose below appropriate d_1 and d_2 choices.

Given a metric $\gamma(\cdot,\cdot)$ on the real line, given two data sequences x^m and y^m; for some positive integer m, let us define,

$$\gamma_m(x^m,y^m) = m^{-1} \sum_{i=1}^{m} \gamma(x_i,y_i) \tag{3}$$

Given some data sequence x^n, let us form a string x of data from repetitions of x^n, and let us denote by μ_{x^n} the empirical measure formed by assigning probability n^{-1} on each string $T^i x$; $i=0,1,\ldots,n-1$, where T indicates one-step shift in time. Given a stochastic process with measure μ, let μ^n denote the n-dimensional measure induced by the measure μ. Given two stationary stochastic processes with measures μ_o and μ, let us denote by $\mathcal{P}(\mu_o^n,\mu^n)$ the class of all joint stationary measures with marginals μ_o^n and μ^n. Let us then define the following Prohorov and rho-bar distances, where $\rho(\cdot,\cdot)$ is a distance on R that is continuous with respect to $\gamma(\cdot,\cdot)$, and it induces bounded values for bounded values of its arguments.

$$\bar{\rho}(\mu_o,\mu_1)=\sup_n \inf_{p^n \varepsilon \mathcal{P}(\mu_o^n,\mu^n)} \int_{R^n x R^n} \rho_n(x^n,y^n)\,dp^n(x^n,y^n) \tag{4}$$

$$\Pi_\gamma(\mu_o^n,\mu^n) \stackrel{\Delta}{=} \inf_{p^n \varepsilon \mathcal{P}(\mu_o^n,\mu^n)} \inf\{\varepsilon : p^n(x^n,y^n : \gamma_n(x^n,y^n)>\varepsilon) \le \varepsilon\} \tag{5}$$

$$\Pi_\gamma(\mu^m_{x^n},\mu^m_{y^n}) = \inf\{\alpha : \#[i : \gamma_m(x_i^{i+m-1},y_i^{i+m-1})>\alpha] \le n\alpha\} \tag{6}$$

$$\Pi_{\Pi_{\gamma,m}}(\mu_o^n,\mu^n) = \inf_{p^n \varepsilon \mathcal{P}(\mu_o^n,\mu^n)} \inf\{\varepsilon : p^n(x^n,y^n : \Pi_\gamma(\mu^m_{x^n},\mu^m_{y^n})>\varepsilon) \le \varepsilon\} \tag{7}$$

We note that the $\bar{\rho}$ distance in (4) is a supremum over Vasershtein distances on n-dimensional restrictions from the processes μ_o, μ, and it bounds from above all the Prohorov distances in (5), for all n. We also note that if γ is a metric, so is the Prohorov distance in (5). This distance is also equivalent to weak convergence. Also, the Prohorov distance in (6) basically measures the distance between two data sequences x^n and y^n, and reflects data outliers directly, and it converges to the Prohorov distance between the measures μ_o^m and μ^m, if the processes μ_o and μ are ergodic, and for $n \to \infty$. Finally, the Prohorov distance in (7) is defined via the metric in (6) on data sequences, and due to the properties of the distance in (6) it reflects weak closeness of the measures μ_o^n and μ^n, for asymptotically large n values. Our d_1 and d_2 selection, for the definition of qualitative robustness in (2), is now reflected by the following definition.

Definition 1

Given the class M_s of stationary processes, given μ_o in M_s, then,

i) Given ℓ, the operation g_ℓ is ρ-robust at μ_o in M_s, iff

Given $\varepsilon > 0$, there exists $\delta > 0$, such that:

$$\mu \varepsilon M_s, \quad \Pi_\gamma(\mu_o^\ell, \mu^\ell) < \delta \to \bar{\rho}(\mu_o \, o g_\ell, \mu \, o g_\ell) < \varepsilon.$$

ii) The sequence $\{g_\ell\}$ of operations is ρ-robust at μ_o, in M_s, iff:

Given $\varepsilon > 0$, there exist integers ℓ_o and m, and some $\delta > 0$, such that:

$$\mu \varepsilon M_s, \; \Pi_{\Pi_{\gamma,m}}(\mu_o^\ell, \mu^\ell) < \delta \to \bar{\rho}(\mu_o \, o g_\ell, \mu \, o g_\ell) < \varepsilon : \forall \ell > \ell_o$$

The d_2 and d_1 selections, represented by definition 1, are important for the time series set up. Indeed, the distance $\bar{\rho}$ is strong, it measures closeness on the total processes, and in the case of parameter estimation it reduces to the marginal Vasershtein distance. Also, as discussed by Papantoni-Kazakos (1981b, 1983), the $\bar{\rho}$ distance also represents closeness of the errors induced by the operation g_ℓ and the processes μ_o and μ. It thus represents closeness in performance, whose specific characteristics are determined by the selection of the metric (or pseudometric) $\rho(\cdot,\cdot)$. When the operation g_ℓ acts on data blocks of finite length ℓ, only the ℓ-dimensional measure of the data process is involved in its performance. Then, the Prohorov distance $\Pi_\gamma(\mu_o^\ell, \mu^\ell)$ is chosen as a measure of contamination, and it basically reflects the probability with which process data may experience contamination or occurrence of outliers, above a certain level. When the operation g_ℓ acts on asymptotically long data sequences, and since there is no uniform upper bound on the Prohorov distances $\Pi_\gamma(\mu_o^\ell, \mu^\ell)$ for nonbounded ℓ values, a more meaningful and tractable contamination measure is the Prohorov distance $\Pi_{\Pi_{\gamma,m}}(\mu_o^\ell, \mu^\ell)$. The latter reflects data contamination, since it utilizes closeness of empirical measures as a metric, and it converges (for $\ell \to \infty$) to closeness of the data processes. We also note that we avoid the problem of possibly nonexisting supremum $\sup_{\ell} \Pi_{\Pi_{\gamma,m}}(\mu_o^{\ell},\mu^{\ell})$, by requiring that for all ℓ above some ℓ_o, the distance $\Pi_{\Pi_{\gamma,m}}(\mu^{\ell},\mu^{\ell})$ be less than δ, in (2). Let us now define two conditions, (A1) and (A2), that may be satisfied by the stationary operation g_{ℓ}:

1. If ℓ given and finite, then given $\varepsilon > 0$, given
$$x^{\ell} \varepsilon R^{\ell}, \exists\ \delta = \delta(\ell, \varepsilon, x^{\ell}) > 0: \quad \gamma_{\ell}(x^{\ell},y^{\ell}) < \delta \longrightarrow \Pi_{\gamma}(g_{\ell,x^{\ell}}, g_{\ell,y^{\ell}}) < \varepsilon \tag{A1}$$
2. Given $\mu_o \;\varepsilon\; M_s$, given $\varepsilon > 0$, $\eta > 0$, $\exists$ integers ℓ_o, m, some $\delta > 0$, and for each $\ell > \ell_o$ some $\Delta^{\ell} \;\varepsilon\; R^{\ell}$ with $\mu_o^{\ell}(\Delta^{\ell}) > 1-\eta$, such that for each $x^{\ell} \;\varepsilon\; \Delta^{\ell}$ with
$$\Pi_{\gamma}(\mu^m_{x^{\ell}},\mu^m_{y^{\ell}}) < \delta \tag{A2}$$
it is implied that,
$$\Pi_{\gamma}(g_{\ell,x^{\ell}},g_{\ell,y^{\ell}}) < \varepsilon$$

We can now express a theorem whose proof is easily verified via the results in Papantoni-Kazakos (1981a, 1981b, 1983).

<u>Theorem 1</u>

Given the class M_s of stationary processes, given μ_o in M_s, then,

i) Given ℓ, let the operation g_{ℓ} satisfy condition (A1). Let $\rho(\cdot,\cdot)$ be continuous with respect to $\gamma(\cdot,\cdot)$, and let it be either bounded, or assume finite values for finite values of its arguments and let then g_{ℓ} map the real line R on a bounded subspace of R. Then, g_{ℓ} is ρ-robust at μ_o in M_s.

ii) Let the sequence $\{g_{\ell}\}$ of operations satisfy condition (A2). Let $\rho(\cdot,\cdot)$ be continuous with respect to $\gamma(\cdot,\cdot)$ and let it be either bounded, or assume finite values for finite values of its arguments and let then g_{ℓ} map the real line R on a bounded subspace of R. Then, $\{g_{\ell}\}$ is ρ-robust at μ_o in M_s.

From theorem 1, we thus conclude that the sufficient conditions derived by Papantoni-Kazakos and Gray (1979), in conjuction with boundness of the operations g_{ℓ}, also guarantee qualitative robustness, as stated by definition 1.

The above formulation of qualitative robustness creates possibilities for the definition and analysis of meaningful performance criteria, for qualitatively robust operations. Such performance criteria are the breakdown points, the sensitivity and the influence curves. The concepts of the breakdown point and the influence curves were introduced by Hampel (1971, 1974, 1978), in the memoryless processes set up. In this set up, the breakdown point in parameter estimation is defined as the largest fraction of the data that may be taken to the boundary of the observation space ($\pm\infty$, in general) without forcing the estimate to the boundary of its own space of values. For i.i.d. processes and in parameter estimation the influence curve of an estimate provides the effect of a single outlier on the estimate, as a function of the value of the outlier. The sensitivity of an estimate at some distribution, and for i.i.d. processes, is defined as the ratio of the distance between induced estimates over the distance between the data distributions that induce them, when the latter distance tends to zero. Clearly, sensitivity has characteristics parallel to derivatives of real functions, and bounded sensitivity is clearly a stronger than qualitative robustness property. Some studies of breakdown points for estimates of autoregressive parameters are provided by Martin and Zeh (1977). The influence curve for parameter estimates of processes with fixed kth-order marginals, has been studied by Martin (1980).

The definition and analysis of the breakdown points, the sensitivity, and the influence curves of qualitatively robust operations for time series is lacking. Yet, those criteria are extremely important for the evaluation of robust operations in time series. They provide measures on the resistance of such operations to various degrees of data contaminations. Furthermore, the direct extension of those criteria, as defined in the set up of i.i.d. processes, is meaningless. For example, considering just percentage of outliers, for the definition of the breakdown point in time series, disregards completely the dependence structure of the nominal as well as of possibly the contaminating processes. Furthermore, in Hampel's definition of the breakdown point, the estimate boundaries are frequently taken to $\pm\infty$. In our formalization of qualitative robustness, as presented in the preceeding discussion, such boundaries are impossible. This is due to the boundness of the rho-bar distance $\bar{\rho}$, in (4). Also, due to our selection of the contamination distance d_1, the definition of the sensitivity of a qualitatively robust operation in time series assumes here different dimensions. Finally, the definition of influence curves in the time series set up is a completely obscure subject at this point. Indeed, restricting the definition to single outliers completely disregards effects due to the dependence structure of the processes at hand. On the other hand, the placement of such outliers is a complicated issue here. One has to search for data patterns that best represent the statistical structure of the nominal process. Regarding the breakdown point and the sensitivity of qualitatively robust operations, we believe that the following definitions are meaningful and consistent with the presented theory of qualitative robustness.

Definition 2

Given the class, M_s, of stationary processes, given μ_o in M_s, given a sequence $\{g_\ell\}$ of operations that is ρ-robust at μ_o in M_s, then,

i) The breakdown point of the sequence $\{g_\ell\}$ at μ_o in M_s, is this constant α^* such that, for every $\alpha > \alpha^*$, there exist integers $m = m(\alpha)$, $\ell_o = \ell_o(\alpha)$, such that,

For $M_\ell(\alpha) = \{\mu \varepsilon M_s, \mu^\ell : \Pi_{\Pi_{\gamma,m}} (\mu_o^\ell, \mu^\ell) < \alpha\}$
the supremum

$$\sup_{\mu \varepsilon M_\ell(\alpha)} \bar{\rho}(\mu, \mu \, og_{\ell k})$$

is independent of α; $\forall \, \ell > \ell_o$.

ii) The sensitivity, $S_\rho(\mu_o, \{g_\ell\})$, of the sequence $\{g_\ell\}$ at μ_o in M_s, is defined as follows,

$$S_\rho(\mu_o, \{g_\ell\}) \triangleq \lim_{\lim_{\ell \to \infty} \Pi_{\Pi_{\gamma,m}} (\mu_o^\ell, \mu^\ell) \to 0} \; \lim_{\ell \to \infty} \frac{\bar{\rho}(\mu_o \, og_\ell, \mu \, og_\ell)}{\Pi_{\Pi_{\gamma,m}} (\mu_o^\ell, \mu^\ell)}$$

; where the integer m is as in definition 1.

We note that both the breakdown point and the sensitivity have been defined asymptotically; that is, for asymptotically long data sequences. This is so because, then, the convergence characteristics of the qualitatively robust operations are reflected by the two criteria. Also, the breakdown point provides now the highest probability with which contamination on a percentage of data blocks, that equals this probability, can be tolerated, without the error induced by the qualitatively robust operation reaching its boundary. This definition of the breakdown point reflects the total statistical structure of the process μ_o, and it allows for bounded error boundaries that are, in general, induced by bounded operations. The sensitivity also reflects now the total statistical structure of the process μ_o. In addition, the requirement for bounded sensitivity imposes now additional, to the sufficient conditions in Papantoni-Kazakos and Gray (1979), restrictions on the qualitatively robust operations. Such restrictions can be found in Papantoni-Kazakos (1983), and they may include uniformly bounded differentiability.

3. Robust Operations in Time Series

In the design of multipurpose, high-quality information processing systems, the corresponding operations must respond in a satisfactory manner, for a whole class of data-generating processes. Qualitatively speaking, this translates to graceful or smooth performance variation or degradation within the overall above class of data processes. This quality is guaranteed by a combination of qualitative robustness theory and the theory of saddle-point games in the domain of stationary stochastic processes. The satisfaction of qualitative robustness at each process in the assumed class of processes provides the necessary continuity conditions for the existence of saddle-point solutions. In addition, the construction of pay off or performance functions with convexity characteristics, and the adoption of convex classes of stochastic processes are also essential elements in the overall process. The challenge of the formulation is particularly enhanced by the involvement of stochastic continuity and convexity measures and the requirement for formulations that are representative of the real underlying problem. When classes of i.i.d. processes are considered, the saddle-point or min-max approaches of Huber (1964, 1973, 1981) are satisfactory, and they generalize easily to additional problems in the i.i.d. set up. When classes of stationary processes with memory are considered, however, straightforward generalizations of Huber's approach do not exist. That should be clear from the fact that, in the time series set up, even the concept of qualitative robustness assumes different and challenging dimensions. The only known to us attempts for game theoretic formalizations in the time series set up, that also incorporate the conditions for qualitative robustness, are offered by Papantoni-Kazakos (1981b, 1983). In those formalizations, the class of considered operations is limited, to satisfy the conditions for qualitative robustness. Then, appropriate performance criteria are determined, conditions for the existence of unique saddle-point solution are developed, and the saddle-point solutions for some specific problems are found. The mathematical formulation of the general problem is summarized by definition 2 and theorem 2 below. The proof of theorem 2 can be derived from parallel derivations in Papantoni-Kazakos (1981b, 1983).

Definition 2

Given a convex class M of stationary processes, that is also compact with respect to the metric $\bar{\gamma}$, then,

i) Given ℓ, given a class S_ℓ of operations, the operation g_ℓ^* in S_ℓ is <u>ρ-robust on $M\text{x}S_\ell$</u>, iff:

g_ℓ^* is ρ-robust at every μ in M, and there exists μ^* in M such that,

$$\bar{\rho}(\mu,\mu\, og_\ell^*) \le \bar{\rho}(\mu^*,\mu^* og_\ell^*) \le \bar{\rho}(\mu^*,\mu^* og_\ell)\,;\ \forall\ \mu\varepsilon M,\ \forall\ g_\ell\ \varepsilon\ S_\ell$$

ii) Given a class S of sequences, $\{g_\ell\}$, of operations, the sequence

$\{g_\ell^*\}$ in S is <u>ρ-robust on MxS</u>, iff:

$\{g_\ell^*\}$ is ρ-robust at every μ in M, and there exist μ^* in M and integer ℓ_o, such that,

$$\bar{\rho}(\mu,\mu\ og_\ell^*) \leq \bar{\rho}(\mu^*,\mu^* og_\ell^*) \leq \bar{\rho}(\mu^*,\mu^* og_\ell) \ ; \ \forall\ \mu\varepsilon M,\ \forall\ g_\ell \varepsilon S_\ell,\ \forall\ \ell > \ell_o$$

<u>Theorem 2</u>

Let M, S_ℓ, and S be as in definition 2. Let, in addition, S_ℓ be a convex class of operations that satisfy condition (A1), and let S be a convex class of sequences of operations that satisfy condition (A2), at every μ in M. Let also $\rho(\cdot,\cdot)$ be continuous with respect to $\gamma(\cdot,\cdot)$, and let it be either bounded, or assuming finite values for finite values of its arguments while then the operations in S_ℓ and S also map the real line R, on a bounded subspace of R. Let the following infima exist,

$$I_{\rho,\ell}(\mu) \stackrel{\Delta}{=} \inf_{g_\ell \,\varepsilon\, S_\ell} \bar{\rho}(\mu,\mu\ og_\ell)$$

$$I_{\rho,\ell_o,\ell}(\mu) \stackrel{\Delta}{=} \inf_{\{g_\ell\} \,\varepsilon\, S} \bar{\rho}(\mu,\mu\ og_\ell);\ \forall\ \ell > \ell_o,\ \text{for some } \ell_o.$$

Then, a unique ρ-robust on MxS_ℓ operation, g_ℓ^*, and a unique ρ-robust on MxS sequence of operations, $\{g_\ell^*\}$, exist. They are such that,

$$I_{\rho,\ell}(\mu^*) \stackrel{\Delta}{=} \sup_{\mu\varepsilon M} I_{\rho,\ell}(\mu) = \bar{\rho}(\mu^*,\mu^*\ og_\ell^*) = \inf_{g_\ell \,\varepsilon\, S_\ell} \bar{\rho}(\mu^*,\mu^*\ og_\ell)$$

$$I_{\rho,\ell_o,\ell}(\mu^*) \stackrel{\Delta}{=} \sup_{\mu\varepsilon M} I_{\rho,\ell_o,\ell}(\mu) = \bar{\rho}(\mu^*,\mu^* og_\ell^*) = \inf_{\{g_\ell\} \,\varepsilon\, S} \bar{\rho}(\mu^*,\mu^* og_\ell);\ \forall\ \ell > \ell_o$$

There exist various alternatives regarding the choice of the classes S_ℓ and S, in theorem 2. An interesting choice evolves from the class H_ℓ of operations, defined below; where $\text{sgn}(x) = \begin{cases} 1 \ ; & x \geq 0 \\ 0 \ ; & x < 0 \end{cases}$

$$H_\ell = \left\{ g_{\ell,x^\ell} : g_{\ell,x_{i+1}^{i+\ell}} = \sum_{k=1}^{\ell+1-m} a_{\ell k}\ \text{sgn}(h_m(x_{i+k}^{i+k+m-1})+v_k) \ ; \ \ell \geq m \right.$$

; where $h_m(x^m)$ a continuous and strictly monotone, with respect to the metric $m^{-1} \sum_{i=1}^{m} \gamma(x_i,y_i)$, scalar and real function, $\{v_k\}$ a sequence of i.i.d. random variables with distribution function $F_v(u)$ that is continuous and such that $F_v(-u)=1-F_v(u);\forall u\varepsilon R$, and $\{a_{\ell k}\}$ some set of real and bounded coefficients $\Big\}$

We note that the class H_ℓ includes operations $h_m(x_{i+k}^{i+k+m-1})$ of sliding (be one) data blocks. As shown by Papantoni-Kazakos (1983), such blocks may induce reduced mean squared errors at all processes in the class M. It can be easily seen that the class H_ℓ satisfies condition (A1), thus it is an appropriate S_ℓ choice. For $\ell \to \infty$, it is sufficient that the set $\{a_{\ell k}\}$ also satisfies the conditions (A3) below, for the $\lim_{\ell \to \infty} H_\ell$ to be an appropriate S choice. The proof of the latter can be found in Papantoni-Kazakos (1981b).

$$a_{\ell j} > 0 \; ; \; \forall_j \; , \; \forall \ell$$

$$\sum_{j=1}^{\ell+1-m} a_{\ell j} = 1 \; : \; \forall \ell \tag{A3}$$

$$\lim_{\ell \to \infty} \sum_{j=1}^{\ell+1-m} a_{\ell j}^2 = 0$$

If condition (A3) is satisfied and if the function $h_m(x^m)$, in class H_ℓ, is also differentiable with respect to the metric $m^{-1} \sum_{i=1}^{m} \gamma(x_i, y_i)$, and has uniformly bounded derivative, then the sensitivity $S_\rho(\mu, \{g_\ell\})$ is bounded at all stationary processes μ (the proof of this can be found in Papantoni-Kazakos (1983)). Such a $h_m(x^m)$ function is the mean average $m^{-1} \sum_i x_i$. In general, the class H_ℓ of operations is convex, with free parameters the set $\{a_{\ell k}\}$ and the function $h_m(x^m)$. The saddle-point game in theorem 2, can be then formulated and solved for various classes, M, of stochastic processes. Some such solutions can be found in Papantoni-Kazakos (1981b, 1983). We point out that the class H_ℓ of operations is applicable to various robust time series problems, such as filtering, smoothing, prediction and interpolation. The specific $\{a_{\ell k}\}$ and $h_m(x^m)$ choices depend on the particular problem and its performance characteristics, as well as on the class, M, of processes considered.

4. References

Beran, R. (1977), "Robust Location Estimates," Annals of Statistics, 5, No. 3, 431-444.

Beran, R. (1978), "An Efficient and Robust Adaptive Estimator of Location," Annals of Statistics, 6, No. 2, 292-313.

Boente, G., Fraiman, R., and Yohai, V., (1982), "Qualitative Robustness for General Stochastic Processes," Dept. of Statistics, Univ. of Washington Seattle, Tech. Report No. 26, Oct.

Cox, Dennis (1978), "Metrics on Stochastic Processes and Qualitative Robustness," Dept. of Statistics, Univ. of Washington Seattle, Tech. Report No. 23.

Hampel, F. R. (1968), "Contributions to the Theory of Robust Estimation," Ph.D. dissertation, Dept. of Statist., University of California, Berkeley.

Hampel, F. R. (1971), "A General Qualitative Definition of Robustness," Annals Math. Stat., 42, 1887-1895.

Hampel, F. R. (1974), "The Influence Curve and its Role in Robust Estimation," J. Amer. Statist. Assoc., 69, 383-394.

Hampel, F. R. (1975), "Beyond Location Parameters: Robust Concepts and Methods," Proceedings of I.S.I. Meeting, 40th Session, Warsaw.

Hampel, F. R. (1978), "Optimally Bounding the Gross-Error-Sensitivity and the Influence of Position in Factor Space," 1978 Proceedings of the A.S.A. Statistical Computing Section, Amer. Stat. Assoc., Washington, DC.

Huber, P. J. (1964), "Robust Estimation of a Location Parameter," Annals Math. Stat., 35, 73-101.

Huber, P. J. (1973), "Robust Regression: Asymptotics, Conjectures and Monte Carlo," Annals of Statist., 1, No. 5, 799-821.

Huber, P. J. (1981), Robust Statistics, Wiley.

Maronna, R. A., and Yohai, V. J., (1981), "Asymptotic Behaviour of General M-Estimates for Regression and Scale with Random Carriers," Z, Wahrscheinlickeitstheorie und verw. Gebiete, 58, 7-20.

Martin, R. D. (1980), "Robust Estimation of Autoregressive Models," in D. R. Brillinger, and G. C. Tias, eds., Direction in Time Series, Haywood CA., Institute of Mathematical Statistics Publication.

Martin, R. D., and Zeh, J. E. (1977), "Determining the Character of Time Series Outliers," Proceedings of the American Statistical Assoc.

Papantoni-Kazakos, P., and Gray, R. M. (1979), "Robustness of Estimators on Stationary Observations," Ann. Probability, 7, 989-1002.

Papantoni-Kazakos, P. (1981a), "Sliding Block Encoders that are Rho-Bar Continuous Functions of their Input," IEEE Trans. Inf. Theory, IT-27, 372-376.

Papantoni-Kazakos, P. (1981b), "A Game Theoretic Approach to Robust Filtering," EECS Dept., Univ. of Connecticut, Tech. Report TR-81-12, Oct.

Papantoni-Kazakos, P. (1983), "Qualitative Robustness in Time Series," EECS Dept., Univ. of Connecticut, Tech. Report UCT/DEECS/TR-83-15, Nov.

TIGHTNESS OF THE SEQUENCE OF EMPIRIC C.D.F. PROCESSES DEFINED FROM REGRESSION FRACTILES

Stephen Portnoy
Division of Statistics
1409 West Green Street
University of Illinois
Urbana, Illinois 61801
U.S.A.

Abstract

The sequence of "regression fractile" estimates of the error distribution in linear models (developed by Bassett and Koenker) is shown to be tight in appropriate metric spaces of distribution functions. This result yields weak convergence to the transformed "Brownian Bridge" process in these spaces, and has obvious application to goodness-of-fit tests for the error distribution. It also suggests that using this estimate may be better than using the empirical distribution of residuals when applying "bootstrap" methods.

1. Introduction

Bassett and Koenker [1982] described how to use regression fractiles to obtain an estimate of the error distribution in linear models which has the same finite dimensional asymptotic distributions as the usual empiric c.d.f. based on a single sample. Later (B-K [1983]), they proved a Glivenko-Cantelli result for this estimate, which from now on will be called the B-K empiric c.d.f. The basic result here is that the sequence of B-K empiric c.d.f. processes $[\sqrt{n}\ (\hat{F}_n - F)]$ is tight (in appropriate metric spaces); and, thus, that this sequence converges to a transformed Brownian Bridge.

This result has obvious application to goodness of fit tests for the error distribution (e.g., Kolmogorov-Smirnov, Cramér-von Mises, etc.). Similar results for the regression fractile processes would also provide an asymptotic normality theory for a wide class of L-type estimators of the regression parameters. However, unlike the case of a single sample, many commonly suggested estimators (e.g., the trimmed least squares estimators of Ruppert and Carroll [1980] or related estimators by Jurečková and Sen [1983]) depend on more than just the regression fractiles (and, thus, would require more complicated settings). Lastly, the B-K empiric c.d.f. has application to the use of Efron's "bootstrap" method of approximating the distribution of arbitrary functions of the regression estimators. Freedman [1981], among others, has suggested using the empiric distribution of the residuals to obtain a "bootstrap" distribution for an estimator of a parametric function of interest. However, the empiric distribution of the residuals will generally <u>not</u> have the same distribution as the sample c.d.f. (a property which, as noted above, will hold for the B-K empiric c.d.f.). As Beran [1982] has shown, the main asymptotic advantage of the bootstrap is the optimality property that it automatically picks up the $1/\sqrt{n}$ correction term to the asymptotic expansion of the distribution of the estimator. He also shows (see remark (c) on p. 218) that this property depends on the empiric c.d.f. having the "right" asymptotic distribution; and, thus, should hold when using the B-K empiric c.d.f. but not when using residuals (at least in non-symmetric cases where the $1/\sqrt{n}$ term does not vanish). A Monte-Carlo simulation experiment is presently being carried out. Although the results of this study will be reported later, preliminary results suggest that there is actually very little difference between the bootstrap distributions (for the distribution of an estimate of a regression parameter) based on these two methods.

The formal framework is as follows: consider a general linear model

(1.1) $$Y = X\beta + R$$

where $Y = (Y_1,\ldots,Y_n)'$, X is a n by $(p+1)$ design matrix with first column all ones and rows $(1\ x_i')$ where $x_i \in R^p$ and

(1.2) $$\sum_{i=1}^{n} x_i = 0,$$

and $(R_1,R_2,\ldots,R_n)$ are are i.i.d. according to c.d.f. F. Assume that F has a bounded density $f(x)$ satisfying $f(x) > 0$ for all x. If necessary, write $w_i' = (1\ x_i')$ for the ith row of X; and also partition the vector $\beta \in R^{p+1}$:

(1.3) $$\beta = (\beta_1 \gamma) \quad \text{with} \quad \gamma \in R^p.$$

Note that F is not identifiable. The empiric c.d.f.s defined below actually estimate the identifiable distribution of $R + \beta_1$. We will assume here that the true value $\beta_1 = 0$, but the general case simply replaces $F(t)$ by $F(t - \beta_1)$. A basic condition on the design matric is the following:

(1.4) $$\frac{1}{n} X'X \to \Sigma \quad (\text{in } R^{p+1})$$

where Σ is a positive definite matrix.

Now for each $\theta \in [0,1]$ define $\hat\beta(\theta) = (\hat\beta_1(\theta),\hat\gamma(\theta))$ to be any vector in R^{p+1} which minimizes

(1.5) $$\sum_{i=1}^{n} \{\theta(Y_i - w_i'\beta)^+ + (1-\theta)(Y_i - w_i'\beta)^-\}$$
$$= \sum_{i=1}^{n} \{\theta I(Y_i > w_i'\beta) + (1-\theta) I(Y_i < w_i'\beta)\} |Y_i - w_i'\beta|$$

where $I(\cdot)$ indicates the indicator function: $I(\cdot)$ is one if its argument is a true statement and equals zero otherwise. K-B [1978] shows that there are breakpoints, $\{\theta_0 = 0,\theta_1,\theta_2,\ldots,\theta_m = 1\}$ such that $\hat\beta(\theta)$ is unique if $\theta_i < \theta < \theta_{i+1}$ and such that $\hat\beta_1(\theta)$ is increasing (almost surely). Thus, a "quantile process" can be obtained by (perhaps) redefining $\hat\beta_1(\theta)$ by continuity from the right at breakpoints. Inverting this process yields the B-K empiric c.d.f.:

(1.6) $$\hat\theta(\beta_1) = \sup\{\theta \in [0,1] : \hat\beta_1(\theta) \le \beta_1\}.$$

(The process $\hat{\theta}$ may sometimes be subscripted by n to emphasize the dependence on the sample size.) Thus, $\hat{\theta}(\beta_1)$ is a non-decreasing, right continuous (step) function on the real line; and, hence, is an element of the space, $\mathcal{D}_0$, of all right continuous functions with left limits. If $\mathcal{D}_0$ is metrized by the topology of "Skorohod" convergence on compact sets, then the processes $\sqrt{n}(\hat{\theta}(\beta_1) - F(\beta_1))$ form a sequence of elements in a metric space and standard tightness definitions can be applied (see Billingsley [1968]). A continuous version of the B-K empiric c.d.f. can also be constructed: consider the graph connecting the points $(\hat{\beta}(\theta_i), \theta_i)$ $(i = 0,1,2,\ldots,m)$ (where $\{\theta_i\}$ are the above breakpoints and $\hat{\beta}(\theta_i)$ are the right limits). Define $\hat{\theta}^*(\beta_1)$ to be this continuous process: then $\sqrt{n}(\hat{\theta}^*(\beta_1) - F(\beta_1))$ form a sequence of elements in $\mathcal{C}_0$, the space of continuous functions metrized by uniform convergence on compact sets. Theorem 2.1 provides tightness for $\{\sqrt{n}(\hat{\theta}_n(\beta_1) - F(\beta_1)$ in $\mathcal{D}_0$ and for $\{\sqrt{n}(\hat{\theta}_n^*(\beta_1) - F(\beta_1))\}$ in $\mathcal{C}_0$. Note that both $\hat{\theta}^*(\beta_1)$ and $\hat{\theta}(\beta_1)$ are zero for $\beta_1 < \hat{\beta}(0)$ and one for $\beta_1 \geq \hat{\beta}(1)$.

Lastly, one more piece of notation is needed: for each subset h of size $p + 1$ from $\{1,2,\ldots,n\}$ let X_h denote the $(p+1)$ by $(p+1)$ matrix with rows $\{w_j' : j \in h\}$ and let Y_h denote the vector in R^{p+1} with coordinates $\{Y_j : j \in h\}$. Then if X_h is non-singular define $\beta_h = X_h^{-1} Y_h$. As K-B [1978] shows, for each θ, $\hat{\beta}(\theta) = \beta_h$ for some subset h of size $(p+1)$ (almost surely) with X_h non-singular.

2. Tightness of the B-K empirical c.d.f. processes

In K-B [1978] it is shown that the regression fractiles must satisfy $\hat{\beta}(\theta) = \beta_h$ if and only if

$$(2.1) \qquad (\theta - 1)\underline{e} \leq \sum_{i=1}^{n} \{I(Y_i < \beta_1 + x_i'\gamma) - \theta I(Y_i \neq \beta_1 + x_i'\gamma)\} w_i X_h^{-1} \leq \theta \underline{e},$$

where h denotes the subset of $(p+1)$ indices fit exactly by $\beta_h = (\beta_1, \gamma)$ as described in Section 1, and $\underline{e} = (1\ 1\ \ldots\ 1)'$. Let

(2.2) $$F_n(\beta) = \frac{1}{n}\sum_{i=1}^{n} I(Y_i < \beta_1 + x_i'\gamma)$$

(2.3) $$H_n(\beta) = \frac{1}{n}\sum_{i=1}^{n} I(Y_i < \beta_1 + x_i'\gamma)x_i .$$

Then (2.1) provides a relationship between $F_n(\beta)$ and the corresponding $\hat{\theta}(\beta_1)$ giving the empiric c.d.f. (see Proposition 2.1). Tightness of the empiric process then follows from tightness of $F_n(\beta)$; but since $F_n(\beta)$ is not exactly a sample c.d.f., some work is needed. Lemma 2.2 proves that $F_n(\beta)$ yields a tight process using Skorohod embedding ideas. However, this argument requires that $\hat{\gamma}(\hat{\theta})$ corresponding to $\hat{\theta}(\beta_1)$ be small in the sense that $\max_i |x_i'\hat{\gamma}| \to 0$ as $n \to \infty$ (uniformly in β_1). (Note: since $\max_i \|x_i\|$ will generally grow with n -- e.g. if $\{x_i\}$ are a sample from some distribution in R^p -- this is stronger than $\|\hat{\gamma}\| \to 0$.) This convergence result is provided by Lemma 2.1; but the only reasonable proof I have requires the rather artificial condition (2.10) which will be discussed in Section 3.

PROPOSITION 2.1: *Assume the condition*

(2.4) $$\max_h \operatorname{tr}(X_h'X_h)^{\frac{1}{2}} = o(\sqrt{n})$$

(2.5) $$\max_h \|\sum_{i\in h} x_i\| = o(\sqrt{n})$$

where h *ranges over the collection of all subsets of size* $(p+1)$ *from* $\{1,2,\ldots,n\}$. *Then*

(2.6) $$F_n(\beta_1, \hat{\gamma}(\hat{\theta}(\beta_1))) - \hat{\theta}(\beta_1) = o(\frac{1}{\sqrt{n}}) \quad \textit{uniformly in } \beta_1,$$

and

(2.7) $$\|H_n(\hat{\beta}(\theta))\| = o(\frac{1}{\sqrt{n}}) \quad \textit{uniformly in } \theta.$$

Proof. From (2.1) it follows directly that $\hat{\beta}(\theta) = \beta_h$ if and only if

$$\text{(2.8)} \qquad \begin{pmatrix} nF_n(\beta_h) - (n-p)\theta \\ \\ \sum_{i=1}^{n} I(Y_i \le \beta_1 + x_i'\gamma)x_i - \theta \sum_{i\notin h} x_i \end{pmatrix} \in X_h A$$

where $A = \{u \in R^{p+1} : \theta - 1 \le u_j \le \theta \text{ for } j = 1,\ldots,p\}$. Since A is contained in the sphere of radius $\sqrt{p+1}$ (uniformly in θ), and the maximum eigenvalue $\lambda_{max}(X_h) \le \le tr(X_h'X_h)^{1/2}$, condition (2.4) immediately implies that each coordinate of any vector in $X_h A$ is $o(\sqrt{n})$ (uniformly in θ as $n \to \infty$). Dividing the first coordinate of (2.8) by n yields $F_n(\hat{\beta}(\theta)) - \theta = o(1/\sqrt{n})$, and (2.6) follows from the definition of $\hat{\theta}(\beta_1)$ (and $\hat{\gamma}(\theta)$). In a similar manner (since $\Sigma x_i = 0$),

$$\| H_n(\hat{\beta}(\theta)) - (\theta/n) \sum_{i\notin h} x_i \| = \| H_n(\hat{\beta}(\theta)) + (\theta/n) \sum_{i\in h} x_i \| = o(1/\sqrt{n}).$$

Thus, (2.7) follows from condition (2.5). □

LEMMA 2.1. *Assume conditions* (2.4) *and* (2.5), *and also assume the following conditions (which will be discussed in Section 3):*

$$\text{(2.9)} \qquad \max_i \|x_i\| = o(n/\log n)^{1/2}$$

(2.10) *for any constant* a *(sufficiently large) there is* $\eta > 0$ *such that for all* $\beta_1 \in [-a,a]$ *and all* γ

$$\gamma' M(\beta_1,\gamma) \ge \eta\, n s(\gamma)$$

where

$$\text{(2.11)} \qquad M(\beta) = \sum_{i=1}^{n} x_i F(\beta_1 + x_i'\gamma)$$

and

$$\text{(2.12)} \qquad s(\gamma) = \min\{\|\gamma\|^2, \|\gamma\|\}.$$

Then (under the hypotheses and definitions of Section 1) for any $a > 0$

$$\sup_{|\beta_1| \le a} \|\hat{\gamma}(\hat{\theta}(\beta_1))\| = O_p(\log n/n)^{\frac{1}{2}} \tag{2.13}$$

and

$$\sup_{|\beta_1| \le a} \max_i |x_i'\hat{\gamma}(\hat{\theta}(\beta_1))| = o_p(1).$$

Proof. From Propositions (2.1) and (2.2) (below)

$$\sup_\theta \frac{1}{n}\|M(\hat{\beta}(\theta))\| = O_p(\log n/n)^{\frac{1}{2}}.$$

Thus, uniformly in θ,

$$(1/n)|\hat{\gamma}'(\theta)M(\hat{\beta}(\theta))| \le \|\hat{\gamma}(\theta)\| O_p(\log n/n)^{\frac{1}{2}}.$$

Insert $\theta = \hat{\theta}(\beta_1)$ and apply condition (2.10). Note that $\hat{\beta}_1(\hat{\theta}(\beta_1)) = \beta_1$. Hence, if $|\beta_1| \le a$ and $\hat{\gamma}$ denotes $\hat{\gamma}(\hat{\theta}(\beta_1))$, (2.10) implies

$$\|\hat{\gamma}\| O_p(\log n/n)^{\frac{1}{2}} \ge \eta \min\{\|\hat{\gamma}\|^2, \|\hat{\gamma}\|\} \quad \text{(in probability)}. \tag{2.14}$$

However, if $\lim\sup\|\hat{\gamma}\|$ were not less than one in probability, (2.14) would be contradicted. Hence, $\min(\|\hat{\gamma}\|^2, \|\hat{\gamma}\|) = \|\hat{\gamma}\|^2$ (in probability), and (dividing by $\|\hat{\gamma}\|$ in (2.14)), the first part of (2.13) follows. The second part now follows immediately from condition (2.9). □

PROPOSITION 2.2. *Under the hypotheses of Lemma 2.1,*

$$\sup_\theta \|H_n(\hat{\beta}(\theta)) - \frac{1}{n} M(\hat{\beta}(\theta))\| = O_p(\log n/n)^{\frac{1}{2}}. \tag{2.15}$$

Proof. From Section 1, $\hat{\beta}(\theta) = \beta_h$ for some subset h. So fix h, let $\beta = \beta_h$, and note that by the proof of Proposition 2.1, condition (2.5) implies

$$H_n(\beta) = \frac{1}{n} \sum_{i \notin h} x_i I(Y_i < w_i'\beta) + o(1/\sqrt{n}). \tag{2.16}$$

Now let

(2.17) $\quad Z_i = x_i I(Y_i < w_i'\beta) - x_i\, F(\beta), \qquad S = \sum_{i \notin h} Z_i.$

Then from (2.16) (and (2.5)) $\|H_n(\beta) - \frac{1}{n} M(\beta) - \frac{1}{n} S\| = o(1/\sqrt{n})$; and (conditional on β_h), for $i \notin h$, $\{Z_i\}$ is independent of β_h and is a sequence of independent, mean zero random vectors.

Now if $W \sim$ binomial $(1,p)$, straightforward expansion shows that there is $\varepsilon > 0$ such that

$$E \exp\{uW\} \leq \exp\{up + u^2\} \quad \text{for} \quad |u| \leq \varepsilon \text{ (uniformly in } p).$$

Thus, the $j\underline{th}$ coordinate of Z_i satisfies

(2.18) $$E \exp\{tZ_i\} \leq \exp\{(x_{ij}t)^2\} \quad \text{for} \quad |x_{ij}t| \leq \varepsilon.$$

Therefore, by the Markov inequality, for $j = 1,\ldots,p$,

(2.19) $$P\{|S_j| \geq c_n\} \leq e^{-tc_n} Ee^{tS_j} + e^{-tc_n} Ee^{-tS_j} \qquad (t > 0)$$

$$\leq 2 \exp\{-tc_n + \sum_{i \notin h} x_{ij}^2 t^2\}.$$

From (1.4) and condition (2.9) there is $b > 0$ such that $\sum_{i \notin h} x_{ij}^2 \geq bn$ $(j = 1,\ldots,p)$. Thus, from (2.19)

$$P\{|S_j| \geq c_n\} \leq 2 \exp\{-tc_n + bnt^2\}$$

$$\leq 2 \exp\left\{-\frac{B^2 \log n}{4b}\right\}$$

where $c_n = B\sqrt{n \log n}$ and we let $t = (B/2b)(\log n/n)^{1/2}$: note that from (2.9) $|x_{ij}| = o(n/\log n)^{1/2}$ so that $|x_{ij}t| \leq \varepsilon$ (for all (i,j) for n large enough). Therefore (since $\|u\| \leq \sqrt{p} \max |u_j|$ for any $u \in R^p$), for any B' there is B with

(2.20) $$P\{\|H_n(\beta) - \frac{1}{n} M(\beta)\| \geq B'(\log n/n)^{1/2}\}$$

$$\leq P\{\|S\| \geq B^*(n \log n)^{1/2}\} \leq \exp\left\{-\frac{B^2 \log n}{4b}\right\}.$$

Lastly, the number of sets, h, of size $(p+1)$ is $\binom{n}{p+1} \leq n^{p+1}$; and, hence

$$\leq \; P\{\sup_\theta \| H_n(\hat{\beta}(\theta)) - \frac{1}{n} M(\hat{\beta}(\theta))\| \geq B'(\log n/n)^{1/2}\}$$

$$\leq P\{\sup_h \| H_n(\beta_h) - \frac{1}{n} M(\beta_h)\| \geq B'(\log n/n)^{1/2}\}$$

$$\leq \; n^{p+1} \exp\left\{- \frac{B^2 \log n}{4b}\right\} \to 0$$

if B is large enough so that $B^2/(4b) > p + 1$. □

LEMMA 2.2. *Suppose Lemma 2.1 holds. Define*

$$U(\delta,a) = \{(\beta,\tilde{\beta}) : |\beta_1| \leq a - \delta, |\beta_1 - \tilde{\beta}_1| \leq \delta, \max_i |x_i'\gamma| \leq \delta, \max_i |x_i'\tilde{\gamma}| \leq \delta\} \tag{2.20}$$

$$\Delta F_n = F_n(\beta) - EF_n(\beta) - [F_n(\tilde{\beta}) - EF_n(\tilde{\beta})]. \tag{2.21}$$

Then under the hypotheses of Section 1, for any ε *and* $a > 0$ *there is* $\delta > 0$ *and* n_0 *such that for* $n \geq n_0$

$$P\{\sup_{U(\delta,a)} |\sqrt{n}\, \Delta F_n| \geq \varepsilon\} \leq \varepsilon.$$

Proof. Using Skorohod embedding, let $X(t), X_1(t), X_2(t), \ldots$ be independent Brownian motions and for each $i = 1,2,\ldots$ define

$$\tau_i(\beta) = \text{time for } X_i(t) \text{ to exit } (-p_i(\beta), 1 - p_i(\beta)) \tag{2.22}$$

where

$$p_i(\beta) = P\{Y_i \leq \beta_1 + x_i'\gamma\} = F(\beta_1 + x_i'\gamma). \tag{2.23}$$

Then

$$\sqrt{n}\, (F_n(\beta) - EF_n(\beta)) \sim \frac{1}{\sqrt{n}} X\left(\sum_{i=1}^{n} \tau_i(\beta)\right). \tag{2.24}$$

Now for each $i = 1,2,\ldots$ define (assuming $p_i(\beta) > p_i(\tilde{\beta})$)

$$V_i(\beta,\tilde{\beta}) = \text{time for } X_i(t) \text{ to exit } (-p_i(\tilde{\beta}), 1 - p_i(\tilde{\beta})) \text{ given } X_i(0) = 1 - p_i(\beta)$$

$$W_i(\zeta) = \text{time for } X_i(t) \text{ to exit } (-p_i(\tilde{\beta}), 1 - p_i(\beta) + \zeta)$$

$$\text{given } X_i(0) = 1 - p_i(\beta)$$

$$+ \text{ time for } X_i(t) \text{ to exit } (-p_i(\tilde{\beta}) - \zeta, 1 - p_i(\beta))$$

$$\text{given } X_i(0) = -p_i(\tilde{\beta})$$

$$W_i^*(\zeta) = \text{twice the time for } X_i(t) \text{ to exit } (-2,\zeta)$$

$$\text{given } X_i(0) = 0.$$

Note that since F is continuous, for $(\beta,\tilde{\beta}) \in U(\delta,a)$ there exists $\zeta = \zeta(\delta)$ such that $|p_i(\beta) - p_i(\tilde{\beta}))| \leq \zeta$ and, in fact, $\zeta(\delta) \to 0$ as $\delta \to 0$. By the above definitions (and continuity of Brownian motion), it is not hard to see that

$$|\tau_i(\beta) - \tau_i(\tilde{\beta})| \leq V_i(\beta,\tilde{\beta}) + V_i(\tilde{\beta},\beta) \leq W_i(\zeta) \leq W_i^*(\zeta).$$

Let

$$T = \min\left\{ \sum_{i=1}^{n} \tau_i(\beta), \sum_{i=1}^{n} \tau_i(\tilde{\beta}) \right\}.$$

Then, using (2.24)

$$\text{(2.25)} \qquad \sup_{U(\delta,a)} |\sqrt{n}\, \Delta F_n| \leq \sup_{U(\delta,a)\cap\{T\leq t\leq T+\Sigma W_i^*(\zeta)\}} \frac{1}{\sqrt{n}} |X(t) - X(T)|.$$

By the strong Markov property, (2.25) is distributed as

$$\text{(2.26)} \qquad \sup_{U(\delta,a)\cap\{0 \leq t \leq \Sigma W_i^*(\zeta)\}} \frac{1}{\sqrt{n}} |X(t)|.$$

But $W_i^*(\zeta)$ decreases to zero as $\zeta \to 0$. So by monotone convergence $EW_i^*(\zeta) \to 0$. Now given ε choose B so that $1 - \Phi(B) \leq \varepsilon/8$. Then choose δ so that $\zeta(\delta)$ is sufficiently small in order that $EW_i^*(\zeta) < \varepsilon^2/(2B^2)$. Since W_i^* are i.i.d., by the Law of Large Numbers, there is n_0 such that for $n \geq n_0$,

$$P\left\{ \sum_{i=1}^{n} W_i^*(\zeta) \leq n\varepsilon^2/B^2 \right\} \geq 1 - \varepsilon/2.$$

Then, from (2.25) and (2.26), for $n \geq n_0$

$$P\left\{\sup_{U(\delta,a)} |\sqrt{n}\,\Delta F_n| \geq \varepsilon\right\} \leq P\left\{\sup_{0\leq t\leq n\varepsilon^2/B^2} \frac{1}{\sqrt{n}} |X(t)| \geq \varepsilon\right\} + \frac{\varepsilon}{2}$$

$$\leq 4\left(1 - \Phi\left(\frac{\sqrt{n}\,\varepsilon}{\sqrt{n}\,\varepsilon/B}\right)\right) + \frac{\varepsilon}{2} \leq \varepsilon. \qquad \square$$

THEOREM 2.1. *Suppose Lemma 2.1 holds. Assume the density* f *has a bounded derivative. Then under the hypotheses of Section 1, the sequence of normalized* B-K *empirical c.d.f. processes* $\{\sqrt{n}(\hat{\theta}_n - F)\}$ *is tight in* $\mathcal{D}_0$*, and the continuous version with* $\hat{\theta}_n^*$ *replacing* $\hat{\theta}_n$ *is tight in* C_0.

Proof. Define

$$V(\delta,a) = \{(\beta_1,\tilde{\beta}_1) : |\beta_1| \leq a - \delta,\ |\beta_1 - \tilde{\beta}_1| \leq \delta\},$$

$$\Delta\hat{\theta} = [\hat{\theta}(\beta_1) - F(\beta_1)] - [\hat{\theta}(\tilde{\beta}_1) - F(\tilde{\beta}_1)].$$

First note that since $\Sigma x_i = 0$ and $f'(u)$ is bounded,

$$(2.27) \quad \frac{1}{n}\sum_{i=1}^{n} F(\beta_1 + x_i'\hat{\gamma}(\hat{\theta}(\beta_1)))$$

$$= F(\beta_1) + \frac{1}{n}\sum_{i=1}^{n} x_i'\hat{\gamma} f(\beta_1) + \frac{1}{2n}\sum_{i=1}^{n} (x_i'\hat{\gamma})^2 f'(\beta_1^*)$$

$$= F(\beta_1) + O(\|\hat{\gamma}\|^2) = F(\beta_1) + O_p(\log n/n)$$

where the fact that $\frac{1}{n}X'X \to \Sigma$ and Lemma 2.1 are used. Thus, (2.27) and Proposition 2.1 imply that

$$(2.28) \qquad |\Delta\hat{\theta} - \Delta F_n| = o_p(1/\sqrt{n})$$

(where ΔF_n has arguments $(\beta_1,\hat{\gamma}(\hat{\theta}(\beta_1)))$ and $(\tilde{\beta}_1,\hat{\gamma}(\hat{\theta}(\tilde{\beta}_1)))$).

Therefore, using Lemma 2.1 again, with probability tending to 1

$$\sup_{V(\delta,a)} \sqrt{n}|\Delta\hat{\theta}| \leq \sup_{\substack{V(\delta,a)\cap\{\max_i|x_i'\hat{\gamma}(\hat{\theta}(\beta_1))| \leq \delta\}\\ \cap\{\max_i|x_i'\hat{\gamma}(\hat{\theta}(\tilde{\beta}_1))| \leq \delta\}}} \sqrt{n}|\Delta\hat{\theta}|$$

$$\leq \sup_{U(\delta,a)} \sqrt{n}|\Delta F_n| = o_p(1).$$

This is a stronger condition than the sufficient condition for tightness in $\mathcal{D}_0$ (see Billingsley [1968, Section 15]). Furthermore, by Theorem 2.2 of B-K [1982] successive jumps in $\hat{\theta}$ are bounded above (almost surely) by $(p+1)/n$. Thus, it follows directly that $\sqrt{n}|\Delta\hat{\theta}^*| \le \sqrt{n}|\Delta\hat{\theta}| + O(1/\sqrt{n})$ (a.s.); and, hence, tightness in C_0 follows immediately (again, see Billingsley [1968]). □

3. On the Conditions on the Design Matrix

Conditions (2.4), (2.5), (2.9), and especially (2.10) appear somewhat artificial and are certainly stronger than the more basic hypotheses of Section 1 (used by Bassett and Koenker). Nonetheless, it will be shown here that these conditions can be expected to hold in almost all common linear models situations.

First note that condition (2.9) $(\max_i \|x_i\| = o(n/\log n)^{\frac{1}{2}})$ is stronger than conditions (2.4) and (2.5):

$$\operatorname{tr}(X_h' X_h)^{\frac{1}{2}} = \left(\sum_{i \in h} \|w_i\|^2\right)^{\frac{1}{2}} \le \sqrt{p}\, \max_i (\|x_i\| + 1)$$

$$\Big\| \sum_{i \in h} x_i \Big\| \le p \max_i \|x_i\| .$$

Clearly condition (2.9) will hold if $\max_{ij} |x_{ij}|$ is bounded (which is the case in ANOVA designs). The other common design situation is when coordinates of the rows $\{x_i'\}$ are random vectors or are simple functions (e.g., polynomials) of random vectors. If $\{x_i\}$ can be assumed to be a sample (i.i.d.) from some distribution in R^p then (2.9) would hold in probability whenever a simple moment condition holds:

PROPOSITION 3.1. *If* $\{x_i : i = 1,2,\ldots,n\}$ *are i.i.d. random vectors in* R^p *with* $E\|x_i\|^{2(1+\delta)} = B < +\infty$ *for some* $\delta > 0$, *then*

$$\max_i \|x_i\| = o_p(n/\log n)^{\frac{1}{2}} .$$

Proof. By Chebychev's inequality, for each $i = 1,\ldots,n$ and any $\varepsilon > 0$,

$$P\{\|x_i\| \geq \varepsilon(n/\log n)^{\frac{1}{2}}\} \leq \frac{B(\log n)^{1+\delta}}{\varepsilon^{2(1+\delta)} n^{1+\delta}} .$$

The result follows since $P\{\|x_i\| \geq A$ for $i = 1,\ldots,n\} \leq$ $\leq nP\{\|x_i\| \geq A\}$. □

Condition (2.10) is somewhat more complicated. The following proposition reduces (2.10) to a somewhat more reasonable (though still quite artificial) condition. We will then show that this condition will generally hold.

PROPOSITION 3.2. *Suppose that for each sufficiently large* $a > 0$ *there are constants* $b > 0$ *and* $d > 0$ *such that*

$$\inf_{\|u\|=1} \sum_{i=1}^{n} |x_i'u| I(b \leq |x_i'u| \leq a) \geq dn. \tag{3.1}$$

Suppose further that F *has a density which is strictly positive everywhere (and assume that* $\Sigma x_i = 0$*). Then for any* a *(sufficiently large) there is* $\eta > 0$ *such that for all* $|\beta_1| \leq a$ *and all* $\gamma \in R^p$

$$\sum_{i=1}^{n} (x_i'\gamma) F(\beta_1 + x_i'\gamma) \geq n\eta \min(\|\gamma\|^2, \|\gamma\|). \tag{3.2}$$

Proof. Let $S(\beta_1,\gamma)$ denote the left-hand side of (3.2). Then

$$S(\beta_1,\gamma) = \sum_{i=1}^{n} (x_i'\gamma)(F(\beta_1 + x_i'\gamma) - F(\beta_1)). \tag{3.3}$$

First consider the case $\|\gamma\| \geq 1$: for any $b > 0$

$$\begin{aligned} S(\beta_1,\gamma) &\geq \sum_{x_i'\gamma \geq b} (x_i'\gamma)(F(\beta_1 + b) - F(\beta_1)) \\ &\quad + \sum_{x_i'\gamma \leq -b} (-x_i'\gamma)(F(\beta_1) - F(\beta_1 - b)) \\ &\geq c_1 \|\gamma\| \sum_{i=1}^{n} \left|x_i' \frac{\gamma}{\|\gamma\|}\right| I(|x_i'\gamma| \geq b) \end{aligned}$$

where $c_1 = \min(F(\beta_1 + b) - F(\beta_1), F(\beta_1) - F(\beta_1 - b)) > 0$ (since the density is stricly positive). Using the fact that $\|\gamma\| \geq 1$,

$$(3.4)\quad S(\beta_1,\gamma) \geq c_1\|\gamma\| \sum_{i=1}^{n} |x_i' \tfrac{\gamma}{\|\gamma\|}| I(a \geq |x_i' \tfrac{\gamma}{\|\gamma\|}| \geq b)$$

$$\geq c_1 d\|\gamma\| n$$

using condition (3.1) directly.

If $\|\gamma\| \leq 1$, then (using the mean value theorem) for $|\beta_1| < a$,

$$S(\beta_1,\gamma) = \sum_{i=1}^{n} (x_i'\gamma)^2 f(\beta_1 + v_i) \qquad (\text{where } |v_i| \leq |x_i'\gamma|)$$

$$\geq c_2 \sum_{i=1}^{n} (x_i'\gamma)^2 I(|x_i'\gamma| \leq a)$$

where $c_2 = \inf\{f(u) : |u| \leq 2a\}$. Thus (using the fact that $\|\gamma\| \leq 1$)

$$(3.5)\quad S(\beta_1,\gamma) \geq c_2\|\gamma\|^2 \sum_{i=1}^{n} (x_i' \tfrac{\gamma}{\|\gamma\|})^2 I(|x_i' \tfrac{\gamma}{\|\gamma\|}| \leq a)$$

$$\geq c_2\|\gamma\|^2 \inf_{\|u\|=1} \Sigma(x_i'u)^2 I(b \leq |x_i'u| \leq a)$$

$$\geq c_2 b\|\gamma\|^2 \inf_{\|u\|=1} \Sigma|x_i'u| I(b \leq |x_i'u| \leq a)$$

$$\geq c_2 bd\|\gamma\|^2 n$$

(using condition (3.1)). Condition (3.2) follows from (3.4) and (3.5). □

If the design is an ANOVA with at least δn observations in each to cell (for some $\delta > 0$), then it is easy to see that condition (3.1) must hold. If $\{x_i\}$ is a sample (i.i.d.) from some distribution in R^p then the argument of Theorem 4.1 in Portnoy [1982] can be used to show that a sufficient condition for (3.1) to hold in probability is the following:

(3.6) for any sufficiently large a there are constants $b > 0$ and $d > 0$ such that

$$\inf_{\|u\|=1} E|x_i'u| I(b \leq |x_i'u| \leq a) \geq d$$

where the expectation is over the distribution of x_i. A sketch of the argument will be given later. First notice, however, that (3.6) holds whenever x_i has a distribution not concentrated on a proper linear subspace of R^p -- a condition which would hold even if some coordinates of x_i are (non-linear) functions of others:

PROPOSITION 3.3. *If* $\{x_i\}$ *are i.i.d. from a distribution not concentrated on a proper linear subspace of* R^p *then condition (3.6) holds.*

Proof. First note that there is $\varepsilon > 0$ such that

$$\inf_{\|u\|=1} P\{|x_i'u| > 0\} \geq \varepsilon. \tag{3.7}$$

For otherwise there would be $\{u_n\}$ such that $P\{|x_i'u_n| > 0\} \to 0$. But, by compactness of the unit sphere, there is a subsequence, $\tilde{u}_n$, and a point u with $\|u\| = 1$ such that $\tilde{u}_n \to u$; and, hence, by Fatou's Lemma, we would have

$$\begin{aligned} 0 = \liminf_{n\to\infty} P(|x_i'\tilde{u}_n| > 0) &= \liminf_{n\to\infty} EI(|x_i'\tilde{u}_n| > 0) \\ &\geq E \liminf_{n\to\infty} I(|x_i'\tilde{u}_n| > 0) \\ &\geq EI(|x_i'u| > 0) = P(|x_i'u| > 0), \end{aligned}$$

which contradicts the hypothesis of the proposition. Thus, choosing a large enough so that $P\{\|x_i\| < a\} > 1 - \varepsilon$, for each u with $\|u\| = 1$,

$$P\{0 < |x_i'u| < a\} > 0. \tag{3.8}$$

Now by way of contradiction, assume (3.6) doesn't hold. Then (as above) there is $u_n \to u$ (in the unit sphere of R^p) and $b_n \to 0$ such that $E|x_i'u_n|I(b_n \leq |x_i'u_n| \leq a) \to 0$. Again, by Fatou's Lemma this would imply

$$\begin{aligned} 0 &= \liminf_{n\to\infty} E|x_i'u_n|I(b_n \leq |x_i'u_n| \leq a) \\ &\geq E \liminf_{n\to\infty} |x_i'u_n|I(b_n \leq |x_i'u_n| \leq a) \\ &\geq E|x_i'u|I(0 < |x_i'u| < a), \end{aligned}$$

which contradicts (3.8). □

Lastly we sketch the argument that (3.6) implies that (3.1) holds in probability (assuming that condition (2.9) holds in probability):

For fixed u (with $\|u\| = 1$), the Markov inequality and (3.6) imply that there is $c > 0$ with

$$P\{\Sigma|x_i'u|I(b + \frac{1}{n} \leq |x_i'u| \leq a - \frac{1}{n}) \leq \frac{1}{2} dn\} \leq e^{-cn}$$

(note that $|x_i'u|I(b + \frac{1}{n} \leq |x_i'u| \leq a - \frac{1}{n})$ is bounded and, hence, has a moment generating function.) Now given u, if $\tilde{u}$ is such that $\|u - \tilde{u}\| \leq 1/n^3$, $\Sigma|x_i'\tilde{u}|I(b \leq |x_i'u| \leq a) \geq \Sigma|x_i'u|I(b + \frac{1}{n} \leq |x_i'u| \leq a - \frac{1}{n}) - B/n$ (for n large enough). Since there are at most An^{3p} balls of radius $1/n^3$ needed to cover the unit sphere in R^p, the probability that the left-hand side of (3.1) is greater than $\frac{1}{3}dn$ is bounded above by $An^{3p}\exp\{-cn\} \to 0$ (as $n \to \infty$).

References

Bassett, G. and Koenker, R. [1982]. An empirical quantile function for linear models with i.i.d. errors, J. Am. Stat. Assoc. 77, 407-15.

__________ [1983]. Convergence of an empirical distribution function for the linear model, to be submitted.

Beran, R. [1982]. Estimating sampling distributions: the bootstrap and competitors, Ann. Statist. 10, 212-25.

Billingsley, P. [1968]. Convergence of Probability Measures, John Wiley and Sons, Inc., New York.

Freedman, D. A. [1981]. Bootstrapping regression models, Ann. Statist. 9, 1218-28.

Jurečková, J. and Sen, P. K. [1983]. On adaptive scale-equivariant M-estimators in linear models, (submitted to Ann. Statist.)

Koenker, R. and Bassett, G. [1978]. Regression quantiles, Econometrika 46, 33-50.

Ruppert, D. and Carroll, R. J. [1980]. Trimmed least squares estimation in the linear model, J. Am. Stat. Assoc. 75, 828-38.

Portnoy, S. [1982]. Asymptotic behavior of M-estimators of p regression parameters when p^2/n is large, (submitted to Ann. Statist.)

ROBUST NONPARAMETRIC AUTOREGRESSION

P.M. Robinson

Department of Mathematics, University of Surrey

1. INTRODUCTION

Throughout we consider a strictly stationary, real-valued, discrete time series $\{X_t; t=1,2,\ldots\}$. The Nadaraya-Watson method of nonparametric regression estimation, and related methods, have been adapted by a number of authors to nonparametric estimation and prediction for time series. Indeed, Watson (1964) applied his estimator to some time series data, and Roussas (1969), Bosq (1980), Doukhan and Ghindes (1980), Pham Dinh Tuan (1981), Robinson (1981), Collomb (1982) have studied asymptotic properties of such estimators and predictors. For some $y \in R^p$ let $\mu(y) = E(X_t | Y_t = y)$, where $Y_t = (X_{t-1},\ldots,X_{t-p})$ and p is a prescribed integer $\geqq 1$. A Nadaraya-Watson type of estimator of $\mu(y)$ is

$$\mu_T(y) = \sum_t X_t K((Y_t - y)/a_T) / \sum_t K((Y_t - y)/a_T), \tag{1}$$

where K is an integrable function on R^p, concentrated round the origin, $a_T > 0$, and throughout Σ_t means $\Sigma_{t=p+1}^T$. The estimator $\mu_T(y)$ is a type of weighted average of the X_t and may be regarded as minimizing the function

$$\sum_t (X_t - \mu)^2 K((Y_t - y)/a_T)$$

when $K \geqq 0$. Related estimators, such as the "predictogram" considered by Bosq (1980), Collomb (1982), or ones replacing K in (1) by nearest neighbour weights (e.g. Stone, 1977), also arise, in effect, from a quadratic loss function.

Evidently $\mu_T(y)$ is highly sensitive to large fluctuations in X_t, particularly when the corresponding Y_t is close to y. Central limit theorems for $\mu_T(y)$ require at least $E(X_t^2) < \infty$, and the asymptotic variance of $n_T^{\frac{1}{2}}\{\mu_T(y) - \mu(y)\}$, for an appropriate norming sequence n_T, is

$$\kappa V(X_t | Y_t = y)/f(y), \tag{2}$$

where f is the probability density function (pdf) of Y_t and $\kappa = \int K^2(y)dy$. One situation in which $\mu_T(y)$ might be formed is when a pth degree autoregression

$$X_t = g(Y_t) + \varepsilon_t$$
$$\varepsilon_t \text{ independent of } \{X_{t-1}, X_{t-2}, \ldots\} \tag{3}$$
$$E(\varepsilon_t) = 0,\ V(\varepsilon_t) = \sigma^2$$

is believed to obtain but the functional form of g - not presumed to be linear - is unknown. Then $V(X_t | Y_t = y) \equiv \sigma^2$; if the innovations ε_t have a long-tailed distribution σ^2 may be very large, making (2) large, while it will be infinite in the case of certain distributions. Bearing in mind the presence in the denominator of (2) of $f(y)$, which will be small for many values of y, and the fact that inevitably $n_T = o(T)$ to avoid asymptotic bias, it seems desirable to seek alternative estimators whose asymptotic variance replaces $V(X_t | Y_t = y)$ by a function which is more likely to be finite, and, when finite, is likely to be smaller than $V(X_t | Y_t = y)$ in the case of long-tailed distributions. We adapt robust M-estimators of a location parameter proposed by Huber (1964) and establish a central limit theorem for such estimators. We do not assume (3) but rather employ the, generally less restrictive, strong mixing condition of Rosenblatt (1956). A further advantage of robust procedures emerges from our work. A condition is needed on the rate of convergence to zero of the mixing coefficients, and the strength of this varies inversely with the strength of moment conditions. Because the usual loss-functions employed in M-estimation lead to a reduction in moment conditions, relative to those needed in the case of quadratic loss, it follows that there are processes $\{X_t\}$ for which (1) does not satisfy the central limit theorem, but certain robust alternatives do.

2. M-ESTIMATORS

Let $\rho(x)$ be a convex function on R^1, so if $K \geqq 0$ the function

$$Q_T(\theta; y) = \sum_t \rho(X_t - \theta) K((Y_t - y)/a_T),$$

is convex in θ. Let $\theta_T(y)$ satisfy

$$Q(\theta_T(y); y) = \min_\theta Q_T(\theta; y).$$

The statistics $\theta_T(y)$ are invariant with respect to location changes in X_t, but not to scale changes; scale-invariant estimators can be constructed in a way analogous to that described by Huber (1964, 1981).

Define $\psi(x) = (d/dx)\rho(x)$, normalised by $\psi(x) = \frac{1}{2}\{\psi(x-0)+\psi(x+0)\}$. We can define $\theta_T(y)$ equivalently as being a zero of

$$\lambda_T(\theta;y) = (a_T^p\ T)^{-1}\sum_t \psi(X_t-\theta)K((Y_t-y)/a_T).$$

We do not assume $\theta_T(y)$ is uniquely defined but we do impose conditions in the next section that ensure the function

$$\gamma(\theta;y) = \int \psi(x-\theta)f(x|y)dx \qquad (4)$$

has a unique zero, where $f(x|y)$ is the conditional pdf of X_t, given that $Y_t = y$; call this zero $\theta(y)$.

Example 1. Let $\rho(x) = \frac{1}{2}x^2$, then $\psi(x) = x$ and $\theta(y) = \mu(y)$. (This is $g(y)$ under (3)).

Example 2. Let $\rho(x) = |x|$, then $\psi(x) = \text{sign}(x)$ and $\theta(y) = \nu(y) \triangleq \text{median}\{X_t|Y_t=y\}$.

Example 3. Let $\rho(x) = \frac{1}{2}x^2$ for $|x| \leqq c$ and $\rho(x) = c|x| - \frac{1}{2}c^2$ for $|x| > c$ (Huber, 1964). Then $\psi(x) = \max[-c,\min(x,c)]$. If $f(x|y)$ is symmetric in x then $\theta(y) = \nu(y)$; if also $\mu(y)$ exists then $\theta(y) = \mu(y)$.

In Section 4 we prove

Theorem. Under the conditions described in Section 3, for distinct $y_i \in R^p$, $i = 1,\dots q$, the variables $(Ta_T^p)^{\frac{1}{2}}\{\theta_T(y_i) - \theta(y_i)\}$ converge in distribution to independent zero-mean normal variables with variances $\sigma^2(y_i)$, where

$$\sigma^2(y) = \kappa\delta(\theta(y);y)/\{\gamma'(\theta(y);y)^2 f(y)\},$$

in which

$$\gamma'(\theta;y) = (\partial/\partial\theta)\gamma(\theta;y),\ \delta(\theta;y) = \int\psi^2(x-\theta)f(x|y)dx.$$

This theorem partly generalizes results of Pham Dinh Tuan (1981), Robinson (1981), who consider estimators of the type (1).

Example 1. (contd.) $\sigma^2(y) = (2)$.

Example 2. (contd.) $\sigma^2(y) = \kappa\{f(\theta(y)|y)^2 f(y)\}^{-1}$.

Example 3. (contd.)

$$\sigma^2(y) = \kappa\{\int_{-c}^{c} x^2 + c^2\int_{-\infty}^{-c} + c^2\int_{c}^{\infty}\}f(x+\theta(y)|y)dx/$$
$$[f(y)\{\int_{-c}^{c} f(x+\theta(y)|y)dx\}^2].$$

By analogy with Huber (1964), $\sigma^2(y)$ is minimized by the choice $\psi(x) = -f'(x|y)/f(x|y)$ when $f(x|y)$ is absolutely continuous. In the context of the autoregressive model (3), $f(x|y)$ is the pdf of ε_t. Here the choice of ψ in Example 1 is optimal if ε_t is normally distributed. The nonparametric estimation approach is motivated by the belief that X_t is non-Gaussian. Though many non-Gaussian nonlinear AR models with Gaussian innovations can be constructed, in practice there may be no reason to believe ε_t in (3) is Gaussian when X_t is non-Gaussian, so (1) might be improved upon by other $\theta_T(y)$, generated by alternative choices of ψ. In fact, as earlier noted, (3) is not assumed and we are considering the general problem of conditional location estimation for non-Gaussian processes. Generalization of our estimators and asymptotic results in various directions is possible. Some discussion of computations is in order at this point. For simplicity we consider only K of the form

$$K(x) = \prod_{i=1}^{p} k(x_i),\ k \geqq 0, \tag{5}$$

where k is an integrable function on R^1, and without further loss of generality we require

$$\int k(x)dx = 1\ . \tag{6}$$

We have $\kappa = (\int k^2(x)dx)^p$.

In general $\theta_T(y)$ is a type of K-weighted location M-estimator. Some simplifications result if the rectangular window is used, $k(x) = 1$ for $|x| < \frac{1}{2}$, $k(x) = 0$ for $|x| \geqq \frac{1}{2}$.

Example 1. (contd.) $\theta_T(y)$ is of arithmetic mean form, $\theta_T(y) = m_T^{-1}\Sigma_t'' X_t$, where the sum is over the set C_T of t such that Y_t is inside the p-dimensional cube of side a_T centred at y, and m_T is the number of elements in C_T.

Example 2. (contd.) $\theta_T(y) = \text{median}\{X_t | t \in C_T\}$.

Example 3. (contd.) $\theta_T(y)$ is a type of Winsorized mean, cf. Huber (1964).

A number of authors (e.g. An and Chen, 1982, Beran, 1975, Bustos, 1982, Gross and Steiger, 1979, Koul, 1977, Lee and Martin, 1982, Martin and Jong, 1976, Portnoy, 1977) have studied asymptotic properties of various M-estimators of location, regression and time series models, in parametric situations. Brillinger (1977) suggested the use of conditional M-estimators in nonparametric nearest neighbour regression, and Stone (1977) established universal consistency of such estimators on the basis of independent observations. Härdle (1982), has established some asymptotic properties of M-estimators for nonparametric regression.

3. ASSUMPTIONS

We require that K satisfy (5), where k is even, satisfies (6), and also

$$|k(x)| \leqq (1+|x|)^{-\max(p,2)-4} .$$

The bandwidth sequence a_T is restricted by

$$a_T^p T \to \infty,\ a_T^{p+4} T \to 0, \text{ as } T \to \infty.$$

Let S be a closed subset of R^p, where y_i is an interior point of S, for $i = 1,\ldots,q$. The following conditions involving y are assumed to hold for all $y \varepsilon S$:

$f(y) > 0$ and $f(y)$ is twice continuously differentiable.

The pdf of (Y_t, X_{t-s}) exists and is bounded for $s = p+1,\ldots,2p$.

The pdf of (Y_t, Y_{t-s}) exists and is bounded uniformly in $s \geqq p$.

The function ψ is monotone increasing and < 0 (> 0) for $x \ll 0$ ($\gg 0$). The function $\gamma(\theta;y)$ (4) satisfies

$$\gamma(\theta(y);y) = 0,$$

$\gamma(\theta;y)$ is differentiable in θ at $\theta(y)$, and $\gamma'(\theta(y);y) < 0$.

$\gamma(\theta;y)$ is continuous in θ on a closed interval containing $\theta(y)$, and is twice continuously differentiable in y.

$\delta(\theta;y)$ is continuous in θ on a closed interval containing $\theta(y)$, and continuous in y.

Now, following Rosenblatt (1956), introduce the strong mixing coefficient

$$\alpha_j = \sup_{A\varepsilon M_{-\infty}^t\ B\varepsilon M_{t+j}^\infty} |P(A\ B) - P(A)P(B)|,\ j > 0,$$

where M_u^v is the σ-field of events generated by X_t, $u \leqq t \leqq v$. We assume <u>either</u> that

$$\sup_x |\psi(x)| < \infty,$$
$$\sum_{j=n}^{\infty} \alpha_j = o(n^{-1}), \text{ as } n \to \infty; \tag{7}$$

or that, for some $\varepsilon > 0$, $\eta > \varepsilon$ and all θ in a neighbourhood of $\theta(y)$,

$$\int|\psi(x-\theta)|^{2+\varepsilon}f(x)dx < \infty,$$
$$\int|\psi(x-\theta)|^{2+\eta}f(x|y)dy < \infty,\ y \in S, \tag{8}$$
$$\sum_{j=n}^{\infty} \alpha_j^{\varepsilon/(2+\varepsilon)} = O(n^{-1}), \quad \text{as } n \to \infty.$$

In case X_t is not bounded conditions (8) are appropriate for Example 1, when they reduce to $E|X_t|^{2+\varepsilon} < \infty$, $E(|X_t|^{2+\eta}|Y_t = y) < \infty$, $y \in S$. Condition (7) is satisfied by Examples 2 and 3, irrespective of the tail behaviour of X_t, when (7) is less restrictive than (8). Thus our conditions indicate that the robust estimators of Examples 2 and 3 satisfy the central limit theorem for some X_t that are too serially dependent and/or too heavy in the tails to allow a central limit theorem in Example 1.

4. PROOF OF THEOREM

Because $k \geqq 0$ and $\psi(x)$ is monotone increasing then $\lambda_T(\theta;y)$ is monotone decreasing in θ for every y. Now $\theta_T(y)$ is not necessarily uniquely defined but as argued by Huber (1964, 1981) it suffices to consider only $\theta_T^*(y) = \sup\{\theta|\lambda_T(\theta;y) > 0\}$. Since $\{\theta_T^*(y) < \theta\} \subset \{\lambda_T(\theta;y) \leqq 0\} \subset \{\theta_T^*(y) \leqq \theta\}$ it remains to show that, for any real $z_1,\ldots,z_q$,

$$P\left(\bigcap_{i=1}^{q}\{\lambda_T(h_{iT};y_i) \leqq 0\}\right) \to \prod_{i=1}^{q}\Phi(z_i), \text{ as } T \to \infty, \tag{9}$$

where Φ is the standard normal distribution function and $h_{iT} = \theta(y_i) + n_T^{-\frac{1}{2}}z_i\sigma(y_i)$, $n_T = a_T^pT$. Define

$$u_{iT} = T^{\frac{1}{2}}E(G_{itT})/\{V(G_{itT})\}^{\frac{1}{2}},\ v_{iT} = T^{-\frac{1}{2}}\sum_t\{G_{itT}-E(G_{itT})\}/\{V(G_{itT})\}^{\frac{1}{2}}$$

where $G_{itT} = \psi(X_t - h_{iT})K((Y_t-y_i)/a_T)$. Then (9) is equivalent to

$$u_{iT} + z_i \to 0 \tag{10}$$

$$\sum_{i=1}^{q}\beta_i v_{iT} \xrightarrow{d} N(0,1), \text{ all } \{\beta_1,\ldots,\beta_q|\sum_{i=1}^{q}\beta_i^2 = 1\} \tag{11}$$

as $T \to \infty$. Our proof of (10) and (11) is facilitated by application of results of Robinson (1981).

Considering (10) first, note that

$$E(G_{itT}) = a_T^p\int\lambda(h_{iT};y_i+a_Tu)K(u)du$$

where $\lambda(\theta,y) = \gamma(\theta;y)f(y)$. Choose $\eta > 0$ then for large enough T_0, $|h_{iT}-\theta(y_i)| \leqq \eta$ for all $T > T_0$. Since $\lambda(\theta;y)$ has uniformly continuous second derivatives in θ on the compact set $A_\eta = \{\theta,y|\theta\varepsilon[\theta(y)-\eta,\theta(y)+\eta],y\varepsilon S\}$, then for small enough η it follows that Lemma 8.8 of Robinson (1981) may be adapted to show that

$$n_T^{\frac{1}{2}}\{a_T^{-p}E(G_{itT})-\lambda(h_{iT};y_i)\} = O(T^{\frac{1}{2}}a_T^{p/2+2}) \tag{12}$$

which $\to 0$ as $T \to \infty$. Also

$$a_T^{-p}V(G_{itT}) = \int\delta(h_{iT};y_i+a_Tu)f(y_i+a_Tu)K^2(u)du - a_T^{-p}E(G_{itT})^2.$$

By Lemmas 8.7, 8.2 of Robinson (1981) and (12),

$$u_{iT}-n_T^{\frac{1}{2}}\lambda(h_{iT};y_i) \;/\{\kappa\delta(h_{iT};y_i)f(y_i)\}^{\frac{1}{2}} \to 0, \text{ as } T \to \infty.$$

Since

$$n_T^{\frac{1}{2}}\lambda(h_{iT};y_i) \sim \lambda'(\theta(y_i);y_i)z_i\sigma(y_i)$$

we have established (10).

We deal with (11) in the case (7), ψ is bounded. In view of what has been established so far, (11) is equivalent to

$$\sum_i\beta_i'n_T^{-\frac{1}{2}}\sum_t\{G_{itT}-E(G_{itT})\} \underset{d}{\to} N(0,1), \tag{13}$$

where $\beta_i' = \beta_i/\{\kappa\delta(\theta(y_i);y_i)f(y_i)\}^{\frac{1}{2}}$. We proceed by verifying the conditions of Lemma 7.1 of Robinson (1981). The LHS of (13) is of the form S_T of that lemma,

$$S_T = \sum_i\sum_t n_T^{-\frac{1}{2}}V_{itT}, \quad V_{itT} = \beta_i'\{G_{itT}-E(G_{itT})\}.$$

Conditions A7.1 and A7.2 of Robinson (1981) are clearly satisfied, whereas

$$a_T^{-p}E(V_{itT}^2) \to \kappa f(\theta(y_i);y_i)f(y_i),$$

as already shown, and

$$a_T^{-p}E(V_{itT}V_{jtT}) \to 0, \; i \neq j,$$

using Lemma 8.6 of Robinson (1981), and distinctness of y_i and y_j. Thus condition A7.3 of Robinson (1981) is verified. Condition A7.4 is checked by use of Lemmas 8.3 and 8.4 of Robinson (1981), to complete the proof of the Theorem in the case of bounded ψ. The proof in the unbounded case (8) follows closely that of Theorem 5.3 of Robinson (1981). For some D, $0 < D < \infty$, we introduce

$$\psi_{itT}' = \psi(X_t - h_{iT})I(|\psi(X_t - h_{iT})| \leq D)$$

$$\psi_{itT}'' = \psi(X_t - h_{iT}) - \psi_{itT}'.$$

As in Theorem 5.3 of Robinson (1981) it may be shown that $V(n_T^{-\frac{1}{2}}\sum_t\psi_{itT}''K((Y_t - y_i)/a_T)) \to 0$ as $D \to \infty$, uniformly in large T. Proceeding as above,

$$\sum_i \beta_i n_T^{-\frac{1}{2}} \sum_t [\psi'_{itT} K((Y_t - y_i)/a_T) - E\{\psi'_{itT} K((Y_t - y_i)/a_T\}]$$

may be shown to be asymptotically normal, and its variance → 1 as D → ∞.

This research was supported by the Social Science Research Council.

REFERENCES

AN, H.-Z. and CHEN, Z.-G. (1982). On convergence of LAD estimates in autoregression with infinite variance. J. Multiv. Anal. 12, 335-345.

BERAN, R. (1977). Adaptive estimates for autoregressive processes. Ann. Inst. Statist. Math. 28, 77-89.

BOSQ, D. (1980). Une methode non parametrique de prediction d'un processus stationnaire. Predictions d'une mesure aleatoire. C.R. Acad. Sci. Paris,t. 290, 711-713.

BRILLINGER, D.R. (1977). Discussion on paper by C.J. Stone. Ann. Statist. 5, 622-623.

BUSTOS, O.H. (1982). General M-estimates for contaminated pth-order autoregressive processes: consistency and normality. Z. Wahr.u. Verw. Geb. 59, 491-504.

COLLOMB, G. (1982). Prediction non parametrique : etude de l'erreur quadratique du predictogramme. C.R. Acad. Sci. Paris,t. 294, 59-62.

DOUKHAN, P. and GHINDES, M. (1980). Estimations dans le processus "$X_{n+1} = f(X_n) + \varepsilon_n$." C.R. Acad. Sci. Paris,t. 291, 61-64.

HÄRDLE, W. (1984) Robust regression function estimation. J. Multiv. Anal., 14, 169 - 180

HUBER, P.J. (1964). Robust estimation of a location parameter. Ann. Math. Statist. 35, 73-101.

HUBER, P.J. (1981). Robust Statistics. Wiley, New York.

KOUL, H.L. (1977). Behaviour of robust estimators in the regression model with dependent errors. Ann. Statist. 5, 681-699.

LEE, C.H. and MARTIN, R.D. (1982). M-estimates for ARMA processes. Tech. Report No.23, Department of Statistics, University of Washington, Seattle, WA.

MARTIN, R.D. AND JONG, J.M. (1976). Asymptotic properties of robust generalized M-estimates for the first order autoregressive parameter. Bell Labs. Tech. Memo, Murray Hill, N.J.

PHAM, D.T. (1981). Nonparametric estimation of the drift coefficient in the diffusion equation. Math. Oper. 12, 61-73.

PORTNOY, S.L. (1977). Robust estimation in dependent situations. Ann. Statist. 5, 22-43.

ROBINSON, P.M. (198). Nonparametric estimators for time series. J. Time Series Anal., 4, 185 - 207

ROBINSON, P.M. (1983). Kernel estimation and interpolation for time series containing missing observations. Preprint.

ROSENBLATT, M. (1956). A central limit theorem and a strong mixing condition. Proc. Nat. Acad. Sci., U.S.A. 42, 43-47.

ROUSSAS, G. (1969). Nonparametric estimation of the transition distribution function of a Markov process. Ann. Math. Statist. 40, 1386-1400.

STONE, C.J. (1977). Consistent nonparametric regression. (With Discussion.) Ann. Statist. 5, 595-645.

WATSON, G.S. (1964). Smooth regression analysis. Sankhyā, Ser. A 26, 359-378.

ROBUST REGRESSION BY MEANS OF S-ESTIMATORS

P. Rousseeuw and V. Yohai

Vrije Universiteit Brussel, CSOO (M 205), Pleinlaan 2,
B-1050 Brussels, Belgium

Departamento de Matemáticas, Facultad de Ciencias Exactas y
Naturales, Ciudad Universitaria, Pabellon 1,
1428 Buenos Aires, Argentina

1. Introduction

There are at least two reasons why robust regression techniques are useful tools in robust time series analysis. First of all, one often wants to estimate autoregressive parameters in a robust way, and secondly, one sometimes has to fit a linear or nonlinear trend to a time series. In this paper we shall develop a class of methods for robust regression, and briefly comment on their use in time series. These new estimators are introduced because of their invulnerability to large fractions of contaminated data. We propose to call them "S-estimators" because they are based on estimators of <u>s</u>cale.

The general linear model is given by $y_i = x_i^t\theta + e_i$ for $i=1,..,n$ where x_i and θ are p-dimensional column vectors, e_i is the error term and y_i is the dependent variable. Our aim is to estimate θ from the data $(x_1,y_1),..,(x_n,y_n)$. The classical least squares estimator of Gauss and Legendre corresponds to

$$\underset{\hat{\theta}}{\text{minimize}} \; \Sigma_{i=1}^{n} r_i^2 \tag{1.1}$$

where the residuals r_i equal $y_i - x_i^t\hat{\theta}$. This estimator is not robust at all, because the occurence of even a single (bad) outlier can spoil the result completely.

In connection with this effect, the <u>breakdown point</u> was invented by Hampel (1971), who gave it a rigourous asymptotic definition. Recently, Donoho and Huber (1983) introduced a simplified version which works on finite samples, like the precursor ideas of Hodges (1967). In this paper we shall use the finite-sample version of the breakdown point (based on replacements). Take any sample X of n data points (x_i, y_i) and any estimator T of the parameter vector θ. Let $\beta(m,T,X)$ be the supremum of $\|T(X')-T(X)\|$ for all corrupted samples X' where any m of the original points of X are replaced by arbitrary values. Then the breakdown point of T at X is defined as

$$\varepsilon_n^*(T,X) = \min\{\frac{m}{n} \, ; \, \beta(m,T,X) \text{ is infinite}\} \quad . \qquad (1.2)$$

In words, ε_n^* is the smallest fraction of contaminated data that can cause the estimator to take on values arbitrarily far from T(X). For least squares, one bad observation can already cause breakdown, so $\varepsilon_n^*(T,X) = \frac{1}{n}$ which tends to 0% when the sample size n becomes large.

The purpose of this paper is to find more robust estimators. A first step came from Edgeworth (1887) who proposed to replace the square in (1.1) by an absolute value, which does not grow so fast. His <u>least absolute values</u> or L_1 criterion is

$$\underset{\hat{\theta}}{\text{minimize}} \; \Sigma_{i=1}^n |r_i| \quad . \qquad (1.3)$$

It turns out that this protects against outlying y_i, but still cannot cope with outlying x_i (called "leverage points"), which have a large effect on the fit. Therefore, we still have $\varepsilon_n^*(T,X) = \frac{1}{n} \to 0\%$.

The next step in this direction is Huber's (1973,1981) <u>M-estimation</u> method, based on the idea of replacing r_i^2 in (1.1) by $\rho(r_i)$, where ρ is a symmetric function with a unique minimum at zero. However, unlike (1.1) or (1.3) this does not yield estimators which are invariant with respect to a magnification of the error scale. Therefore, Huber estimates the scale parameter simultaneously (which corresponds to "Proposal 2" of Huber (1964) for the location problem):

$$\Sigma_{i=1}^n \psi(r_i/\hat{\sigma}) \, x_i = 0 \qquad (1.4)$$

$$\Sigma_{i=1}^n \chi(r_i/\hat{\sigma}) = 0 \qquad (1.5)$$

where ψ is the derivative of ρ . (Finding the simultaneous solution of the system of equations (1.4) and (1.5) is not trivial, and in practice one uses some iteration scheme, which makes it hard to use non-monotone ψ-functions.) Motivated by minimax asymptotic variance arguments, Huber proposed to use the function $\psi(u)=\min(k,\max(u,-k))$ where k is some constant, usually around 1.5 . As a consequence, such M-estimators are statistically more efficient than L_1 at a central model with Gaussian errors. However, again $\varepsilon_n^*(T,X) = \frac{1}{n}$ because of the possibility of leverage points.

Because of this vulnerability to leverage points, generalized M-estimators ("GM-estimators" for short) were considered. Their basic idea is to bound the influence of outlying x_i , making use of some weight function w. Mallows (1975) proposed to replace (1.4) by

$$\Sigma_{i=1}^{n} w(x_i)\psi(r_i/\hat{\sigma})\ x_i = 0 \quad ,$$

whereas Schweppe (see Hill 1977) suggested to use

$$\Sigma_{i=1}^{n} w(x_i)\psi(r_i/(w(x_i)\hat{\sigma}))\ x_i = 0 \quad .$$

Making use of influence functions, good choices of ψ and w were made (Hampel 1978, Krasker 1980, Krasker and Welsch 1982, Ronchetti and Rousseeuw 1982). However, there are some problems finding suitable defining constants, making the application of these estimators to real data a nontrivial task. Moreover, it turns out that all GM-estimators have a breakdown point of at most $1/(p+1)$, where p is the dimension of x_i (Maronna, Bustos and Yohai 1979, Donoho and Huber 1983). In fact, some numerical experiments by the first author indicate that most commonly used GM-estimators do not reach $1/(p+1)$ at all. Several other estimators have been proposed by Theil (1950), Brown and Mood (1951), Sen (1968), Jaeckel (1972) and Andrews (1974), but they all have a breakdown point less than 30% in the case of simple regression (p=2).

All this raises the question whether robust regression with high breakdown point is at all possible. The affirmative answer was given by Siegel (1982) who proposed the repeated median with a 50% breakdown point. Indeed, 50% is the best we can expect (for larger amounts of contamination, it becomes impossible to distinguish the "good" and the "bad" parts of the data). However, the repeated median is not affine equivariant, by which we mean that it depends on the choice of the coordinate axes of the x_i . The least median of squares (LMS) technique

$$\underset{\hat{\theta}}{\text{minimize}}\ \text{median}(r_i^2) \tag{1.6}$$

(Rousseeuw 1982, based on an idea of Hampel 1975) also possesses a 50% breakdown point and is affine equivariant, but has the disadvantage of converging like $n^{-1/3}$. Finally, the least trimmed squares (LTS) estimator (Rousseeuw 1983) is given by

$$\underset{\hat{\theta}}{\text{minimize}}\ \Sigma_{i=1}^{h}\ (r^2)_{i:n} \tag{1.7}$$

where h is the largest integer $\leq \frac{n}{2}+1$ and $(r^2)_{1:n} \leq \ldots \leq (r^2)_{n:n}$ are the ordered squared residuals. The LTS also has a 50% breakdown point, is affine equivariant, and even converges like $n^{-1/2}$ (Rousseeuw 1983). On the other hand, the computation of its objective function (for fixed p) takes O(n log n) steps (because of the ordering), compared to only O(n) steps for the LMS. The rest of the paper is devoted to the construction of a 50% breakdown, affine equivariant estimator with rate $n^{-1/2}$, with an O(n) objective function and a higher asymptotic efficiency than the LTS.

Our aim is to find simple high-breakdown regression estimators which share the flexibility and nice asymptotic properties of M-estimators. Looking at (1.4) and (1.5), we see that the function χ needed for the estimate $\hat{\sigma}$ is independent of the choice of the function ψ needed for $\hat{\theta}$. On the other hand, a completely different approach would be to take the scale of the residuals as the central notion, and to derive the estimate $\bar{\theta}$ from it.

We will consider one-dimensional estimators of scale defined by a function ρ satisfying

(R1) ρ is symmetric, continuously differentiable and $\rho(0)=0$;

(R2) there exists $c > 0$ such that ρ is strictly increasing on $[0,c]$ and constant on $[c,\infty)$.

For any sample $\{r_1,\ldots,r_n\}$ of real numbers, we define the scale estimate $s(r_1,\ldots,r_n)$ as the solution of

$$\frac{1}{n}\Sigma_{i=1}^{n} \rho(r_i/s) = K \tag{1.8}$$

where K is taken to be $E_\Phi[\rho]$, where Φ is the standard normal. (If there happens to be more than one solution to (1.8), we put $s(r_1,\ldots,r_n)$ equal to the supremum of the set of solutions; if there is no solution to (1.8), then we put $s(r_1,\ldots,r_n) = 0$.)

<u>DEFINITION</u> Let $(x_1,y_1),\ldots,(x_n,y_n)$ be a sample of regression data with p-dimensional x_i . For each vector θ , we obtain residuals $r_i(\theta)=y_i-x_i^t\theta$ of which we can calculate the dispersion $s(r_1(\theta),\ldots,r_n(\theta))$ by (1.8), where ρ satisfies (R1) and (R2). Then the <u>S-estimator</u> $\hat{\theta}$ is defined by

$$\underset{\hat{\theta}}{\text{minimize}}\ s(r_1(\theta),\ldots,r_n(\theta)) \tag{1.9}$$

and the final scale estimator is

$$\hat{\sigma} = s(r_1(\hat{\theta}),\ldots,r_n(\hat{\theta})) \quad . \tag{1.10}$$

We have decided to call $\hat{\theta}$ an S-estimator because it is derived from a <u>s</u>cale statistic in an implicit way, like the derivation of R-estimators from <u>r</u>ank statistics. S-estimators are clearly affine equivariant, and we shall see later on that they possess a high breakdown point and are asymptotically normal.

An example of a ρ-function for (1.8) is

$$\begin{aligned} \rho(x) &= \frac{x^2}{2} - \frac{x^4}{2c^2} + \frac{x^6}{6c^4} \qquad \text{for } |x| \le c \\ &= \frac{c^2}{6} \qquad \text{for } |x| \ge c \quad , \end{aligned} \tag{1.11}$$

the derivative of which is Tukey's biweight function:

$$\psi(x) = x\left(1-\left(\frac{x}{c}\right)^2\right)^2 \qquad \text{for } |x| \le c$$
$$= 0 \qquad \text{for } |x| \ge c \quad . \qquad (1.12)$$

Another possibility is to take a ρ corresponding to the function $\tilde{\psi}$ proposed by Hampel, Rousseeuw and Ronchetti (1981). In general, $\psi(x) = \rho'(x)$ will always be zero for $|x| \ge c$ because of condition (R2); such ψ-functions are usually called "redescending".

2. The breakdown point of S-estimators

Let us start by considering a function ρ for which (R1) and (R2) hold, and also

$$\text{(R3)} \quad \frac{E_\Phi[\rho]}{\rho(c)} = \frac{1}{2} \quad .$$

This is not impossible: in the case of (1.11) it is achieved by taking $c \simeq 1.547$. Let us now look at the scale estimator $s(r_1,\ldots,r_n)$ which is defined by (1.8).

LEMMA 1. For each ρ satisfying (R1) to (R3) and for each n there exist positive constants α and β such that the estimator s given by (1.8) satisfies $\alpha\text{median}(|r_1|,\ldots,|r_n|) \le s(r_1,\ldots,r_n) \le \beta\text{median}(|r_1|,\ldots,|r_n|)$.

Proof. Verification shows that these inequalities hold with $\alpha = \frac{1}{c}$ and $\beta = 1/(\rho^{-1}[\rho(c)/(n+1)])$ for n odd, respectively $\beta = 2/(\rho^{-1}[2\rho(c)/(n+2)])$ for n even. When either $\text{median}(|r_1|,\ldots,|r_n|)$ or $s(r_1,\ldots,r_n)$ is zero, the other is zero too.

From now on, we work with regression data $\{(x_1,y_1),\ldots,(x_n,y_n)\}$. We assume that all observations with $x_i = 0$ have been deleted, because they give no information on θ . Moreover, no more than half of the points (x_i,y_i) should lie on a vertical subspace (here, a vertical subspace is one containing (0,1).)

LEMMA 2. For any ρ satisfying (R1) to (R3), there always exists a solution to (1.9).

<u>Proof.</u> Making use of Lemma 1, this follows from the proof of Theorem 1 of (Rousseeuw 1982) where the result was essentially given for minimization of median$(|r_1|,\ldots,|r_n|)$.

We shall say the observations are in <u>general position</u> when any p of them give a unique determination of θ . In the case of simple regression (p=2), this means that no two points may coincide or determine a vertical line.

<u>THEOREM 1.</u> An S-estimator constructed from a function ρ satisfying (R1) to (R3) has breakdown point

$$\varepsilon_n^* = ([\tfrac{n}{2}]-p+2)/n$$

at any sample $\{(x_1,y_1),\ldots,(x_n,y_n)\}$ in general position.

<u>Proof.</u> Follows from Theorem 2 of (Rousseeuw 1982) using Lemma 1.

The breakdown point only depends slightly on n, and for $n \to \infty$ we obtain $\varepsilon^* = 50\%$, the best we can expect. A nice illustration of this high resistance is given by the following result.

<u>COROLLARY.</u> If there exists some θ such that at least $[\frac{n}{2}]+p$ of the points satisfy $y_i = x_i^t\theta$ <u>exactly</u> and are in general position, then $\hat{\theta} = \theta$ whatever the other observations are.

For example, in the case of simple regression this implies that when 12 out of 20 points lie exactly on a non-vertical straight line, then this line will be found.

If condition (R3) is replaced by $E_\Phi[\rho]/\rho(c) = \lambda$ where $0 < \lambda \leq \frac{1}{2}$, then the corresponding S-estimator has a breakdown point tending to $\varepsilon^* = \lambda$ for $n \to \infty$. In the case of (1.11), values of c larger than 1.547 yield better asymptotic efficiencies at a Gaussian central model, but smaller breakdown points.

3. Examples

In order to demonstrate the usefulness of S-estimators, let us look at an example with a large fraction of outliers. The Belgian Statistical Survey (edited by the Ministry of Economy) gives, among other things, the total number of international phone calls made. These data are listed in Table 1 (in tens of millions), and seem to show an upward trend over the years. However, this time series contains heavy contamination from 1964 till 1969 (and probably the 1963 and 1970 values are also affected). In Table 1, we marked the spurious values with (*). Actually, it turns out that the discrepancy was due to the fact that in this period another recording system was used, which only gave the total number of minutes of these calls!

Table 1.

x	y	x	y
50	0.44	62	1.61
51	0.47	63	2.12(*)
52	0.47	64	11.9(*)
53	0.59	65	12.4(*)
54	0.66	66	14.2(*)
55	0.73	67	15.9(*)
56	0.81	68	18.2(*)
57	0.88	69	21.2(*)
58	1.06	70	4.30(*)
59	1.20	71	2.40
60	1.35	72	2.70
61	1.49	73	2.90

If someone would not look carefully at these data and just apply the least squares method in a routine way, he would obtain $y = 0.5041x - 26.01$ which corresponds to the dotted line in Figure 1. This dotted line has been attracted very much by the 1964-1969 values, and does not fit the good or the bad data points. In fact, some of the good values (such as the 1972 one) yield even larger LS residuals than some of the bad values! Now let us apply the S-estimator corresponding to the function (1.11) with $c = 1.547$. This yields $y = 0.1121x - 5.42$ (plotted as a solid line in Figure 1), which neglects the outliers and gives a good approximation to the other points. (This is not to say that a linear fit is necessarily

Figure 1.

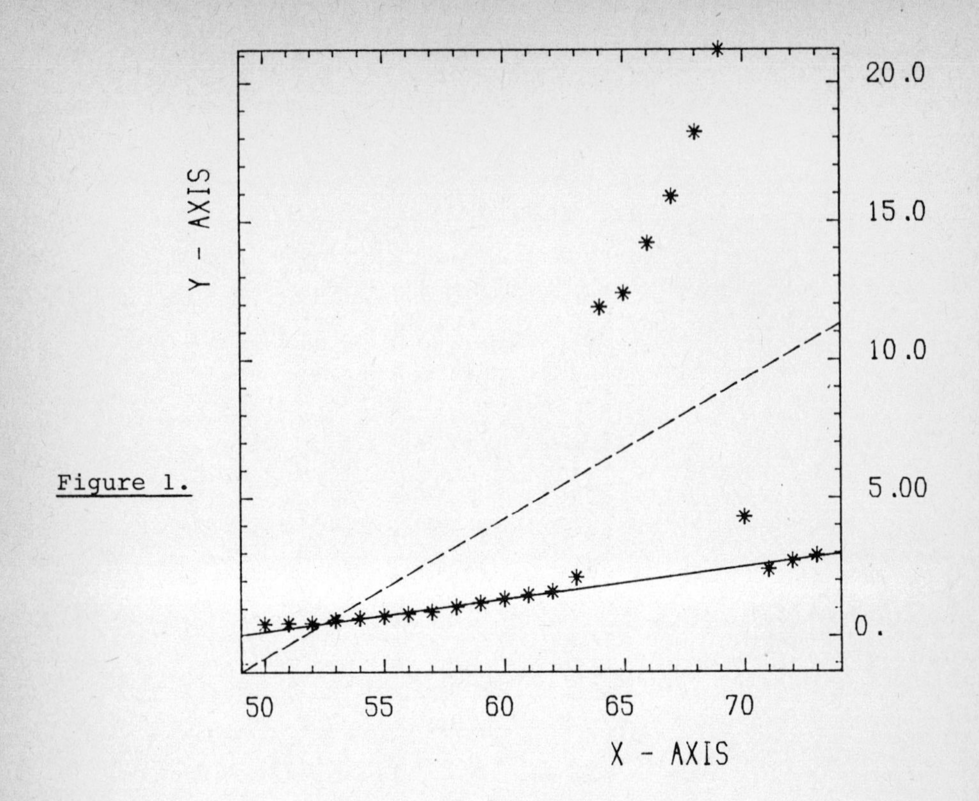

the best model, because collecting more data might reveal a more complicated kind of relationship.)

Let us now look at an example from astronomy. The data in Table 2 form the Hertzsprung-Russell diagram of the star cluster CYG OB1, which contains 47 stars in the direction of Cygnus. Here, x is the logarithm of the effective temperature at the surface of the star, and y is the logarithm of its light intensity. These numbers were given to us by Claude Doom (personal communication) who extracted the raw data from Humphreys (1978) and performed the calibration according to Vansina and De Greve (1982).

The Hertzsprung-Russel diagram itself is shown in Figure 2. It is a scatterplot of these points, where the log temperature x is plotted from right to left. Our eye sees two groups of points: the majority which seems to follow a steep band, and the four stars in the upper right corner. These parts of the diagram are well-known in astronomy: the 43 stars are said to lie on the <u>main sequence</u>, whereas the four

Table 2.

x	y	x	y
4.37	5.32	4.56	5.74
4.26	4.93	4.56	5.74
4.30	5.19	4.46	5.46
3.84	4.65	4.57	5.27
4.26	5.57	4.37	5.12
3.49	5.73	4.43	5.45
4.48	5.42	4.01	4.05
4.29	4.26	4.42	4.58
4.23	3.94	4.42	4.18
4.23	4.18	3.49	5.89
4.29	4.38	4.29	4.22
4.42	4.42	4.49	4.85
4.38	5.02	4.42	4.66
4.29	4.66	4.38	4.90
4.22	4.39	3.48	6.05
4.38	4.42	4.56	5.10
4.45	5.22	3.49	6.29
4.23	4.34	4.62	5.62
4.53	5.10	4.45	5.22
4.53	5.18	4.43	5.57
4.38	4.62	4.45	5.06
4.50	5.34	4.45	5.34
4.55	5.54	4.45	4.98
4.42	4.50		

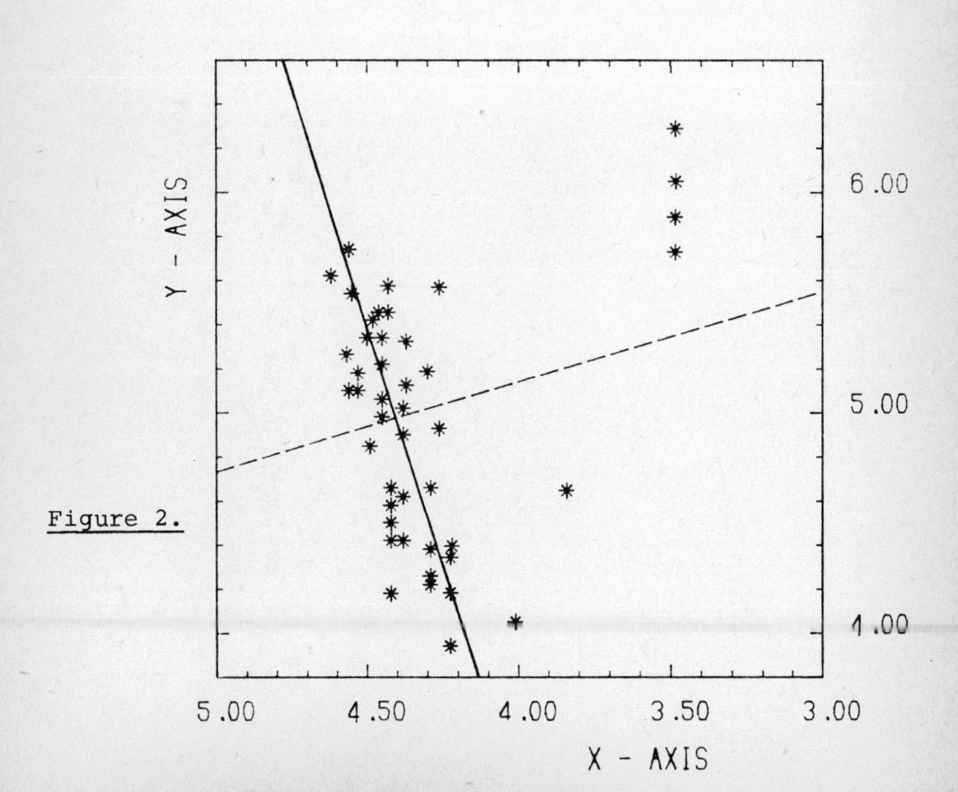

Figure 2.

remaining stars are called giants. Application of our S-estimator to this data yields the solid line $y = 3.289x - 9.59$, which fits the main sequence nicely. On the other hand, the least squares solution $y = -0.408x + 6.77$ corresponds to the dotted line in Figure 2, which has been pulled away by the four giant stars (which it does not fit well either).

4. Asymptotic behaviour of S-estimators

In section 2 we saw that S-estimators do not break down easily when the data are contaminated. Of course, we also want our estimators to behave well when the data are not contaminated, that is, when they satisfy the classical assumptions. Therefore, let us look at the asymptotic behaviour of S-estimators at the central Gaussian model, where (x_i, y_i) are i.i.d. random variables satisfying

$$y_i = x_i^t \theta_0 + e_i \quad , \tag{4.1}$$

x_i follows some distribution H, and e_i is independent of x_i and distributed like $\Phi(e/\sigma_0)$ for some $\sigma_0 > 0$ (here, Φ is the standard normal cdf).

THEOREM 2 (CONSISTENCY). Let ρ be a function satisfying (R1) and (R2), with the derivative $\rho' = \psi$. Assume that
(i) $\psi(u)/u$ is nonincreasing for $u > 0$;
(ii) $E_H[\|x\|] < \infty$, and H has a density.
Let (x_i, y_i) be i.i.d. according to the model (4.1), and let $\hat{\theta}_n$ be a solution of (1.9) for the first n points, and $\hat{\sigma}_n = s(r_1(\hat{\theta}_n),..,r_n(\hat{\theta}_n))$. Then

$$\hat{\theta}_n \to \theta_0 \quad \text{a.s.}$$

$$\hat{\sigma}_n \to \sigma_0 \quad \text{a.s.}$$

Proof. This follows from Theorems 2.2 and 3.1 of Maronna and Yohai (1981) because S-estimators satisfy the same first-order necessary conditions as M-estimators. Indeed, let θ be any p-dimensional parameter vector. By definition, we know that $s(\theta)=s(r_1(\theta),\ldots,r_n(\theta))$ is larger than or equal to $\hat{\sigma}_n = s(\hat{\theta}_n)$. Keeping in mind that $s(\theta)$ satisfies $\frac{1}{n}\Sigma_{i=1}^{n}\rho(r_i(\theta)/s(\theta))=K$ and that $\rho(u)$ is nondecreasing in $|u|$, it follows that always $\frac{1}{n}\Sigma_{i=1}^{n}\rho(r_i(\theta)/\hat{\sigma}_n) \geq K$. At $\theta = \hat{\theta}_n$, this becomes an equality. Therefore, $\hat{\theta}_n$ minimizes $\frac{1}{n}\Sigma_{i=1}^{n}\rho(r_i(\theta)/\hat{\sigma}_n)$. (This fact cannot be used for determining $\hat{\theta}_n$ in practice, because $\hat{\sigma}_n$ is fixed but unknown.) Differentiating with respect to θ, we find $\frac{1}{n}\Sigma_{i=1}^{n}\psi(r_i(\theta)/\hat{\sigma}_n)x_i = 0$. If we denote ρ-K by χ we conclude that $(\hat{\theta}_n,\hat{\sigma}_n)$ is a solution of the system of equations (1.4)-(1.5). Unfortunately, these equations cannot be used directly because there are infinitely many solutions (ψ is redescending) and the familiar iteration procedures easily end in the wrong place if there are leverage points. (This means we still have to minimize (1.9) with brute force in order to actually compute the S-estimate in a practical situation.) But anyway, the fact that $(\hat{\theta}_n,\hat{\sigma}_n)$ satisfies (1.4)-(1.5) is sufficient to apply the results of Maronna and Yohai (1981) to S-estimators. This ends the proof.

THEOREM 3 (ASYMPTOTIC NORMALITY) Let $\theta_0 = 0$ and $\sigma_0 = 1$ for simplicity. If the conditions of Theorem 2 hold and also

(iii) ψ is differentiable in all but a finite number of points, $|\psi'|$ is bounded and $\int\psi' d\Phi > 0$;

(iv) $E_H[xx^t]$ is nonsingular and $E_H[\|x\|^3] < \infty$, then

$$n^{1/2}(\hat{\theta}_n-\theta_0) \xrightarrow{\mathcal{L}} N(0, E_H[xx^t]^{-1}\int\psi^2 d\Phi/(\int\psi' d\Phi)^2)$$

$$n^{1/2}(\hat{\sigma}_n-\sigma_0) \xrightarrow{\mathcal{L}} N(0, \int(\rho(y)-K)^2 d\Phi(y)/(\int y\psi(y)d\Phi(y))^2) .$$

Proof. This follows from Theorem 4.1 of Maronna and Yohai (1981).

As a consequence of this theorem, we can compute the asymptotic efficiency e of an S-estimator at the Gaussian model as $e = (\int\psi' d\Phi)^2/\int\psi^2 d\Phi$. Table 3 gives the asymptotic efficiency of the S-estimator corresponding to (1.11), for different values of the breakdown point ε^*.

We note that taking c=2.560 yields a value of e which is larger than that of L_1 (for which e is about 64%), and gains

Table 3.

ε^*	e	c	K
50%	28.7%	1.547	.1995
45%	37.0%	1.756	.2312
40%	46.2%	1.988	.2634
35%	56.0%	2.251	.2957
30%	66.1%	2.560	.3278
25%	75.9%	2.937	.3593
20%	84.7%	3.420	.3899
15%	91.7%	4.096	.4194
10%	96.6%	5.182	.4475

us a breakdown point of 30%. In practice, we do not recommend the estimators in the table with a breakdown point smaller than 25%. In fact, it seems like a good idea to apply the c=1.547 estimator because of its 50% breakdown point, and to make up for its low efficiency by computing a one-step M-estimator (Bickel 1975) from this first solution, with a re-descending and more efficient ψ . Such a two-stage procedure inherits the 50% breakdown point from the first stage, and the high asymptotic efficiency from the second.

5. Outlook

The computation of S-estimators is not at all easy. In fact, there is a relation to the projection pursuit technique (Friedman and Tukey 1974) for the analysis of multivariate data. To see this, let us consider the (p+1)-dimensional space of the (x,y)-data. In this space, linear models are defined by $(x,y)(-\theta,1)^t=0$ for some p-dimensional vector θ . The definition of S-estimators amounts to the following: for any vector θ , we consider the projection on the y-axis in the direction orthogonal to $(-\theta,1)$, and we select the $\hat{\theta}$ for which the projected sample has minimal dispersion $s(\theta)$. (In fact, we could generalize this easily to orthogonal regression by taking any nonzero (p+1)-vector and by projecting on this vector itself instead of on the y-axis.) This means that S-estimators belong to the highly computer-intensive part of statistics, just like projection pursuit and the bootstrap (Efron 1982). In Table 4 we give a schematic overview of criteria in affine equivariant regression.

Table 4: Schematic Overview

Criterion	Method	Computation	ε^*
Best Linear Unbiased	LS	explicit	0
Minimax Variance	M	iterative	0
Bounded Influence	GM	iterative (harder)	$\leq 1/(p+1)$
High Breakdown	S	projection pursuit	constant, up to 1/2

At present, we have a portable Fortran program for the computation of S-estimators in simple regression, which is based on the LMS program of Rousseeuw (1982). We are also working on a program for multiple regression, the computation time of which is feasable up to around $p = 10$.

An interesting project (which hopefully will be started at the University of Washington in 1984) is to make a Monte Carlo comparison of the available methods for robust regression in different situations. This study would be the natural sequel to the work of Andrews et al (1972). For the S-estimators, this would also be a good occasion to study the behaviour of one-step improvements, both one-step M and one-step reweighted LS.

Looking at Table 3, there appears to be a tradeoff between breakdown point and efficiency for S-estimators. Probably one could obtain higher values of e for the same ε^* by using ρ-functions based on hyperbolic tangent estimators (Hampel, Rousseeuw and Ronchetti 1981). In fact, we wonder what is the maximal efficiency e, given a certain value of ε^*.

Our definition of S-estimators (1.9) could easily be generalized by allowing other types of dispersion measures s. If we consider all permutation invariant and scale equivariant measures s (that is, $s(r_{\pi(1)},\ldots,r_{\pi(n)}) = s(r_1,\ldots,r_n)$ and $s(\lambda r_1,\ldots,\lambda r_n) = |\lambda| s(r_1,\ldots,r_n)$ for all permutations π and factors λ), then also the following methods are S-estimators: least squares; least absolute deviations, least p-th power deviations (Gentleman 1965), the method of Jaeckel (1972), least median of squares (Rousseeuw 1982) and least trimmed squares (Rousseeuw 1983).

S-estimators could also be used for robust analysis of variance, even in the general linear model. Instead of comparing sums of squares (with respect to the reduced model and the full model) in order to obtain an F-statistic, we could compare the scale estimators (1.10) which are obtained from S-estimators. By (1.9), we are sure that $\hat{\sigma}$ for the reduced model is at least as large as $\hat{\sigma}$ for the full model. Making use of a similar reasoning, Doug Martin (personal communication) proposes to define robust sequential partial correlations by means of S-estimators. In both cases, it is probably not possible to find the exact finite sample distribution of these statistics in an analytical way. However, one could resort to small-sample asymptotics (Field and Hampel 1982) or simulation in order to determine critical values for hypothesis testing.

Acknowledgments The research of the first author was supported by the Belgian National Science Foundation, and that of the second author by Office of Naval Research Contract N000/4-82-K-0062. Most of the work leading to this paper was done while the authors were sharing the hospitality of Doug Martin, at the Statistics Department of the University of Washington. We also wish to thank Annick Leroy for assistance with the programming, and Frank Hampel and Werner Stahel for useful suggestions.

References

ANDREWS, D.F. (1974), "A Robust Method for Multiple Linear Regression," Technometrics, 16, 523-531.

ANDREWS, D.F., BICKEL, P.J., HAMPEL, F.R., HUBER, P.J., ROGERS, W.H., and TUKEY, J.W. (1972), Robust Estimates of Location: Survey and Advances, Princeton University Press.

BICKEL, P.J. (1975), "One-Step Huber Estimates in the Linear Model," Journal of the American Statistical Association, 70, 428-434.

BROWN, G.W., and MOOD, A.M. (1951), "On Median Tests for Linear Hypotheses," Proceedings 2nd Berkeley Symposium on Mathematical Statistics and Probability, 159-166.

DONOHO, D.L., and HUBER, P.J. (1983), "The notion of Breakdown Point," in A Festschrift for E.L. Lehmann, Wadsworth.

EDGEWORTH, F.Y. (1887), "On Observations Relating to Several Quantities," Hermathena, 6, 279-285.

EFRON, B. (1982), The Jackknife, the Bootstrap and other Resampling Plans, SIAM Monograph No. 38, Society for Industrial and Applied Mathematics.

FIELD, C.A., and HAMPEL, F.R. (1982), "Small-Sample Asympto-

tic Distributions of M-estimators of Location," Biometrika, 69, 29-46.
FRIEDMAN, J.H., and TUKEY, J.W. (1974), "A Projection Pursuit Algorithm for Exploratory Data Analysis," IEEE Transactions on Computers, C-23, 881-889.
GENTLEMAN, W.M. (1965), "Robust Estimation of Multivariate Location by Minimizing p-th power Deviations," Ph.D. dissertation, Princeton University, and Bell Tel. Laboratories memorandum MM65-1215-16.
HAMPEL, F.R. (1971), "A General Qualitative Definition of Robustness," Annals of Mathematical Statistics, 42, 1887-1896.
HAMPEL, F.R. (1975), "Beyond Location Parameters: Robust Concepts and Methods," Bulletin of the International Statistical Institute, 46, 375-382.
HAMPEL, F.R. (1978), "Optimally Bounding the Gross-Error-Sensitivity and the Influence of Position in Factor Space," 1978 Proceedings of the ASA Statistical Computing Section, 59-64.
HAMPEL, F.R., ROUSSEEUW, P.J., and RONCHETTI, E. (1981), "The Change-of-Variance Curve and Optimal Redescending M-estimators," Journal of the American Statistical Association, 76, 643-648.
HILL, R.W. (1977), "Robust Regression when there are Outliers in the Carriers," unpublished Ph.D. dissertation, Harvard University.
HODGES, J.L. Jr. (1967), "Efficiency in Normal Samples and Tolerance of Extreme Values for some Estimates of Location," Proceedings 5th Berkeley Symposium on Mathematical Statistics and Probability, Vol. 1, 163-168.
HUBER, P.J. (1964), "Robust Estimation of a Location Parameter," Annals of Mathematical Statistics, 35, 73-101.
HUBER, P.J. (1973), "Robust Regression: Asymptotics, Conjectures and Monte Carlo," Annals of Statistics, 1, 799-821.
HUBER, P.J. (1981), Robust Statistics, New York: Wiley.
HUMPHREYS, R.M. (1978), "Studies of Luminous Stars in Nearby Galaxies. I. Supergiants and O stars in the Milky Way," The Astrophysical Journal Supplement Series, 38, 309-350.
JAECKEL, L.A. (1972), "Estimating Regression Coefficients by Minimizing the Dispersion of the Residuals," Annals of Mathematical Statistics, 5, 1449-1458.
KRASKER, W.S. (1980), "Estimation in Linear Regression Models with Disparate Data Points," Econometrica, 48, 1333-1346.
KRASKER, W.S., and WELSCH, R.E. (1982), "Efficient Bounded-Influence Regression Estimation," Journal of the American Statistical Association, 77, 595-604.
MALLOWS, C.L. (1975), "On Some Topics in Robustness," unpublished memorandum, Bell Tel. Laboratories, Murray Hill.
MARONNA, R.A., and YOHAI, V.J. (1981), "Asymptotic Behaviour of General M-estimates for Regression and Scale with Random Carriers," Zeitschrift für Wahrscheinlichkeitstheorie und verwandte Gebiete, 58, 7-20.
MARONNA, R.A., BUSTOS, O., and YOHAI, V.J. (1979), "Bias- and Efficiency Robustness of General M-estimators for Regression with Random Carriers," in Smoothing Techniques for Curve Estimation, eds. T. Gasser and M. Rosenblatt, New York: Springer Verlag, 91-116.
RONCHETTI, E., and ROUSSEEUW, P.J. (1982), "Change-of-Variance Sensitivities in Regression Analysis," Research Report

No. 36, Fachgruppe für Statistics, ETH Zürich.
ROUSSEEUW, P.J. (1982), "Least Median of Squares Regression," Research Report No. 178, Centre for Statistics and Operations Research, VUB Brussels.
ROUSSEEUW, P.J. (1983), "Multivariate Estimation with High Breakdown Point," Research Report No. 192, Centre for Statistics and Operations Research, VUB Brussels.
SEN, P.K. (1968), "Estimates of the Regression Coefficient Based on Kendall's Tau," Journal of the American Statistical Association, 63, 1379-1389.
SIEGEL, A.F. (1982), "Robust Regression Using Repeated Medians," Biometrika, 69, 242-244.
THEIL, H. (1950), "A Rank-Invariant Method of Linear and Polynomial Regression Analysis," I, II and III. Nederlandsche Akademie van Wetenschappen Proceedings Serie A, 53, 386-392, 521-525, and 1397-1412.
VANSINA, F., and DE GREVE, J.P. (1982), "Close Binary Systems Before and After Mass Transfer," Astrophysics and Space Science, 87, 377-401.

ON ROBUST ESTIMATION OF PARAMETERS FOR AUTOREGRESSIVE MOVING AVERAGE MODELS

Pham Dinh Tuan
Université de Grenoble, Laboratoire IMAG
BP 68, 38402 St Martin d'Hères, France

SUMMARY

A simplified version of the AM estimate introduced by Martin in connection with robust estimation for autoregressive moving average model with additive outliers is investigated. It is seen that the estimate is not consistent under the ideal model, so a similar procedure is introduced and studied. Simulation results in the simple case of the first order autoregressive model show that the estimates are very robust against the additive outlier model, but not quite robust against the innovation outlier model.

1. INTRODUCTION

Robust methods for time series have not been received much attention. The most well known robust estimates of parameters of the autoregressive moving average model (ARMA) model are the analogues of Huber's M estimate and its generalisation GM estimate. They have been studied in some details in the autoregressive case by Denby and Martin (1979) and Martin (1979a,1980). Recently, another approach for obtaining robust estimates for the ARMA model with additive outlier (AO) has been proposed by Martin (1981). The procedure, referred to as the approximate maximum (AM) likelihood procedure, has an intuitive appeal, but not much is known about its performance and a theoretical study seems to be intractable. In this paper, we introduce some simplifications to make the procedure easier to implement and we study the estimate in the simple case of the first order autoregressive model by Monte-Carlo method. It is seen from theoretical considerations, supported by simulation results, that the estimate is not consistent under the ideal model. This is not serious from the practical point of view since in small sample, it only results in a small bias. Still, it is an undesirable feature. Therefore, we shall introduce a similar procedure which should eliminate the bias. It involves the use of a "robust" prediction filter which censors "bad" data before computing the prediction. The new estimate is also studied by simulation and is compared with the original one.

2. THE AM PROCEDURE

We here describe briefly the procedure proposed by Martin (1981). Let the AO model be

(2.1) $$Y_t = X_t + V_t$$

where V_t is a sequence of i.i.d. random variables with a long tail distribution such that $\gamma = P(V_t \neq 0)$ is small and X_t is a Gaussian ARMA process and hence can be represented in state form

(2.2) $$X_t = H\,Z_t + \mu$$

$$Z_t = A\,Z_{t-1} + B\,e_t$$

where e_t are i.i.d. normal variates with zero mean and variance σ^2 and H, A, B, are appropriate row, square, collumn matrices respectively, with HB = 1. Note that the above model becomes the innovation outlier (IO) model if $V_t = 0$ and e_t is not normal but has a long tail distribution. Now, let $f_t(y;\, Y_1, \ldots, Y_{t-1})$ be the conditional density of Y_t given $Y_1, \ldots, Y_{t-1}$, then the log likelihood function of the model based on a sample of size n is

$$\sum_{t=1}^{n} \log f_t(Y_t;Y_1,\ldots,Y_{t-1}) .$$

The exact form of the above density is however usually impossible to obtain, so the idea is to approximate the conditional distribution of Z_t given $Y_1, \ldots, Y_{t-1}$ by a Gaussian distribution of mean $\hat{Z}_{t;t-1}$ and covariance matrix M_t. One can then obtain the following updating equation for $\hat{Z}_{t;t-1}$ and M_t (Masreliier, 1976, Martin, 1979b, 1981) :

(2.3)
$$\hat{Z}_{t;t} = \hat{Z}_{t;t-1} + M_t H' \psi_t(Y_t;Y_1,\ldots,Y_{t-1}) ,$$
$$P_t = M_t - \psi'_t(Y_t;Y_1,\ldots,Y_{t-1}) M_t H'H M_t ,$$
$$M_t = A P_t A' + BB'\sigma^2 ,$$
$$\hat{Z}_{t+1;t} = A \hat{Z}_{t;t}$$

where $\psi_t(y;Y_1,\ldots,Y_{t-1})$ and $\psi'_t(y;Y_1,\ldots,Y_{t-1})$ are the first and second derivatives of $-\log f_t(y;Y_1,\ldots,Y_{t-1})$ with respect to y and $\hat{Z}_{t;t}$ and P_t denote the expected value and the covariance matrix of the conditional distribution of Z_t given $Y_1, \ldots, Y_t$. Now, it still remains the problem of computing the convolution of the distribution of V_t and the Gaussian distribution of mean $H \hat{Z}_{t;t-1} + \mu$ and variance $H M_t H'$. However, it is not unreasonable to approximate this convolution density by $g[(Y_t - H\hat{Z}_{t;t-1} - \mu)/s_t]$ where g is a certain function depending on the distribution of V_t only and s_t is a scale factor. Following the usual M-estimate approach, we shall replace $-\log g$ by $\log s_t + \log \rho$ where ρ is some good robustifying rho function, and since γ is small we take s_t^2 to be HM_tH'. The procedure thus consists of minimising

(2.4)
$$\sum_{t=1}^{n} \log s_t + \sum_{t=1}^{n} \rho\left(\frac{Y_t - H \hat{Z}_{t;t-1} - \mu}{s_t}\right)$$

where $s_t^2 = HM_tH'$ and $\hat{Z}_{t;t-1}$ and M_t are computed recursively by (2.3) with $\psi_t(Y_t;Y_1,\ldots,Y_{t-1}) = \psi(r_t/s_t)/s_t$, $\psi_t'(Y_t;Y_1,\ldots,Y_{t-1}) = \psi'(r_t/s_t)/s_t^2$, $r_t = Y_t - H\hat{Z}_{t;t-1} - \mu$ and ψ, ψ' being the first and second derivatives of ρ. The appealing feature of the procedure is that in the updating equation for $\hat{Z}_{t;t-1}$, large residuals are downweighed. Specifically if $\psi(x) = x$ for $|x| < c$ and $0 < \psi(x)/x < 1$ otherwise, then the expected value of X_t given $Y_1,\ldots, Y_{t-1}$ is Y_t if $|r_t| < cs_t$ and is pulled toward to $H\hat{Z}_{t;t-1} + \mu$ otherwise, which agrees with the intuitive idea that y_t corresponds to a contaminated value of the process if r_t is large.

The minimisation of (2.4) is rather complicated. Therefore, we introduce some simplifications. We first observe that in case of the ideal model where $\psi(x) = x$, if we start the recursion (2.3) with $M_1 = BB'\sigma^2$, then $M_t = BB'\sigma^2$ for all t, since $HB = 1$. Ofcourse the above starting value is not the correct one, but one can show that for any starting value M_t converges to $BB'\sigma^2$ so it does not much matter which starting value we use. Now, to deal with the AO model, ψ could not be the identity function so that M_t, s_t would be data dependent and cannot be made constant, but if $\psi(x) = x$ except for large x, one can expect that s_t^2 stays close to σ^2 most of the time. Write $s_t^2 = \sigma^2\alpha_t^2$ and treat α_t as independent of σ^2, we get the following estimating equations

$$(2.5) \qquad \sum_{t=1}^{n} \frac{1}{s_t} \frac{\partial r_t}{\partial \theta} \psi(\frac{r_t}{s_t}) = 0$$

$$(2.6) \qquad \sum_{t=1}^{n} [1 - \frac{r_t}{s_t} \psi(\frac{r_t}{s_t})] = 0$$

where θ denotes the vector of the autoregressive and moving average coefficients and $\partial r_t/\partial\theta$ denotes the vector of partial derivatives of r_t with respect to θ. Equation (2.6) has however the undesired feature that the term $(r_t/s_t)\psi(r_t/s_t)$ is unbounded which would result in a nonrobust estimate. Therefore, following the usual robustifying procedure, we replace this term by $V(r_t/s_t)$ where V is a positive bounded function such that $E\,V(X) = 1$ for a standard normal variate X. We shall take $V(x) = K \min(x^2, c^2)$, K being an appropriate constant. Then (2.6) becomes

$$\sigma^2 = \frac{K}{n} \sum_{t=1}^{n} \min\left(\frac{r_t^2}{\alpha_t^2}, c^2\sigma^2\right). \tag{2.6'}$$

The above procedure, however, may not yield consistent estimate under the ideal model. This can be best seen in the case of the first order autoregressive model of mean zero ($\theta = a = A$, $H = B = 1$, $\mu = 0$) with Huber psi function ($\psi(x) = x$ if $|x| < c$, $= c\,\mathrm{sign}(x)$ otherwise). Indeed, if the estimates $\hat{a}$, $\hat{\theta}$ converge to their true values, then one could expect that the time average

$$\frac{1}{n}\left[\sum_{t=1}^{n} \frac{1}{\alpha_t} \frac{\partial r_t}{\partial a} \psi\left(\frac{r_t}{\sigma\alpha_t}\right)\right]_{a=\hat{a},\theta=\hat{\theta}}$$

converges to the expected value of $(\partial r_t/\partial a)\psi(r_t/\sigma\alpha_t)/\alpha_t$ (the distribution of this random variable can be considered, up to a transient effect, as independent of t). Hence the above expected value is zero. Now, an explicit computation shows that the conditional expectation of the above random variable, given $Y_1, \ldots, Y_{t-1}$, equals 0 when $|r_{t-1}| < c\sigma\alpha_{t-1}$ and equals

$$-\frac{1}{\alpha_t}\left[2aY_{t-2} + c\sigma\,\mathrm{sign}(e_{t-1})\right] E\,\psi\left[\frac{X + a\{e_{t-1}/\sigma - c\,\mathrm{sign}(e_{t-1})\}}{(1+a^2)^{1/2}}\right]$$

when $|r_{t-1}| < c\sigma\alpha_{t-1}$, $|r_{t-2}| < c\sigma\alpha_{t-2}$, where X denotes a standard normal variate. It can be seen that the integral of the above expression on the set $\{|r_{t-1}| < c\sigma\alpha_{t-1}, |r_{t-2}| < c\sigma\alpha_{t-2}\}$ is not zero and has the same sign as $-a$, since e_{t-1} is greater than c in absolute value on this set. Since the probability that both $|r_{t-1}|$ and $|r_{t-2}|$ exceed their thresholds is very small, it is very likely that the above random variable has non zero expectation so that the estimate is not consistent. Simulation results in section 4 clearly exhibit this nonconsistency. By a heuristic augument using the Taylor expansion, it can be seen that for finite sample, the estimate has a small bias of the same sign as a.

3. A NEW PROCEDURE

To obtain consistency, we introduce the following procedure. The observations are censored, yielding a random sequence of subsets D_n, defined recursively as follows. D_1 consists of the point Y_1 and D_t is the semi-interval $[\tilde{X}_{t;t-1} + c\sigma\alpha_t, \infty)$ or $(-\infty, \tilde{X}_{t;t-1} - c\sigma\alpha_t]$ which contains Y_t or consists of the point Y_t if none of the above interval contains Y_t, $\sigma^2\alpha_t^2$ being the conditional variance of X_t given that $X_1 \in D_1, \ldots, X_{t-1} \in D_{t-1}$ and $\tilde{X}_{t;t-1}$ is the unique value of m satisfying :

$$E\{\psi(\frac{X_t - m}{\sigma\alpha_t}) | X_1 \in D_1, \ldots, X_{t-1} \in D_{t-1}\} = 0, \ \psi(x) = \text{sign}(x) \min(|x|, c).$$

The procedure again consists of solving for (2.5), (2.6') with r_t now equals to $Y_t - \tilde{X}_{t;t-1}$. Clearly, by construction, the random variable $\psi(r_t/\sigma\alpha_t)$ is uncorrelated with any random variable depending only on $D_1, \ldots, D_{t-1}$ and hence with $\partial r_t/\partial\theta$. The procedure, however, requires the computation of $\tilde{X}_{t;t-1}$ which is practically impossible. Fortunately, one can approximate $\tilde{X}_{t;t-1}$ very closely by $\hat{X}_{t;t-1}$, the conditional expectation of X_t given that $X_1 \in D_1, \ldots, X_{t-1} \in D_{t-1}$. This is justified by the fact that c is large and that the conditional distribution of Z_t given that $X_1 \in D_1, \ldots, X_{t-1} \in D_{t-1}$ is nearly normal. The last point is best illustrated in the first order autoregressive case. If D_{t-1} is reduced to a point, then the above distribution is exactly normal; if it is not but D_{t-2} is, then this is the distribution of a^2Y_{t-2} plus or minus $\sigma u_{a,c}$ according to as D_{t-1} is an upper or lower interval, where $u_{a,c}$ is distributed as the conditional distribution of $aN_1 + N_2$ given that $N_1 > c$, N_1, N_2 being independent standard normal variates. Numerical compupation shows that the distribution of $u_{a,c}$ is surprisingly close to normal (see Appendix). In the general case, let U_t be distributed like the conditional distribution of Z_t given that $X_1 \in D_1, \ldots, X_{t-1} \in D_{t-1}$, then U_{t+1} is distributed as the convolution of a Gaussian distribution with the conditional distribution of AU_t given that $HU_t + \mu \in D_t$. Let G be a matrix such that (H G) is non singular and GU_t is uncorrelated with HU_t, then the above conditioning would mainly alter the distribution of HU_t, but after convoluting we would again obtain a distribution close to Gaussian as is seen before. This suggests approximating the distribution of U_t by a Gaussian distribution of mean $\hat{Z}_{t;t-1}$ and covariance matrix M_t, say. Then one can obtain the following updating equation for $\hat{Z}_{t;t-1}$ and M_t :

(3.1) $$\hat{Z}_{t+1;t} = A\,\hat{Z}_{t;t}\,,$$

$$\hat{Z}_{t;t} = \hat{Z}_{t;t-1} + \hat{z}_t\,,$$

$$M_{t+1} = A\,P_t\,A' + BB'\sigma^2$$

where $\hat{z}_t$ and P_t are the conditional expectation and covariance matrix of z_t given that $Hz_t + H\hat{Z}_{t;t-1} + \mu \in D_t$, z_t being a Gaussian vector with mean 0 and covariance matrix M_t. A direct computation gives :

(3.2) $$\hat{z}_t = M_tH'(Y_t - H\hat{Z}_{t;t-1} - \mu)/s_t^2 \quad \text{if } |Y_t - H\hat{Z}_{t;t-1} - \mu| \le cs_t,$$

$$= M_tH'm_c\,\text{sign}(Y_t - H\hat{Z}_{t;t-1} - \mu) \quad \text{otherwise},$$

(3.3) $$P_t = M_t - M_tHH'M_t/s_t^2 \quad \text{if } |Y_t - H\hat{Z}_{t;t-1} - \mu| \le cs_t,$$

$$= M_t - (1-v_c)M_tHH'M_t/s_t^2 \quad \text{otherwise}$$

where $s_t^2 = HM_tH'$ and m_c, v_c are the conditional expectation and variance of a standard normal variate given that it is greater than c. The procedure again consists of solving for (2.5), (2.6') with $r_t = Y_t - H\hat{Z}_{t;t-1} - \mu$, $s_t = \sigma\alpha_t$. We refer to it as the Censor procedure.

Instead of censoring, one might just reject the "doubtful" data. Define the random sequence subsets D_t by : D_1 consists of the point Y_1, D_t consists of the point Y_t if $|Y_t - \hat{X}_{t;t-1}| < c\sigma\alpha_t$ and of all the real numbers otherwise, where $\hat{X}_{t;t-1}$, $\sigma^2\alpha_t^2$ are the the conditional expectation and variance of X_t given that $X_1 \in D_1, \ldots, X_{t-1} \in D_{t-1}$. By symetry, $\psi(Y_t - \hat{X}_{t;t-1})/\sigma\alpha_t$, under the ideal model, is uncorrelated with any random variable function of $D_1, \ldots, D_{t-1}$ only. The procedure consisting of solving for (2.5), (2.6') with $r_t = Y_t - \hat{X}_{t;t-1}$ would again lead to consistent estimates. Note that $\hat{X}_{t;t-1} = H\hat{Z}_{t;t-1} + \mu$ and can be updated by (3.2), (3.3) with $m_c = 0$, $v_c = 1$. The procedure will be refered to as the Rejection procedure.

4. SIMULATIONS RESULTS

To solve (2.5). (2.6') we use the Newton Rhapson and the fixed point iteration. We consider only the fist order zero mean autoregressive case, starting the recursion (2.3) at $t = 2$ with Y_1 as starting value of $\hat{Z}_{1;1}$ and dropping the first term in the sums in (2.5), (2.6') and using $K/(n-1)$ instead of K/n. The iteration requires computing $u_t = \partial r_t/\partial a$ and $\partial(u_t/\alpha_t)/\partial a$ but terms involving the last quantity can be neglected. By treating α_t as independent of a, u_t can be updated as follows : $u_{t+1} = Y_t$ if $|r_t| < c\sigma\alpha_t$, $= au_t + \hat{Z}_{t;t}$ otherwise. As starting values for a and σ, we use the Least squares estimate and (1/.6745) times the median absolute deviation of the residuals.

In a first simulation study, we check the consistency of the estimate. A first order autoregressive process with $a = .5$ is generated and we compute the Least squares, Huber, Censor and Rejection estimates, using the first $n = 100, 200, \ldots, 5000$

observations. The results are plotted in figure 1. The Least Squares, Censor and Rejection estimates seem to converge to the right value, but the Huber estimate is consistently too high. For n = 5000, the formers are .513, .513 and .509 respectively while the latter is .544 .

In a second simulation study, we look at the performance of the procedure under the ideal and various $AO(\gamma,\sigma^2)$ and $IO(\gamma,\sigma^2)$ models, using 250 replications. Here $AO(\gamma,\sigma^2)$ means the AO model with V_t having the discrete contaminated normal distribution $(1-\gamma)\ \delta_o + \gamma\, N(0,\sigma^2)$ where δ_o denotes the Dirac distribution with mass at 0 and $N(0,\sigma^2)$ the normal distribution of mean 0 and variance σ^2, and $IO(\gamma,\sigma^2)$ means the IO model with contaminated normal distribution $(1-\gamma)\ N(0,1) + \gamma\, N(0,\sigma^2)$ for the innovations. Table 1 gives the mean square efficency (MSE) of the estimates relative to the Least Squares estimate for a = .5 and .9 and for sample sizes n = 50 and 200. The Huber and Censor estimate are seen to be quite robust against the A0 model. For n = 200 and a = .5, the Huber estimate has rather low MSE under the ideal model. This is because the estimate is biased, as seen in table 2 (the first number in each entry of this table refers to the mean while the second refers to the standard deviation). There is a somewhat strange result : the MSE of the Huber estimate is higher than 1 for a = .9 under the Ideal model. A closer examination shows that the distribution of the Least Squares in this case is rather skew with a long tail in the lower end (this seems to come from the fact that the estimate rarely exceeds 1, but the true value of a is quite close to 1, and that the asymptotic distribution of the estimate is not normal when a = 1. Thus the estimate is biased toward 0 and has higher variance than its theoretical asyptotic value. Since the Huber estimate tends to be higher than the Least squares estimate (when a > 0), its distribution tends to have higher means and shorter tail, which results in a reduction of bias and variance. Thus by a happy coincidence, the Huber procedure outperforms the other procedures when a is close to 1.

Model	n	a = .5			a = .9		
		Huber	Censor	Rejection	Huber	Censor	Rejection
Ideal model	50	.90	.98	.90	1.07	.94	.92
	200	.71	.96	.90	1.20	.96	.78
AO(.2,4)	50	1.16	1.17	1.13	1.75	1.36	1.41
	200	1.23	1.30	1.18	2.01	1.38	1.26
AO(.2,9)	50	1.57	1.54	1.44	2.73	2.07	2.15
	200	2.24	2.18	2.02	3.74	2.48	2.21
AO(.2,36)	50	3.06	2.91	2.53	8.29	6.08	3.35
	200	9.63	8.12	7.65	21.2	12.9	3.84
AO(.2,100)	50	4.56	4.26	3.39	17.3	12.5	3.14
	200	22.0	17.5	15.5	72.5	44.4	9.53
IO(.5,9)	50	.97	1.07	.73	1.31	1.12	.33
	200	.63	.89	.60	1.27	1.13	.26
IO(.5,36)	50	.70	.73	.43	1.32	1.15	.09
	200	.35	.47	.40	1.22	.73	.06
IO(.05,100)	50	.46	.45	.23	1.23	.76	.03
	200	.18	.20	.29	.86	.33	.02

Table 1 : Mean squares efficiency of estimates

It can be seen that the estimates are not robust against the IO model. This is a rather discouraging result comparing with the GM estimate which is robust both at the AO and the IO models (Birch and Martin, 1982). However, the poor performance of the estimate at the IO model can be corrected. In fact, in the Censor and Rejection procedures, we have used the Huber psi function in (2.5), (2.6) with the <u>same</u> cutoff value c as the one in (3.2), (3.3), but for the argument in section 3 to be valid, we only need that psi has a cutoff value c' not exceeding c (or more

Model	n	a = .5								a = .9							
		L.S.		Huber		Censor		Rejection		L.S.		Huber		Censor		Rejection	
Ideal model	50	.487	.118	.511	.125	.482	.119	.483	.124	.866	.086	.874	.085	.864	.088	.866	.090
	200	.502	.058	.532	.060	.500	.059	.499	.061	.892	.035	.900	.033	.891	.036	.891	.040
AO(.02,4)	50	.462	.126	.514	.121	.480	.120	.481	.122	.858	.084	.881	.068	.869	.074	.873	.074
	200	.469	.060	.518	.058	.484	.057	.485	.060	.876	.054	.896	.029	.884	.032	.886	.035
AO(.02,9)	50	.439	.139	.507	.121	.473	.119	.483	.125	.839	.103	.880	.070	.865	.076	.875	.078
	200	.442	.069	.514	.058	.479	.057	.486	.061	.860	.043	.895	.030	.882	.033	.888	.037
AO(.02,36)	50	.376	.176	.500	.123	.465	.121	.489	.135	.773	.174	.877	.071	.862	.079	.895	.118
	200	.349	.104	.508	.058	.471	.057	.491	.065	.786	.089	.893	.031	.878	.034	.908	.048
AO(.02,100)	50	.320	.203	.495	.127	.459	.125	.488	.147	.690	.247	.875	.074	.859	.082	.935	.180
	200	.251	.127	.506	.059	.467	.058	.498	.071	.668	.152	.891	.031	.877	.034	.952	.073
IO(.05,9)	50	.488	.126	.532	.124	.507	.122	.484	.147	.869	.077	.864	.070	.897	.075	.856	.138
	200	.493	.059	.542	.061	.513	.061	.488	.075	.888	.032	.903	.030	.898	.032	.882	.065
IO(.05,36)	50	.493	.114	.565	.120	.549	.124	.475	.173	.873	.067	.896	.063	.901	.067	.840	.231
	200	.495	.057	.577	.060	.558	.062	.484	.091	.890	.032	.915	.026	.923	.032	.871	.133
IO(.05,100)	50	.497	.103	.597	.115	.587	.125	.467	.209	.876	.064	.905	.061	.917	.076	.824	.370
	200	.498	.054	.614	.059	.604	.064	.482	.099	.892	.031	.926	.023	.946	.032	.872	.206

Table 2 : Mean (first entry) and standard deviation (second entry) of estimates

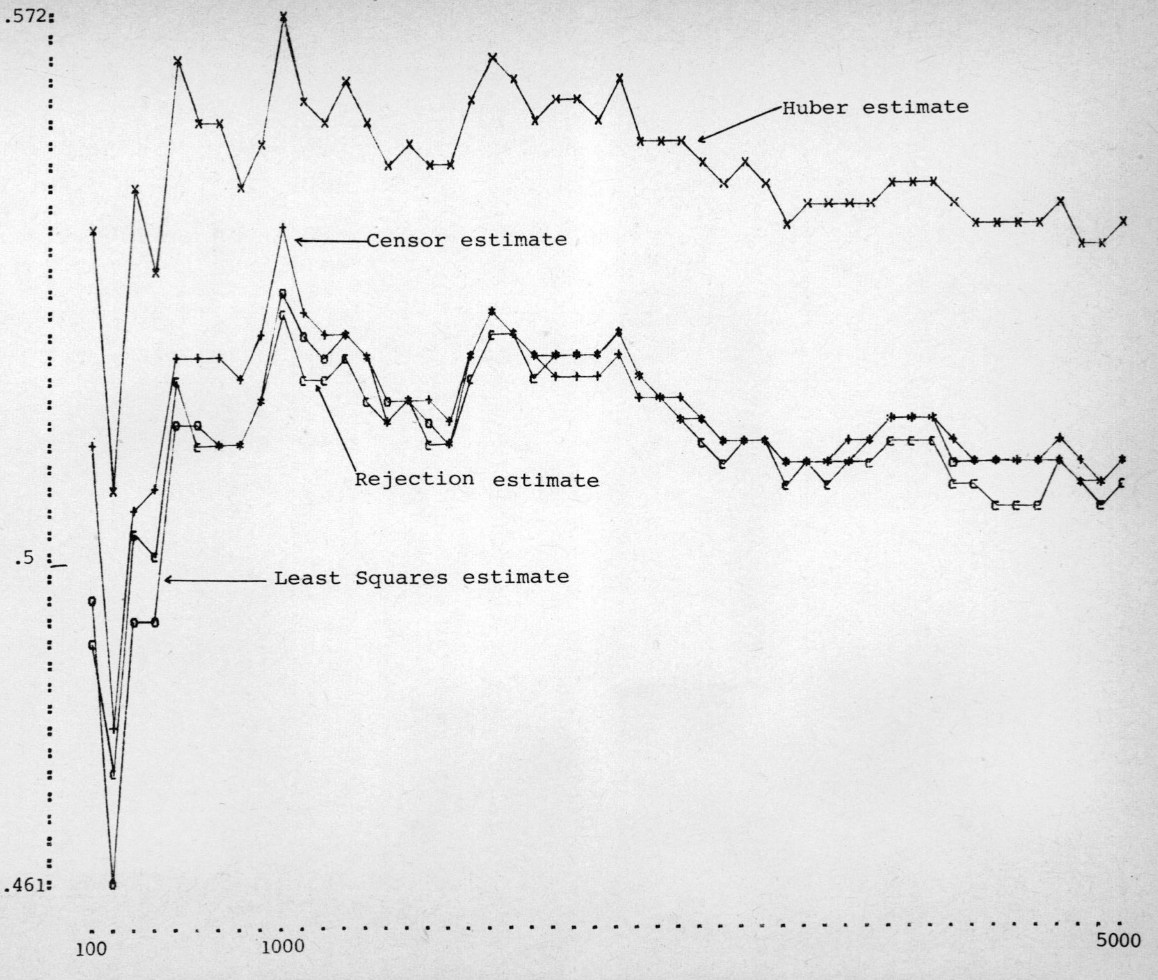

Figure 1 : Estimates of parameter as function of sample size.

generally ψ is constant on $(-\infty,-c]$ and on $[c,\infty)$). Thus, we may improve the efficiency toward the IO model by taking high c and low c' (if $c = \infty$, we get the M-estimate). Our preliminary simulations seem to indicate that this can be done with very little loss of efficiency toward the AO model. This observation is very interesting in the case of the Rejection procedure, since in this case c should be rather high to avoid rejecting too much data so we need a low c' to achieve robustness. The poor performance of the Rejection procedure in our simulation might come from the fact that we have used the same c and c'.

APPENDIX : Distribution of the variable $u_{a,c}$ of section 3

We have computed by numerical integration the cumulative distribution of the variable $u_{a,c}$, normalised to have mean 0 and variance 1. For c = 1.5, a = .5, .9, we get :

x	-3.0	-2.0	-1.8	-1.6	-1.4	-1.2	-1.0	-0.8	-0.6
a = .05	.001	.023	.036	.055	.081	.115	.158	.211	.274
a = .09	.002	.024	.037	.056	.081	.115	.158	.210	.272

x	-0.4	-0.3	-0.2	-.01	0	0.1	0.2	0.3	0.4
a = .5	.344	.381	.420	.459	.499	.539	.579	.617	.655
a = .9	.341	.379	.417	.457	.497	.537	.576	.615	.653

x	0.6	0.8	1.0	1.2	1.4	1.6	1.8	2.0	3.0
a = .5	.725	.788	.841	.885	.919	.945	.964	.977	.999
a = .9	.725	.788	.842	.886	.921	.947	.966	.978	.999

The above distribution is nearly symetric and is quite close to normal. Its median differs from its mean by .002 for c = 1.5, a = .5 and .007 for c = 1.5, a = .9. Also the value m for which $\psi(x-m)$ integrates to zero is .001 for c = 1.5, a = .5 and .004 for c = 1.5, a = .9.

REFERENCES

BIRCH, J.B., MARTIN, R.D. (1982) Confident interval for robu estimate of the first order autoregressive parameter. J. Time Series Ana. 3, 206-220.

DENBY, L., MARTIN, R.D. (1979) Robust estimation of the first order autoregressive parameter. J. Amer. Statist. Assoc. 74, 140-46.

MARTIN, R.D. (1979a) Robust estimation of time series auto-regressions. In Robustness in Statistics, R.L. Launer and G. Wilkinson eds., New-York : Academic Press.

MARTIN, R.D. (1979b) Approximate conditional mean smoother an interpolator. In smoothing Techniques for Curves estimation, Th. Gasser and M. Rosenblatt eds., Heidelberg : Springer-Ver

MARTIN, R.D. (1980) Robust estimation for autoregressive mode In Direction in Time Series, D.R. Brillinger and Tiao eds., Inst. Math. Statist. Publication.

MARTIN, R.D. (1981) Robust methods for time series. In Applie Time Series Analysis II, D.F. Findley ed., New-York : Accadem Press.

MASRELLIER, C.J. (1975) Approximate non Gaussian filtering wit linear state and observation relation. IEEE Trans. Automatic Control AC-22, 361-71.